AN

ELEMENTARY TREATISE

ON

MECHANICS,

EMBRACING THE

THEORY OF STATICS AND DYNAMICS,

AND ITS APPLICATION TO

SOLIDS AND FLUIDS.

Prepared for the Under-graduate Course in the Wesleyan University.

BY AUGUSTUS W. SMITH, LL.D.,
PROFESSOR OF MATHEMATICS AND ASTRONOMY.

NEW YORK:
HARPER & BROTHERS, PUBLISHERS.
1874.

PREFACE.

The preparation of the present treatise was undertaken under the impression that an elementary work on analytical Mechanics, suited to the purposes and exigencies of the course of study in colleges, was needed. This impression is the result of long experience in teaching, and a fair trial of all known American works and reprints designed for such use.

It can scarcely be necessary, at this period, to assign at length the reasons for adopting the analytical methods of investigation. Whether the object be intellectual discipline, or a knowledge of facts and principles, or both, the preference must be given to the modern analysis. It affords a wider field for the exercise of judgment, calls more fully into exercise the inventive powers, and taxes the memory less with unimportant particulars, thus developing and strengthening more of the mental faculties, and more equably by far than the geometrical methods. It is more universal in its application, shorter in practice, and far more fruitful in results. It is, indeed, the only method by which the student can advance beyond the merest rudiments of the science, without an expense of time and energy wholly disproportioned to the ends accomplished—the only method by which he can acquire a self-sustaining and a progressive power.

As the hope of furnishing to the student some *additional* facilities for a pleasant and profitable prosecution of this branch of study was the motive for undertaking the preparation of this manual, it will be proper to refer to some of the specific objects had in view. The most formidable obstacles to the acquisition of any branch of science are generally found at the very outset. It has therefore been a specific object to introduce the subject by giving distinctness to the elementary truths, dwelling upon them till they are rendered familiar, adopting the simplest mode of investigation and proof consistent with rigor of demonstration, and avoiding all reference to the metaphysics of the science as out of place in a work designed for beginners. At every successive stage of advancement the student is required to review the ground passed over by the use of the principles learned, in the solution of examples which will require their application, and test at once his knowledge of them and his ability to apply them. Having examined in this way each division of forces as classified, more general methods are introduced, and their application illustrated by numerous problems. These methods are often employed in particular cases when others less general would be shorter; for it is a readiness in their use and a familiarity with their application that gives to the student his power over more difficult and complicated questions.

To adapt the work to the exigencies of the recitation-room, the whole is divided, at the risk, perhaps, of too much apparent formality, into distinct portions or propositions, suitable for an individual exercise. In each case the object to be accomplished is distinctly proposed, or the truth to be established is briefly and clearly enunciated. The student has thus a definite object before him when called upon to recite, acquires a convenient formulary of words by which to quote and apply his arguments, and the more clearly conceives and marks his own progress.

In this institution the mathematical and experimental courses are assigned to different departments. It is not, however, for this reason alone that I have purposely abstained from swelling the volume with diffuse verbal explanations, and the introduction of experimental illustrations. The experienced teacher will always have at command an abundance of matter of this kind to meet every emergency, and can adapt the kind and mode of illustration to the specific difficulty that arises in the mind of his pupil. There is an objection to the introduction of such matter, growing out of a tendency, in some at least, to become the passive recipients of their mental aliment. These mental dyspeptics loathe that which costs them labor, and, satisfied with the stimulus of inflation, seize upon the lighter portions, and neglect that which alone can impart vigor to their mental constitution. Whoever caters for this class of students must share in the responsibility of raising up a race effeminate in mind, if not in body.

The investigations are limited to forces in the same plane, except in the case of parallel forces. This limit, while it is sufficiently ample to include the most important topics of terrestrial mechanics, and to embrace an interesting field of celestial mechanics, is fully sufficient to occupy the time which the crowded course of study in our colleges will admit of being appropriated to it. It may, without marring the integrity of the work, be very much reduced by the omission of all those portions in which the integral calculus is employed; and still further, if desired, by omitting all that relates to the principle of virtual velocities, and its application to the mechanical powers. A copious analysis is given in the contents, convenient for frequent and rapid reviews, and suggestive of questions for examination.

In this work no claim is advanced to originality. The materials have been sought and freely taken from all available sources. Nearly all the matter, in some form, is found in almost every author consulted, and credit could not be given in every case, if, in an elementary work designed as a text-book, it were desirable to do so. In the portion on Statics, I am most indebted to the excellent introductory work of *Professor Potter*, of University College, London. The chapter on Couples is substantially taken from *Poinsot's Elemens de Statique.* The questions, generally simplified in their character, are mostly taken from *Walton's Mechanical and Hydrostatical Problems*, and *Wrigley and Johnston's Examples.* The works more especially consulted are those of *Poisson*, *Francœur*, *Gregory*, *Whewell*, *Walker*, *Moseley*, and *Jamieson.* Something has been taken from each, but modified to suit the specific object kept constantly in view—the preparation of a manual which should be simple in its character, would most naturally, easily, and successfully induct the student into the elementary principles of the science, and prepare him, if so disposed, to prosecute the study further, without the necessity of beginning again and studying entirely new methods. How far I have succeeded must be left to the decision of others, especially of my co-laborers in this department of instruction.

WESLEYAN UNIVERSITY,
Middletown, Conn., Jan., 1849.

CONTENTS.

STATICS.

CHAPTER I.

CHAPTER V.

CHAPTER VII.

CHAPTER VIII.

DYNAMICS.

CHAPTER VI.

HYDROSTATICS.

HYDRODYNAMICS.

HYDROSTATIC AND HYDRAULIC INSTRUMENTS.

ELEMENTARY MECHANICS.

INTRODUCTION.

1. MECHANICS is the Science which treats of the laws of Equilibrium and Motion. It is subdivided into Statics, Dynamics, Hydrostatics, and Hydrodynamics.

Statics treats of the necessary relations in the intensities and directions of forces, in order to produce equilibrium of solid bodies.

Dynamics treats of the effects of forces on solid bodies when motion is produced.

Hydrostatics investigates the conditions of equilibrium in fluid bodies.

Hydrodynamics investigates the effects of forces on fluids when motion results.*

2. *Force* is that which produces or tends to produce motion or change of motion.

The consideration of the *nature* of force does not belong to the present subject. Mechanics is concerned only with the effects of force as exhibited in the production of *motion* or *rest*.

3. The *effect* of a force depends on, 1st, its Intensity; 2d, its Direction; and, 3d, its Point of Application.

The *Intensity* of a force may be measured, statically, by the pressure it will produce, or by the weight which will counterpoise it; dynamically, by the quantity of motion it will produce. By assuming for a *unit of force* that force which is counter poised by a known weight, the intensity or magnitude of any other force will be expressed by the numerical ratio which its counterpoise will bear to the counterpoise of the unit of force.

* Theoretic Statics and Dynamics are those branches of theoretic Mechanics which treat of the effects of forces applied to material points or particles regarded as without weight or magnitude. Static and Dynamic Somatology would then embrace the application of theoretic Statics and Dynamics to bodies of definite form and magnitude, both solid and fluid.

In the same manner, by fixing on any *line* to represent the unit of force, any other force will be represented by the line which bears to the linear unit the same ratio which the force in question bears to the unit of force.

The *Direction* of a force is the line which a material point, acted on by that force, would describe were it perfectly free.

The point of application of a force is that point in its line of direction on which the force acts.

4. As the magnitudes, directions, and points of lines are all determinable by the principles of Analytical Geometry, so forces, of which lines are the appropriate representatives, come under the dominion of the same principles. The application of these principles to the determination of the laws of equilibrium and motion, considered as the effects of forces, constitutes *analytical mechanics.*

5. Forces may act on a point either by *pushing* or *pulling* it. As these are readily convertible, the one into the other, without affecting the intensity, direction, or point of application of the forces themselves, all forces will be regarded as pulling unless otherwise expressly stated.

For convenience, we shall call those forces which have a common point of application *concurring forces,* and those which act along the same line toward the same parts, *conspiring* forces.

6. A *body* is an assemblage of material points. The material points, or elementary particles, are connected together in various ways, according to the nature of the body.

A body is said to be *rigid* when the relative position of its particles remains unchanged by the action of forces upon it.

In *flexible* and *elastic* bodies the relative positions of the particles change by the action of forces.

STATICS.

CHAPTER I.

ON THE COMPOSITION AND EQUILIBRIUM OF CONCURRING FORCES.

7. Prop. *Two equal forces applied to the same point in exactly opposite directions are in equilibrium.*

For no reason can be assigned why motion should take place in the direction of one force which will not equally apply to the other.

8. Prop. *Two concurring forces forming an angle with each other can not be in equilibrium.*

If possible, let the two forces P and Q acting at A, and making, by their directions, the angle PAQ, be in equilibrium. Apply to the point A the force P_1, equal and opposite to Q. Since, by hypothesis, the forces P and Q are in equilibrium, the point A will move in the direction of P_1. But P_1 and Q, being equal and opposite (*Art.* 7), will be in equilibrium, and hence the point A must move in the direction of P. Therefore the point A must move in two directions at once, which is absurd.

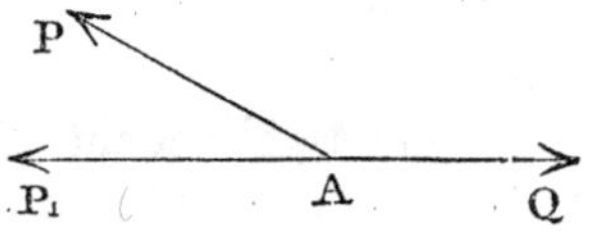

P
P_1 A Q

9. Definition. The *resultant* of two or more forces is a force which singly will produce the same mechanical effect as the forces themselves jointly.

The original forces are called *components.*

Cor. In all statical investigations the components may be replaced by their resultant, and *vice versa.*

10. Prop. *The resultant of several conspiring forces is a single force equal to their sum, and acting in the same direction.*

This is obvious from the admitted mode of measuring forces.

11. Prop. *The resultant of two unequal forces acting in opposite directions is a single force equal to their difference, and acting in the direction of the larger.*

For the smaller will obviously be expended in annulling in the larger a quantity equal to itself, and thus leave an effective balance equal to the excess of the larger over the smaller, and acting in the direction of the larger.

12. Prop. *The resultant of any number of forces acting in the same right line is equal to their algebraic sum.*

By *Art.* 10, the resultant of each system of conspiring forces is equal to their sum, and by *Art.* 11, the resultant of these two resultants is equal to their difference with the sign of the greater prefixed.

13. Prop. *In a system of points invariably connected, any point in the direction of a force may be taken as the point of application.*

If the force P be applied at the point B, and to any other point A in its direction two other forces be applied, each equal to it, but opposite to each other, since B, by hypothesis, is invariably connected with A, one of these forces (*Art.* 7) will be in equilibrium with P. There will remain, therefore, a single effective force equal to P and applied at A. In the same manner it may be transferred to any other point in its direction.

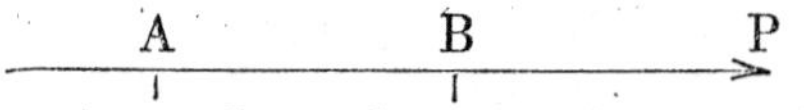

14. Prop. *The resultant of several concurring forces in one plane lies in the same plane.*

For if we suppose the resultant to lie out of the plane of the forces on one side, we may always conceive a line symmetrically situated on the other side of the plane; and since no reason can be assigned why it should be in one of these lines rather than in the other, it can be in neither of them, unless we admit the absurd consequence, that it is in both; in other words, that a system of forces has two resultants.

15. Prop. *The resultant of two equal concurring forces is in*

the direction of a line bisecting the angle formed by the components.

For no reason can be assigned why it should tend to one side rather than the other of this line.

16. Cor. When the forces are unequal, it is obvious that the direction of the resultant will make a less angle with the larger force than with the smaller, and the greater the disparity in the forces, the smaller will be this angle.

17. Prop. *If all the forces of a system, while their directions are preserved, are increased or diminished in any ratio, their resultant, without changing its direction, will be varied in the same ratio; and if the components were previously in equilibrium, they will remain so in whatever ratio their intensities be varied.*

Let P_1, P_2, P_3 . . . . . be any concurring forces whatever. Since (*Cor., Art.* 9) we can replace them by their resultant R, if we double or treble each of the components, this will only double or treble the resultant R, without changing its direction, because, in so doing, we only add to the system one or more systems precisely equal to the first.

In like manner, if we reduce the original components to one half or one third of their former intensities, the resultant will still preserve its direction, but will become one half or one third as large as before; since to double or treble these is only to double or treble their resultant, and thus reproduce the original system.

If the forces are in equilibrium, by varying the magnitudes of all in the same ratio, we only add to or suppress from the system other systems already in equilibrium.

18. Prop. *Three equal concurring forces, inclined at angles of* 120°, *will be in equilibrium.*

Since each force is inclined in the same angle to the directions of the other two, any reason that can be assigned why either one will prevail, will apply with equal force to show that each of the others will prevail.

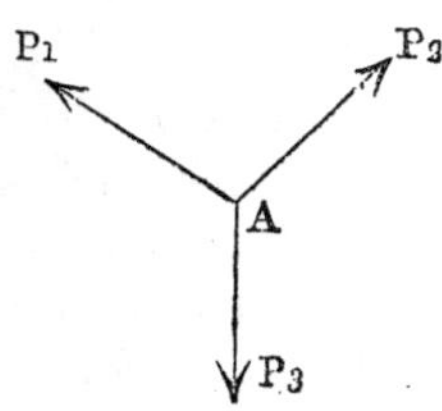

19. Prop. *Two equal concurring forces, inclined at an angle of* 120°, *have for their resultant a force which will be represented in magnitude and direction by the diagonal of a rhombus constructed on the lines representing the components.*

Let AD and AB represent the two forces P_1 and P_2, making the angle BAD=120°. To the point A apply a force $P_3=P_1$ or P_2, and making, with P_1 and P_2, angles of 120°. By *Art.* 18, these three forces will be in equilibrium. Now since P_1 and P_2 equilibrate P_3, their resultant must also. Therefore produce FA to C, making AC=AF; and since AC=R equilibrates P_3, it will be the resultant of P_1 and P_2. Join DC and BC, and ABCD will be a rhombus, of which AC is the diagonal. For since CAD=60°, and AC=AD, the triangle CAD is isosceles; and since each angle is equal to 60°, it is equilateral. ∴ DC=AD=AB. In the same manner it may be shown that CB=AD, and hence the figure is a rhombus.

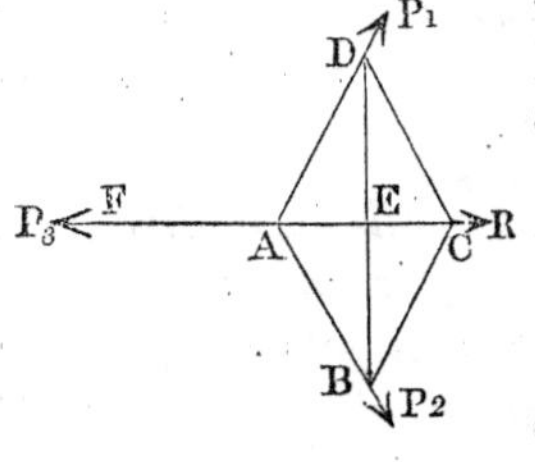

Cor. Denoting the angle BAC by a, we have AE=AB cos. a=P cos. a, and the resultant AC=R=2P cos. a.

20. Prop. *If three forces are in equilibrium, each will be equal and opposite in direction to the resultant of the other two.*

For, replacing either two by their resultant, we shall have two forces in equilibrium. If they be not opposite in direction, by *Art.* 8, they can not be in equilibrium; and if they be not equal, by *Art.* 11, motion will ensue.

PARALLELOGRAM OF FORCES.

21. Prop. *If two forces be represented in magnitude and direction by the two adjacent sides of a parallelogram, the diagonal will represent their resultant in magnitude and direction.*

First. The *direction* of the diagonal is that of the resultant

1°. When the forces are *equal*, this is obvious from *Art.* 15.

2°. Let us assume, for the present, that it is also true for the two systems of *unequal* forces P and Q, and P and R; then it will be true for the forces P and Q+R. Let P and Q act at A in the directions AD and AB respectively, and be represented in magnitude by these lines. Suppose the force R to act at B (*Art.* 13), a point in its direction, and to be represented in magnitude by BC. Complete the parallelogram ADFC, draw BE parallel to AD, and the diagonals AE, AF, and BF. Then P at A, in AD, and Q at A, in AB, have by hypothesis, their resultant in the direction AE, and (*Art.* 13) may be supposed to act at E. Replacing this resultant by its components acting in their original directions, we have a force P acting at E in the direction BE, and a force Q acting at E in the direction EF. Transfer P to B and Q to F, without changing their directions. Then P at B, in BE, and R at B, in BC, will also, by hypothesis, have their resultant in the direction BF, which may be supposed to act at F. We now have all the forces acting at F, and this without disturbing their effect upon the point A, supposed to be invariably connected with F. Hence, if the assumption be correct, F is a point in the direction of the resultant of the forces P and Q +R. But the assumption is correct when Q and R are each equal to P (*Art.* 15). Therefore, the proposition for the *direction* of the resultant is true for P and 2P. Again, making Q =2P and R=P, it is true for P and 3P, and so for P and *n*P. Also, putting *n*P for P, Q=P, and R=P, it will be true for *n*P and 2P, and so on for *m*P and *n*P (*m* and *n* being positive integers), or for all commensurable forces.

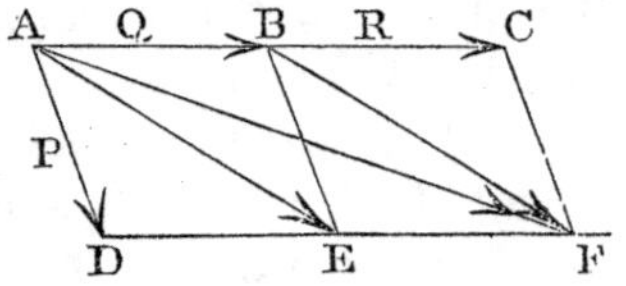

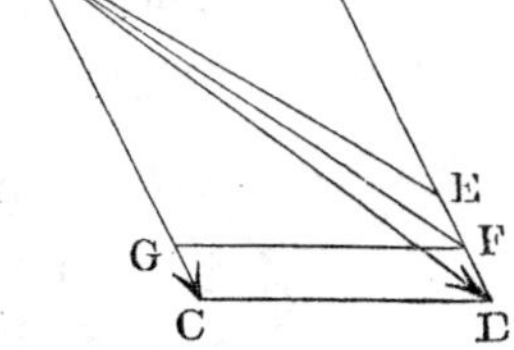

3°. When the forces are incommensurable. Let AB and AC represent the forces. Complete the parallelogram, and draw the diagonal AD. AD will represent the direction of the resultant. If not, let some other line, as AE, be its direction. Divide AB into a number of equal parts less than DE, and on AC take as many of these parts as possible. Since AC and AB are incommensurable, there will be a remainder GC less than DE. Draw GF parallel to AB, and join AF. AB and AG representing two commensurable forces, AF will represent the direction of their resultant. But (*Art.* 16) the resultant of AB and AC will make a less angle with AC than the resultant of AB and AG does. ∴ AE can not be the direction of the resultant of AB and AC; and similarly, it can be shown that no other direction than AD can be that of the resultant.

Second. The diagonal will represent the *magnitude* of the resultant. Let P, Q, and R be three forces in equilibrium, and AD, AB, AF represent their magnitudes respectively. Complete the parallelograms BD and DF, and draw the diagonals AC and AE. AC being the direction of the resultant of P and Q, must, by *Art.* 20, be in the same straight line with AF, and is therefore parallel to DE. In the same manner, AE being the direction of the resultant of P and R, will be in the direction of BA produced, and therefore parallel to CD. Hence CAED is a parallelogram, and CA is equal to DE, which is equal to AF, by construction. But, by *Art.* 20, AF represents the magnitude of the resultant of P and Q. Hence CA, the diagonal of the parallelogram constructed on the lines AD and AB, represents the resultant of P and Q, in *magnitude* as well as direction.

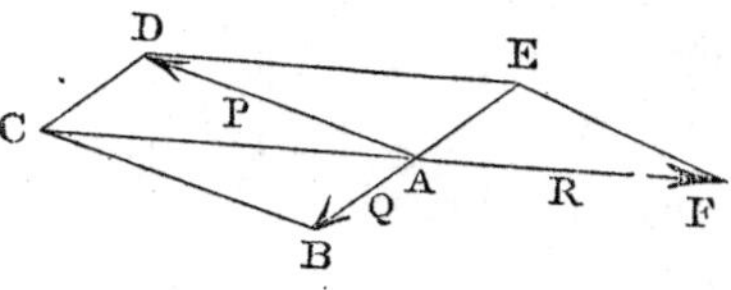

TRIANGLE OF FORCES.

22. Prop. *If three concurring forces are in equilibrium, and a triangle be formed by lines drawn in their directions, the sides of the triangle, taken in order, will represent the forces. Conversely, if the forces can be represented by the sides of a triangle, taken in order, they will be in equilibrium.*

Let the forces P, Q, and R be in equilibrium, and be represented by AB, AD, and AE respectively. Produce EA, draw BC parallel to AD, and complete the parallelogram. AC will represent the resultant of P and Q (*Art.* 21); and since R equilibrates P and Q, it must be equal and opposite to their resultant. ∴ EA = AC. And since BC is equal and

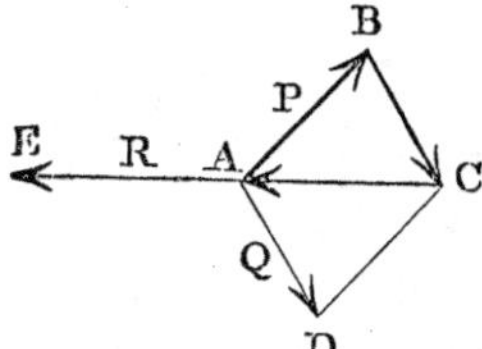

also parallel to AD, that is, has the same direction, the three sides of the triangle ABC taken in order represent the three forces P, Q, and R.

Conversely. If the three sides AB, BC, and CA of the triangle ABC, taken in order, represent the direction and magnitude of the three forces P, Q, and R, they will be in equilibrium. Draw from A a line parallel to BC, and from C a line parallel to AB, meeting the former in D. The resultant of the two forces P and Q, represented by AB and BC, or AB and AD, is equal to, and in the direction of the diagonal AC; that is, equal and opposite in direction to the force R. Hence P, Q, and R are in equilibrium.

23. Cor. Hence a given force may be resolved into two component forces, acting in given directions. Also into two others, one of which is given in magnitude and direction.

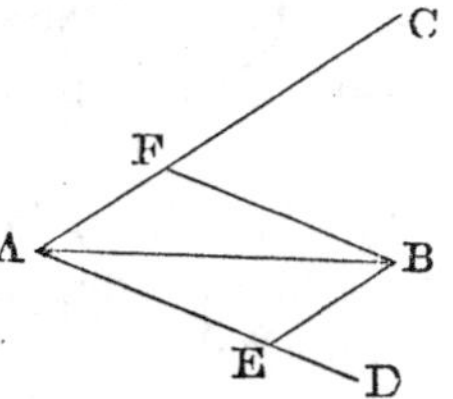

1°. Let AB be the given force, and AC, AD the given directions; that is, making known angles with AB. From B draw BF, and BE parallel to AD and AC. Then AF and AE are the two components (*Art.* 21) acting in the given directions AC and AD.

2°. To resolve AB into two others, one of which shall be in the direction AC, and be equal to AF. Draw FB, and complete the parallelogram. AE will be the other component.

When the directions of the components are arbitrary, their valuation will be most easily effected by assuming these directions at right angles. If the angle CAD be right, then AE =AB cos. BAE, and AF=AB cos. BAF.

POLYGON OF FORCES.

24. Prop. *If any number of concurring forces be represented in magnitude and direction by the sides of a polygon, taken in order, they will be in equilibrium.*

Let the sides of the polygon ABCDEA represent the magnitudes and have the directions respectively of the forces P P_2, P_3, P_4, and P_5. These forces will be in equilibrium.

Draw the diagonals AC and AD. By *Art.* 22, AC, the third side of the triangle ABC, represents a force equivalent to the two forces AB and BC. AC is therefore the resultant of P_1 and P_2, and may be substituted for them. In the same manner, AD is equivalent to the forces AC and CD, or to P_1, P_2, and P_3; and AE to the forces AD and DE, or to the forces P_1, P_2, P_3, and P_4. That is, AE is the resultant of the four forces P_1, P_2, P_3, and P_4, and acts in the direction AE. But P_5 is represented in magnitude by EA, and acts in the direction EA. Hence P_5 is equal and opposite to the resultant of the other four, or the forces are in equilibrium.

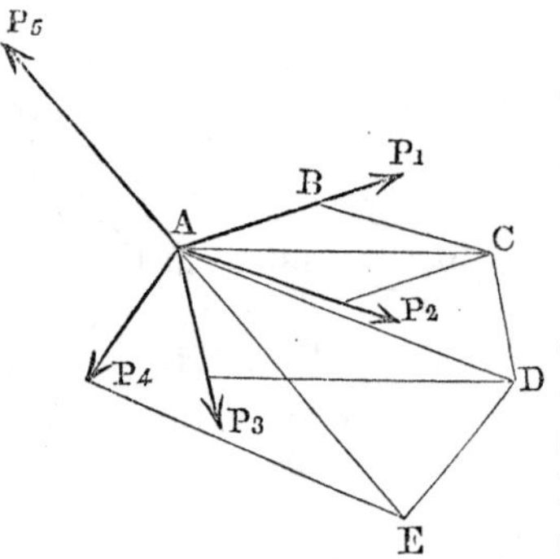

Cor. The proposition is true of forces which do not lie all in one plane. For the proof is independent of this supposition.

25. Scholium. When *three* forces are in equilibrium, *any* three lines, taken parallel to their directions, will form a triangle, the sides of which respectively will represent the *relative* magnitudes of the forces; but when there are four or more forces this will not hold, since the relation which subsists between the sides and angles of triangles does not obtain in polygons of more than three sides.

26. Prop. *To find*, graphically, *the resultant of any number of forces, acting at different points in the same plane.*

Let the forces P_1, P_2, P_3 have their points of application at A, B, and C. Producing the directions of the forces P_1 and P_2 till they meet in D, construct the parallelogram AB on the lines representing them, and the diagonal will represent their resultant R_1. Producing R_1 until it meets the direction of P_3 in F, and constructing the parallelogram on the lines representing R_1 and P_3, the diagonal will represent their resultant

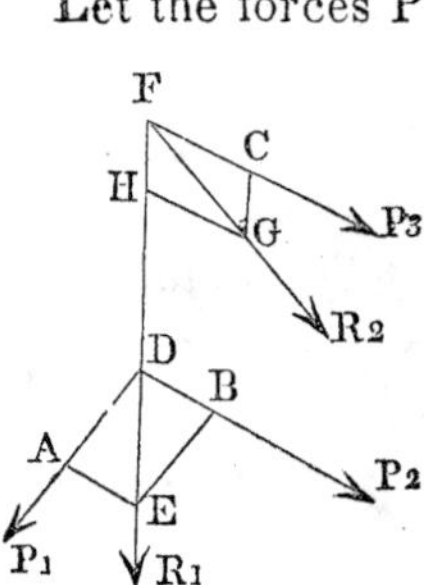

R_2, or the resultant of P_1, P_2, and P_3. By the same process we may find the resultant of any number of forces. If any two are parallel, we must first compound one of these with a third, whose direction is inclined to it. If all are parallel, the case will belong to *parallel forces.*

PARALLELOPIPED OF FORCES.

27. Prop. *If three concurring forces lying in different planes be represented in magnitude and direction by the three edges of a parallelopiped, then the diagonal will represent their resultant in magnitude and direction; and conversely, if the diagonal represents a force, it is equivalent to three forces represented by the edges of the parallelopiped.*

Let the three edges AB, AC, AD of the parallelopiped represent the three forces. Then AE, the diagonal of the face ACED, represents the resultant of the forces AD and AC. Compounding this with the third force, represented by AB, we have AF, the diagonal of the parallelogram AEFB, for the resultant of AE and AB, or of the forces AC, AD, AB.

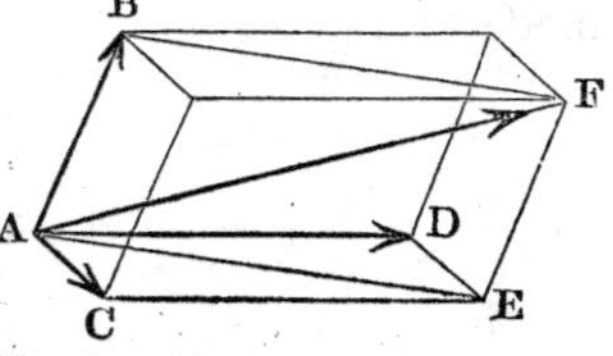

Reciprocally, the force AF is equivalent to the components AB, AE, or to the components AB, AC, and AD.

28. Prop. *If three forces are in equilibrium, they are proportional each to the sine of the angle made by the directions of the other two.*

By *Art.* 22 *and figure,*

$$P : Q : R = AB : BC : CA,$$

and by trig.,

$$= \sin. BCA : \sin. CAB : \sin. ABC;$$

$$= \sin. CAD : \sin. CAB : \sin. ABC;$$

or because $\sin. A = \sin. (180° - A)$,

$$= \sin. DAE : \sin. EAB : \sin. BAD;$$

$$= \sin. \widehat{QR} : \sin. \widehat{PR} : \sin. \widehat{PQ}.$$

29. Prop. *If* P *and* Q *be two concurring forces,* θ *the angle made by their directions, and* R *their resultant, then* $R^2 = P^2 + Q^2 + 2PQ \cos. \theta$.

By trig. we have, from *fig.*, *Art.* 22,

$$AC^2=AB^2+BC^2-2AB.BC \cos. ABC,$$

and $\cos. ABC=-\cos. BAD=-\cos. \theta.$

$$\therefore \quad R^2=P^2+Q^2+2PQ \cos. \theta.$$

30. Def. The *moment* of a force about any point is the product of the force into the perpendicular let fall from that point on the direction of the force. The point is called the *origin of moments.*

The moment of a force measures the tendency of the force to produce rotatory motion about a fixed point.

31. Prop. *The moment of the resultant of two forces equals the algebraic sum of the moments of the components.*

1°. When the origin of moments falls without the angle made by the forces.

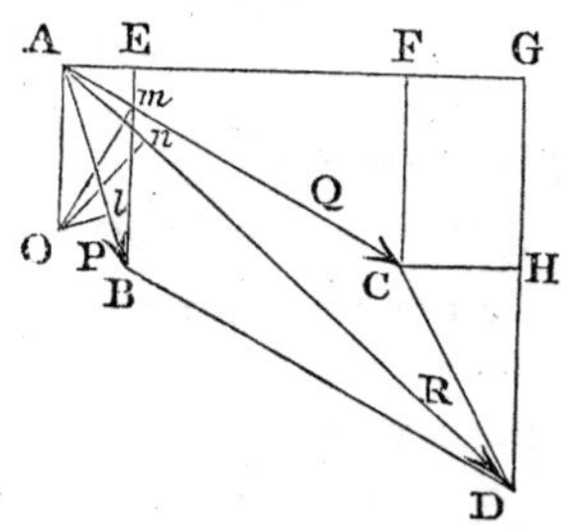

Let AB, AC represent the two forces P and Q, and AD the diagonal of the parallelogram constructed on AB and AC, their resultant R. Take any point, O, for the origin of moments; join OA and draw AG perpendicular to OA; draw Ol, Om, On respectively perpendicular to AB, AC, AD, and BE, CF, DG perpendicular to AG; also CH parallel to AG.

The triangles OlA, OmA, OnA are respectively similar to the triangles AEB, AFC, AGD.

Whence $AE : AB=Ol : OA$, or $AE=\frac{AB.Ol}{OA}$,

$AF : AC=Om : OA$, or $AF=\frac{AC.Om}{OA}$,

$AG : AD=On : OA$, or $AG=\frac{AD.On}{OA}$.

But the triangles ABE and CDH being equal, AE=CH=FG.

$$\therefore \quad AE+AF=AG,$$

or $AB.Ol+AC.Om=AD.On$, or designating Ol, Om, and On by p, q, and r respectively, $P.p+Q.q=R.r$.

2°. When the origin of moments is taken within the angle

In this case the moments of the forces tend to produce rotation in opposite directions. Assuming one direction as positive, to distinguish them we must regard the other as negative. Let the positive direction be that of the hands of a watch; then $Q.Om$ will be positive, and $P.Ol$ negative.

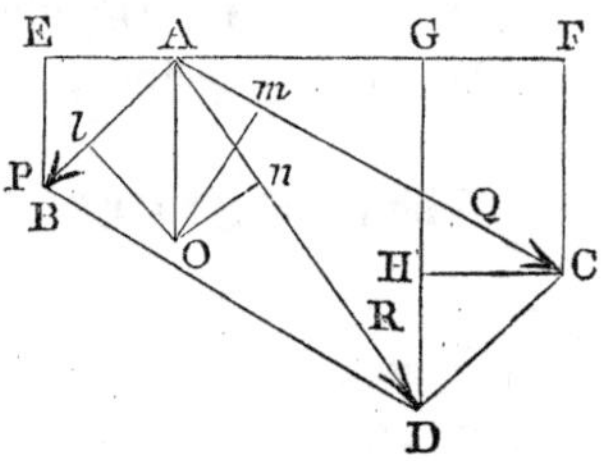

The proof is the same in this case, except that $AG=AF-AE$,

and $$AC.Om-AB.Ol=AD.On,$$

or $$Q.q-P.p=R.r.$$

32. Cor. 1. *The moment of the resultant of any number of concurring forces in the same plane, is equal to the algebraic sum of the moments of the components.*

By compounding the resultant R with a third force, we should obtain a like result. In the same manner, the proposition may be extended to any number of forces.

33. Cor. 2. If the origin of moments be a fixed point, and taken in the direction of the resultant, On will become zero, and

$$P.Ol=Q.Om, \text{ or } P.p=Q.q;$$

that is, while the fixed point O, by its resistance, counteracts the resultant force, the forces P and Q will be in equilibrium about that point, since their moments, tending to cause rotation in opposite directions, are equal.

34. Cor. 3. If several forces are in equilibrium, the resultant force R is zero, and the moment of the resultant $R.r=0$. Hence the moments in one direction balance those in the opposite direction, and there is no tendency to motion either of translation or rotation.

35. EXAMPLES.

1. When the component forces are P and Q, and the angle made by their directions θ; what is the magnitude of the resultant R when $\theta=0$ and $\theta=\pi$?

Ans. $(P+Q)$ and $(P-Q)$.

2. Show that R is greatest when $\theta=0$, least when $\theta=\pi$, and intermediate for intermediate values of θ.

3. When P=Q and $\theta=60°$, find R.

Ans. $R=P\sqrt{3}$.

4. When P=Q and $\theta=135°$, find R.

Ans. $R=P\sqrt{2-\sqrt{2}}$.

5. When the three concurring forces $3m$, $4m$, and $5m$ are in equilibrium, find the angle $\widehat{3m,4m}$.

Ans. 90°.

6. If P=Q and $\theta=120°$, find R.

Ans. R=P.

7. If P=6, Q=11, and $\theta=30°$, find the magnitude of R, and of the angles $\widehat{P,R}$ and $\widehat{Q,R}$.

Ans. R=16.47, $\widehat{P,R}=19°.30'$, $\widehat{Q,R}=10°.30'$.

8. Apply the proof of the polygon of forces to the case of five equal forces represented by the sides of a regular pentagon taken in order.

9. A cord is tied round a pin at the fixed point A, and its two ends are drawn in different directions by the forces P and Q. Find the angle $\widehat{P,Q}=\theta$, when the pressure on the pin is $R=\frac{P+Q}{2}$.

Ans. $\text{Cos.}\ \theta=\frac{2PQ-3(P^2+Q^2)}{8PQ}$.

10. A cord, whose length is $2l$, is tied at the points A and B in the same horizontal line, whose distance is $2a$; a smooth ring upon the cord sustains a weight w: find the force T of tension in the cord.

Ans. $T=\frac{w}{2\sqrt{1-\frac{a^2}{l^2}}}$.

11. Given the four concurring forces $P_1=1$, $P_2=2$, $P_3=3$, $P_4=4$, and the angles $\widehat{P_1,P_3}=90°$, $\widehat{P_2,P_4}=90°$, and $\widehat{P_1,P_2}=60°$. Find the magnitude of the resultant R, and its direction $\widehat{P_1,R}$.

Ans. R=6.889 and $\widehat{P_1,R}=102°.16'$

CHAPTER II.

PARALLEL FORCES.

HAVING considered the subject of forces which have a common point of application, or which are reducible to this condition, we next proceed to consider forces which act on different points, connected together in some invariable manner, as in rigid bodies, and whose directions are parallel.

36. PROP. *The resultant of two parallel forces, acting at the extremities of a rigid right line, is parallel to the components, equal to their algebraic sum, and divides the right line, or the right line produced, into parts reciprocally proportional to the forces.*

1°. When the forces act in the same direction.

Let the parallel forces P and P_1 act at A and B. At these points apply the equal and opposite forces P' and P'_1; these will not disturb the system. P and P' at A will have a resultant P'', and P_1 and P'_1 at B, a resultant P''_1. The directions of P'' and P''_1 will meet in some point D, at which we may suppose them to act.

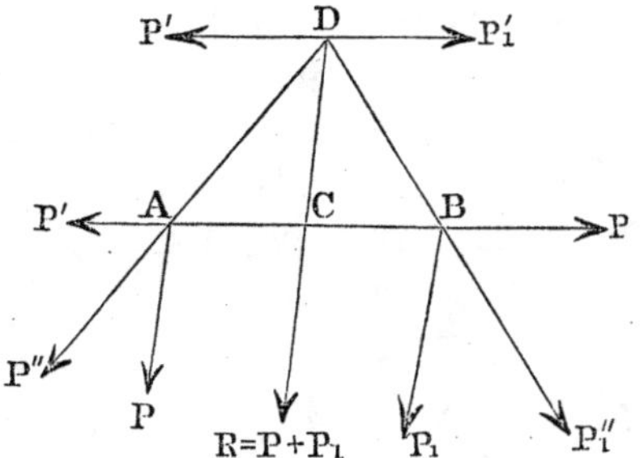

Replacing P'' by P and P', and P''_1 by P_1 and P'_1, we now have, acting at D, the four forces P, P', P_1, and P'_1, of which P' and P'_1 are equal and opposite, while P and P_1 act in the direction DC, and have a resultant $R=P+P_1$ (*Art.* 10). To determine the point C in the line AB, where R acts, since the sides of the triangle ACD have the directions respectively of the forces P, P', and their resultant P'' (*Art* 22), we have,

$$P : P' = DC : CA,$$

and similarly from triangle BCD,

$$P'_1 : P_1 = BC : DC.$$

Compounding these ratios, we have, since $P = P'_1$,

$$P : P_1 = BC : CA, \text{ or } P.AC = P_1.BC.$$

2°. When the forces act in opposite directions.

Let P and P_1 ($P_1 > P$) be two parallel forces, acting at A and B in opposite directions, and apply to these points, as before, the equal and opposite forces P' and P'_1. The resultant P'' of P and P' will meet the resultant P''_1 of P_1 and P'_1 in some point D since (*Art.* 16) P''_1 makes a less angle with P_1 than P'' does with P. We shall then have, as before, the four forces P, P_1, P', and P'_1, acting at D, of which P' and P'_1 are equal and opposite, while P and P_1 are opposite and unequal, and have a resultant $R = P_1 - P$, acting at any point in its direction, as C in AB produced.

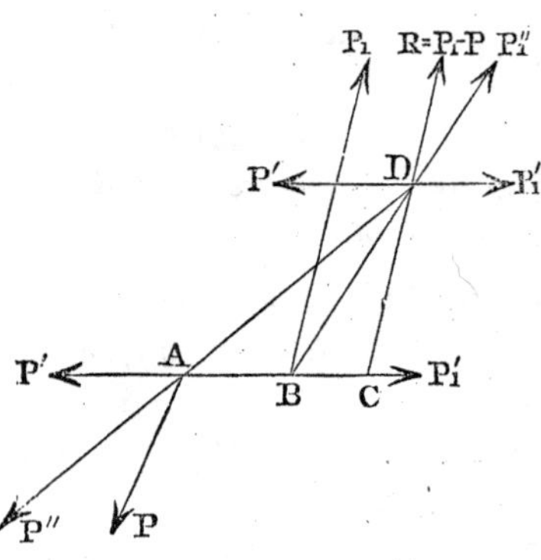

To determine the point C, we have (*Art.* 22), from the triangles ACD and BCD, as in *Case* 1°,

$$P : P' = DC : CA$$

and $\qquad P'_1 : P_1 = BC : CD;$

compounding $P : P_1 = BC : CA$, or $P.AC = P_1.CB$.

37. Def. When AB is perpendicular to the direction of the forces, AC and BC are called the *arms* of the forces, and the products P.AC and P_1.CB are the *moments* of the forces about the point C.

38. Cor. If to the point C we apply a force R_1, equal and opposite to R, the forces P, P_1, and R_1 will be in equilibrium about that point. For the resultant R, passing through C, will be counteracted by R_1, so there can be no motion of translation. And if we draw through C a line perpendicular to the

direction of the forces, calling the parts intercepted by P and P_1 respectively p and p_1; from the similar triangles, thus formed, we should have

$$BC : CA = p_1 : p.$$

$$\therefore\ P : P_1 = p_1 : p, \text{ or } P.p = P_1 p_1.$$

Hence the moments of P and P_1, which measure the tendency to rotation in opposite directions, being equal, there can be no motion of rotation.

√ 39. Prop. *To determine the point of application of the resultant in terms of the components and distance between their points of application.*

1°. When the forces act in the same direction. By *Art.* 36 *and figure*, we have

$$P.AC = P_1.BC = P_1.(AB - AC),$$

or
$$AC = \frac{P_1}{P + P_1}.AB;$$

and
$$P.AC = P_1.BC = P(AB - BC),$$

or
$$BC = \frac{P}{P + P_1}.AB.$$

2°. When the forces act in opposite directions, we have

$$P.AC = P_1.BC = P_1.(AC - AB),$$

or
$$AC = \frac{P_1}{P_1 - P}.AB;$$

and similarly,
$$BC = \frac{P}{P_1 - P}.AB.$$

√ 40. Cor. 1. When the forces act in *opposite* directions, the resultant lies without the components and on the side of the larger.

√ 41. Cor. 2. When the forces are *equal* and *opposite*, we have

$$R = P_1 - P = 0,$$

or there will be no motion of translation.

Also,
$$AC = \frac{P_1}{P_1 - P}.AB = \frac{P_1}{0}.AB = \infty.$$

Hence an equilibrium can not be produced except by the application of an infinitely small force at a point whose distance is infinite; that is, two equal and parallel forces, acting in opposite directions, can have no single resultant.

42. Def. Such forces are called a *statical couple.* Their effect is, tendency to rotatory motion only, and all tendency to rotatory motion can be referred to forces forming such *couples.*

43. Prop. *To find the resultant of any number of parallel forces acting at any points in the same plane.*

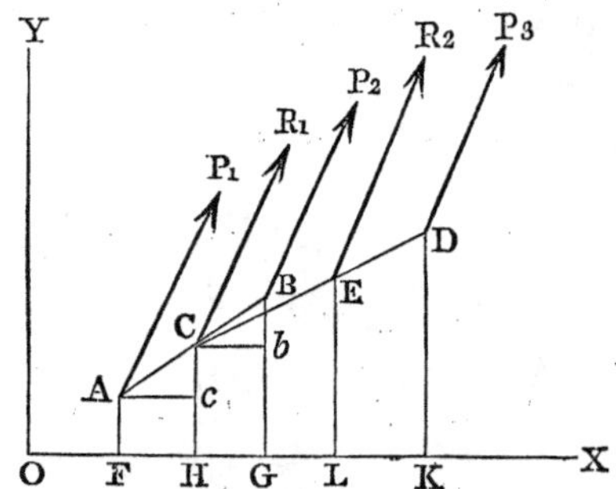

1°. Let the parallel forces $P_1, P_2, P_3 \ldots P_n$ all act in the *same* direction. Draw any two rectangular axes OX, OY, and let

$x_1\ y_1$	be the co-ordinates of		A,	the point of application of			P_1
$x_2\ y_2$	"	"	B,	"	"	"	P_2
$x_3\ y_3$	"	"	D,	"	"	"	P_3
....			...				...
$x_n\ y_n$	"	"	N,	"	"	"	P_n
also,							
$x'\ y'$	"	"	C,	"	"	"	R_1
$x''\ y''$	"	"	E,	"	"	"	R_2
....			...				...
$\bar{x}\ \bar{y}$	"	"	R,	"	"	"	R.

Draw A*c* and C*b* parallel to OX. The triangles AC*c* and CB*b* are similar, and give

$$AC : BC = Cc : Bb.$$

By *Art.* 36, $$R_1 = P_1 + P_2,$$

and $P_1 : P_2 = BC : AC = Bb : Cc$, or $P_1.Cc = P_2.Bb$,

$$P_1.(CH-AF)=P_2.(BG-CH),$$

$$(P_1+P_2)CH=P_1AF+P_2.BG;$$

or

$$R_1.y'=P_1y_1+P_2.y_2.$$

Compounding R_1 with P_3, we shall find the second resultant,

$$R_2=R_1+P_3=P_1+P_2+P_3,$$

and

$$R_2.EL=R_1.CH+P_3.DK,$$

$$R_2.EL=P_1AF+P_2.BG+P_3DK;$$

or

$$R_2.y''=P_1y_1+P_2y_2+P_3y_3.$$

By continuing the same process, we should find, ultimately,

$$R=P_1+P_2+P_3+\ldots\ldots P_n \qquad (a),$$

$$R\bar{y}=P_1y_1+P_2y_2+P_3y_3+\ldots\ldots P_ny_n \qquad (c).$$

By drawing lines from A, C, B, E, D, &c., parallel to OX, we may find, similarly,

$$R\bar{x}=P_1x_1+P_2x_2+P_3x_3+, \&c. \ldots\ldots +P_nx_n \qquad (b).$$

2°. When some of the forces act in opposite directions. To these the negative sign must be prefixed.

If R_1, the resultant of P_1 and P_2, as found in the previous case, be compounded with P_3 acting at D, in a direction opposite to that of P_1 and P_2, by joining CD and producing it (*Art.* 40), in the direction of the greater force, say R_1, we have R_2, the second resultant.

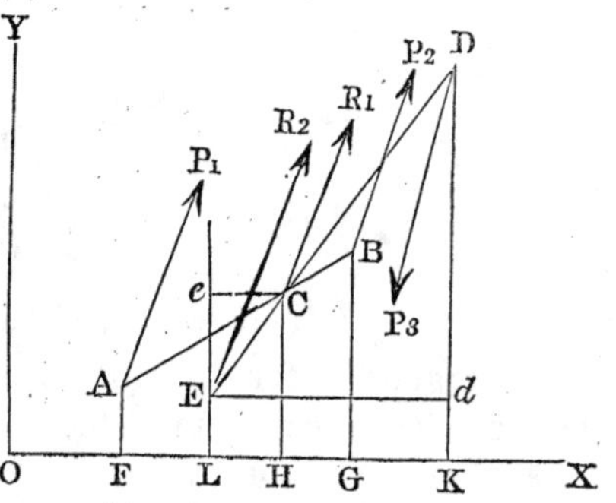

$R_2=R_1-P_3$, and E being its point of application, $R_1.EC = P_3.ED$. Drawing Ce, Ed, parallel to OX, and meeting DK in d and LE produced in e, the triangles ECe, EDd are similar; and

$$EC : DE = Ee : Dd.$$

$$\therefore R_1.Ee=P_3.Dd;$$

or

$$R_1(CH-EL)=P_3.(DK-EL),$$

or $$(R_1 - P_3).EL = R_1.CH - P_3.DK,$$

or $$R_2.EL = P_1.AF + P_2 BG - P_3.DK,$$

or $$R_2.y'' = P_1 y_1 + P_2 y_2 - P_3 y_3,$$

and so on for any other forces acting in directions opposite to P_1 and P_2.

3°. When the point of application of either force lies in any other angle than YOX, to its co-ordinates must be given their appropriate sign.

The ordinate y_2 of B, the point of application of P_2, will be negative. Draw Aa, Cb parallel to OX.

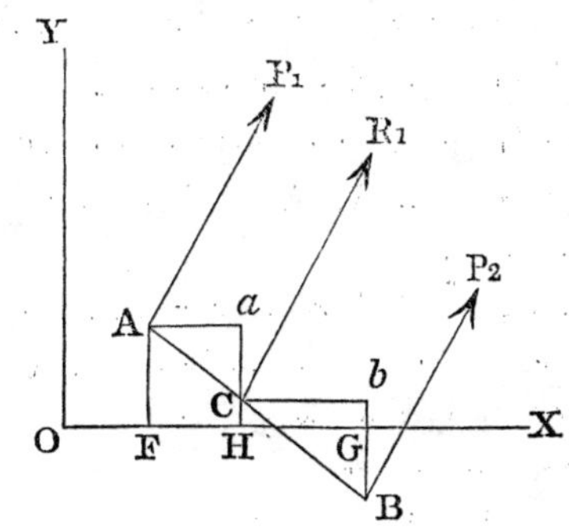

$$P_1.CA = P_2.CB.$$

By sim. triangles, $$P_1.Ca = P_2.Bb;$$

or $$P_1(AF - CH) = P_2(CH + GB),$$

or $$(P_1 + P_2).CH = P_1.AF - P_2.BG,$$

or $$R_1 y' = P_1 y_1 - P_2 y_2.$$

∴ The general formulæ (*a*), (*b*), (*c*) will apply to all cases, by giving the proper signs to the forces and to the co-ordinates of their points of application.

These formulæ are more concisely written by using the Greek letter Σ as the sign of summation, P to represent either force, and x, y its co-ordinates; thus,

$$R. = \Sigma(P) \qquad (a'),$$

$$R.\bar{x} = \Sigma(P.x) \qquad (b'),$$

$$R.\bar{y} = \Sigma(P.y) \qquad (c').$$

When, therefore, the magnitudes of the forces and the co-or

dinates of their points of application are given, (*a*) will give the magnitude of the resultant, and (*b*) and (*c*) the co-ordinates $\bar{x}, \bar{y}$ of its point of application.

44. Def. The point whose co-ordinates are $\bar{x}, \bar{y}$ is called *the center of parallel forces.* Its position depends on the magnitude of the forces and the co-ordinates of their points of application, but is independent of their common direction; for, by turning the forces around their respective points of application, at the same time preserving their parallelism, it will be seen that the position of the point E is not thereby affected.

45. Prop. *The moment of the resultant of any number of parallel forces, acting in the same plane, is equal to the algebraic sum of the moments of the components.*

Since the origin of the co-ordinate axes and their direction are arbitrary (*see figure of Art.* 43), suppose the axis OX to be drawn perpendicular to the direction of the forces, and the forces produced to intersect it. Then P_1x_1, P_2x_2, &c., will represent the moments of P_1, P_2 respectively, and $R\bar{x}$ the moment of the resultant.

46. Def. *The moment of a force with reference to a plane* is the product of the intensity of the force by the perpendicular let fall from the point of application upon the plane. Thus P_1x_1 is the moment of the force P_1, in reference to the plane passing through OY perpendicular to OX, and P_1y_1 is the moment of the force P_1, with reference to the plane through OX perpendicular to OY.

47. Prop. *To determine the conditions of equilibrium of any number of parallel forces.*

Suppose the forces in the figure (*Art.* 43) all turned round their points of application so as to become perpendicular to the plane of YOX. They will no longer be in the same plane, but as their intensity is not thus changed, nor their points of application, (a'), (b'), (c') will still hold. Then,

1°. We must obviously have $R=0$ in (a'), or

$$P_1+P_2+P_3+, \&c.=0.$$

2°. If this value of R be substituted in (b'), we have

$$\bar{x}=\frac{\Sigma.Px}{R}=\frac{\Sigma.Px}{0}.$$

Therefore $\bar{x}$ will be infinite unless $\Sigma.Px=0$. But when $\bar{x}$ is infinite, we have a *couple* (*Art.* 41), and, consequently, there can be no equilibrium. Hence we must also have $\Sigma.Px=0$, or

$$P_1x_1+P_2x_2+P_3x_3+, \&c.=0.$$

And in the same manner it will appear that $\Sigma.Py=0$ in case of an equilibrium. But $\Sigma.Px$ and $\Sigma.Py$ are the moments of the forces in reference to two planes parallel to their directions.

Hence the necessary conditions of equilibrium are,

1°. *The sum of the forces must be equal to zero.*

2°. *The sum of their moments, with reference to two planes parallel to their directions, must each be equal to zero.*

48. Cor. 1. If $R=0$, but $\Sigma.Px$ and $\Sigma.Py$ are not equal to zero, there will be no motion of translation, but simply a motion of rotation.

For if R_1 be the resultant of all the positive forces, and $\bar{x}_1$ the abscissa of its point of application, R_2 the resultant of all the negative forces, and $\bar{x}_2$ its abscissa,

then $$R_1\bar{x}_1-R_2\bar{x}_2 \gtrless 0;$$

or, since $R_1=R_2$, $$R_1(\bar{x}_1-\bar{x}_2) \gtrless 0.$$

But R_1 being finite, $\bar{x}_1-\bar{x}_2$ must also be finite.

That is, the points of application of R_1 and R_2 are not the same, and a tendency to motion of rotation around the axis of Y exists.

And $$\Sigma.Py \gtrless 0$$

gives, in like manner, a tendency to motion around the axis of x.

49. Cor. 2. If the equilibrium subsist, one force, as P_1, must be equal and opposite to the resultant of all the others.

50. Cor. 3. If the equilibrium subsist with one direction, it will subsist whatever be the common direction of the forces.

51. EXAMPLES.

1. Two parallel forces, acting in the same direction, have their magnitudes 5 and 13, and their points of application A

and B 6 feet distant. Find the magnitude of their resultant, and its point of application C.

Ans. $R=18$,
$AC=4\frac{1}{3}$,
$BC=1\frac{2}{3}$.

2. Find the resultant, and its point of application, when the forces in the last question act in opposite directions.

Ans. $R=8$,
$AC=9\frac{3}{4}$,
$BC=3\frac{3}{4}$.

3. If two parallel forces, P and Q, act in the same direction at A and B, and make an angle θ with AB, find the moment of each about the point of application of their resultant.

Ans. $\frac{P.Q}{P+Q}AB \sin. \theta$.

4. If the weights 1, 2, 3, 4, and 5 lbs. act perpendicularly to a straight line at the respective distances of 1, 2, 3, 4, and 5 feet from one extremity, required their resultant and its point of application.

Solution. From equations (*a*) and (*b*) we have

$$R=\Sigma.P=1+2+3+4+5=15 \text{ lbs.}$$

And, taking the extremity of the line for the origin of co-ordinates,

$$R\bar{x}=15.\bar{x}=\Sigma.Px=1\times1+2\times2+3\times3+4\times4+5\times5=55.$$

$\therefore\ \bar{x}=3\frac{2}{3}$ feet.

5. Let the weights 4, -7, 8, and -3 lbs. act perpendicularly to a straight line at the points A, B, C, and D, so that $AB=5$ feet, $BC=4$ feet, and $CD=2$ feet; find the resultant and its point of application E.

Ans. $R=2$ lbs.,
$AE=2$ feet.

6. Let three forces which, if concurring, would be in equilibrium, act each in the side of the triangle which represents them in magnitude and direction. Show that they are equivalent to a statical couple.

CHAPTER III.

THEORY OF COUPLES.

52. Def. A *statical couple* consists of two equal and parallel forces acting in opposite directions at different points of a body.

53. Def. The *arm* of a couple is the perpendicular distance between the directions of the forces.

54. Def. The *moment* of a couple is the product of the arm by one of the forces.

55. Prop. *A couple may be turned round in any manner in its own plane without altering its statical effect.*

Let P_1ABP_2 be the original couple. Suppose the arm AB turned around any point in it to the position *ab*. Apply the equal and opposite forces P_3 and P_4 perpendicularly to *ab* at *a*, and similarly P_5 and P_6 at *b*, and let each be equal to P_1 or P_2. These forces, being in equilibrium, will not disturb the system. The two equal forces P_1 at A, and P_4 at *a*, may be regarded as acting at their point of intersection E, and will have a resultant in the direction CE bisecting the angle P_1EP_4. Similarly, the resultant of P_2 and P_6 at D will be an equal force in the direction CD. These forces, being equal and opposite, may be removed; that is, we may remove from the system the forces P_1, P_2, P_4, and P_6, and we have remaining the forces P_3 and P_5 at *a*

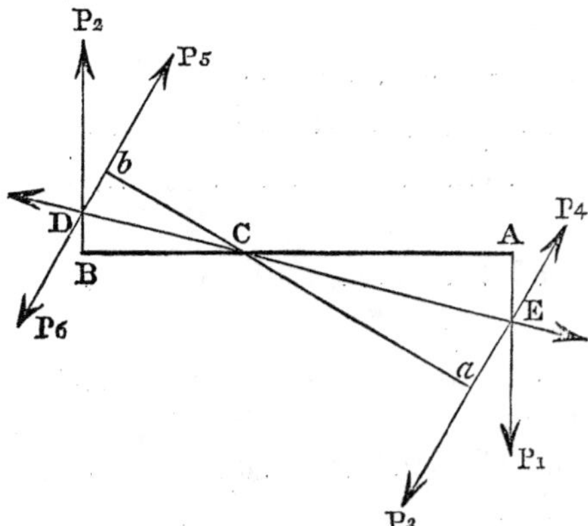

and b, forming the couple P_3abP_5, which is the same as if we had turned the original couple round the point C until its arm came to the position ab.

56. Prop. *A couple may be removed to any position in its own plane, the parallelism of its arm being preserved, without changing its statical effect.*

Let the arm AB of the couple P_1ABP_2 be removed in its own plane to the parallel position ab, and let the forces P_3, P_4, P_5, P_6, each equal to the original forces, be applied perpendicularly to ab, at the extremities a and b, in opposite pairs.

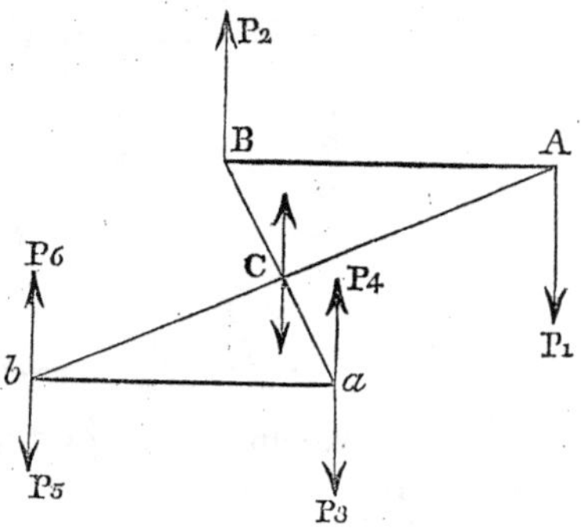

Join Ab and aB. These lines will bisect each other in C. P_1 at A and P_5 at b will have a resultant $2P_1$ at C, parallel to the original direction. Similarly, P_2 at B and P_4 at a have a resultant $=2P_2$ at C, opposite to the former; these will consequently balance each other, and may therefore be removed, or the forces P_1, P_2, P_4, P_5 may be removed, and we have remaining the couple P_3abP_6, equivalent to the original couple, removed parallel to itself in its own plane.

57. Prop. *A couple may be removed into any other plane, parallel to the original one, without altering its statical effect, the parallelism of its arm being preserved.*

Let the arm AB of the original couple be transferred from its own plane MN parallel to itself to ab in the parallel plane RS. Let forces be applied at a and b, as in the preceding proposition, each being equal to P_1 or P_2. Join Ab and aB. These lines will bisect each other in C. The forces P_1 and P_6 will have a resultant $=2P_1$ at C, and P_2 and P_4 a resultant $=2P_2$ at C. These resultants will be equal and opposite and may therefore be removed without disturbing the statical effect of the system. We have then remaining the couple

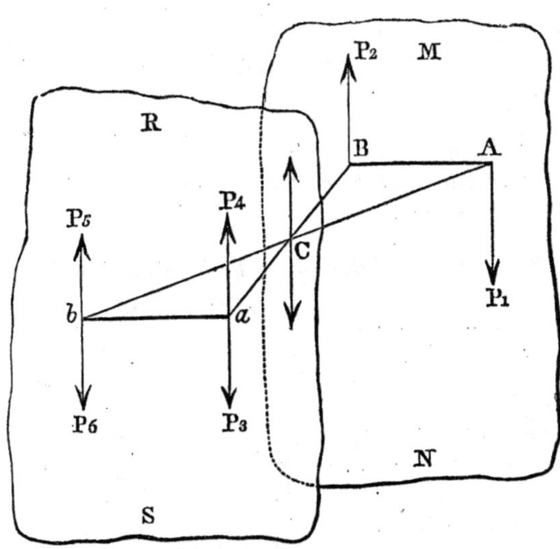

P_3abP_5, equivalent to the original couple, and transferred to the plane RS.

58. Prop. *All statical couples are equivalent to each other whose planes are parallel and moments equal.*

Let P_1ABP_2 and Q_1DEQ_2 be any two couples whose planes are parallel and moments equal. If they are not in the same plane, by the foregoing propositions let the latter be transferred to the plane of the former, and moved in that plane until the extremities D and A of their arms coincide, and DE take the position AC. Apply at C, perpendicularly to AC, two equal and opposite forces Q_3 and Q_4, each equal to Q_1. Resolve P_1 into two forces P' and Q_1, so that $P_1 = P' + Q_1$, or $P_1 - Q_1 = P'$.

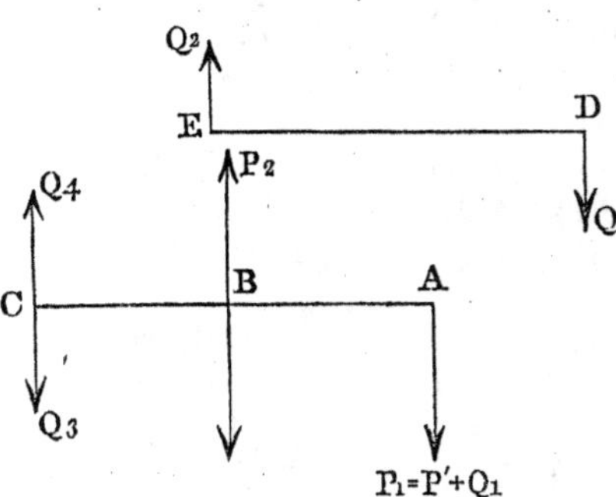

Then, by hypothesis,

$P_1.AB = Q_1.DE = Q_1.AC = Q_1.AB. + Q_1.BC = Q_1.AB + Q_3.BC$

and $(P_1 - Q_1)AB = P'.AB = Q_3.BC,$

or the resultant of P′ and Q_3 (*Art.* 36) passes through B. It is also equal and opposite to P_2. Hence the forces P′, Q_3, and P_2 may be removed, and there will remain the couple Q_1ACQ_4, equivalent to P_1ABP_2.

59. Prop. *Any statical couple may be changed into another, which shall be equivalent and have an arm of given length.*

Let P_1p_1 be the moment of any couple, of which P_1 is one of the forces and p_1 its arm, and let p be the given arm. Find a fourth proportional P'_1 to p, p_1, and P_1; or take $p : p_1 = P_1 : P'_1$, which gives $P'_1p = P_1p_1$. Hence, by *Art.* 58, their moments being equal, and in the same plane, the couples are equivalent.

60. Def. The *axis of a couple* is a line perpendicular to the plane of the couple.

If the length of the axis be taken proportional to the moment of the couple, and drawn above its plane when the couple is positive or tends to produce rotatory motion in the direction of the hands of a watch, and below its plane, when negative or tends to produce motion in the opposite direction, then the axis will completely represent the couple in position, intensity, and sign.

61. Cor. By considering the previous propositions, it will be obvious that the axis of a couple, as thus defined and limited, may be removed parallel to itself, to any position within the body acted on by the couple.

62. Def. The *resultant* of two or more couples is one which will produce the same effect singly as the couples themselves jointly.

63. Prop. *The moment of the resultant of two or more couples in the same or parallel planes, equals the algebraic sum of the moments of the component couples.*

Let P_1, P_2, P_3, &c., be the forces; p_1, p_2, p_3, &c., their arms respectively. The couples may all be removed into one

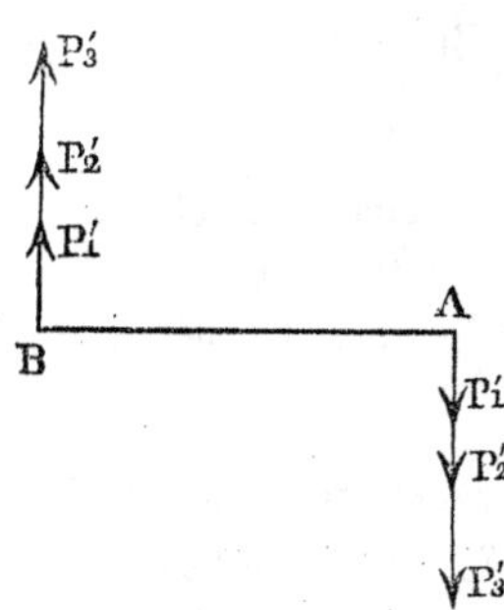

plane (*Art.* 57), turned round in that plane (*Art.* 55), moved in it (*Art.* 56), and their arms changed (*Art.* 59) to a common arm, while their moments remain the same. Let p be the common arm AB; P'_1, P'_2, P'_3, &c., the forces; so that $P'_1p=P_1p_1$, $P'_2p=P_2p_2$, $P'_3p=P_3p_3$, &c. Now the forces P'_1, P'_2, P'_3 at A are equivalent to a force

$$P'_1+P'_2+P'_3+, \&c., =\frac{P_1p_1}{p}+\frac{P_2p_2}{p}+\frac{P_3p_3}{p}+, \&c.$$

And the forces at B are equal to the same sum. We have then for the moment of the resultant couple,

$$(P'_1+P'_2+P'_3+, \&c.)\,AB=(P'_1+P'_2+P'_3+)p,$$
$$=P_1p_1+P_2p_2+P_3p_3+, \&c.$$

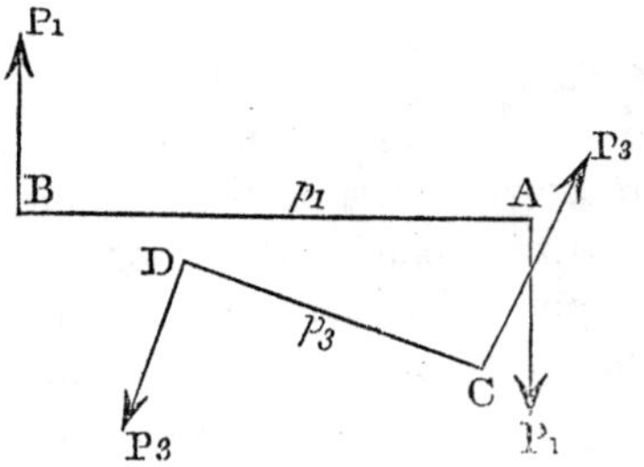

If either of the original couples, as P_3p_3, tend to produce motion in the opposite direction, its sign will be negative. The sum of the forces will be

$$P'_1+P'_2-P'_3+, \&c.,$$

and the moment of the resultant couple

$$(P'_1+P'_2-P'_3+, \&c.)p=P_1p_1+P_2p_2-P_3p_3+, \&c.$$

64. Cor. The axis of the resultant couple will be equal to the algebraic sum of the axes of the component couples.

Schol. The composition of couples in the same or parallel planes, by means of their axes, is therefore analogous to the composition of *conspiring* forces.

65. Prop. *To find the resultant of two couples in different planes inclined to each other.*

Let each couple be turned round and moved in its own plane until its arm coincide with the common intersection of their planes, and let the arms be made coincident and of the same length, the moments of the couples remaining the same. Let AB be this common arm, and PABP, QABQ the couples. Complete the parallelograms on the lines representing P and Q, and the diagonals will represent their resultants R, which may be substituted for them. Hence we have the resultant couple RABR equivalent to the component couples PABP and QABQ.

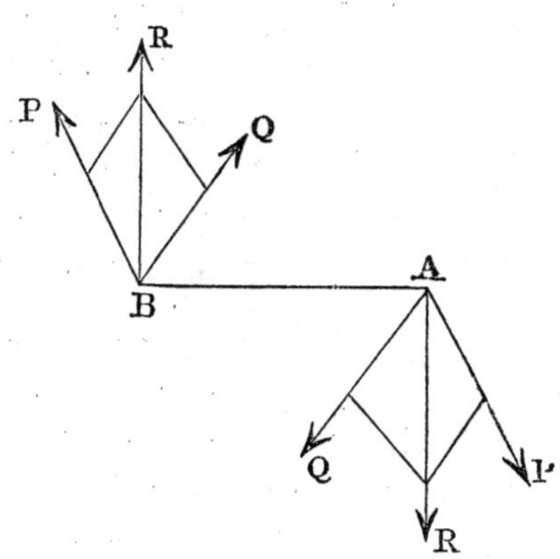

66. Cor. The diagonal of the parallelogram, constructed on the lines representing the axes of the component couples, will represent the axis of a couple equivalent to them.

Let OK and OL be the axes of the couples, and θ the angle made by the planes of the couples. Since the axes are perpendicular to these planes, the angle made by the axes will be θ. We have, therefore, θ=PAQ=LOK.

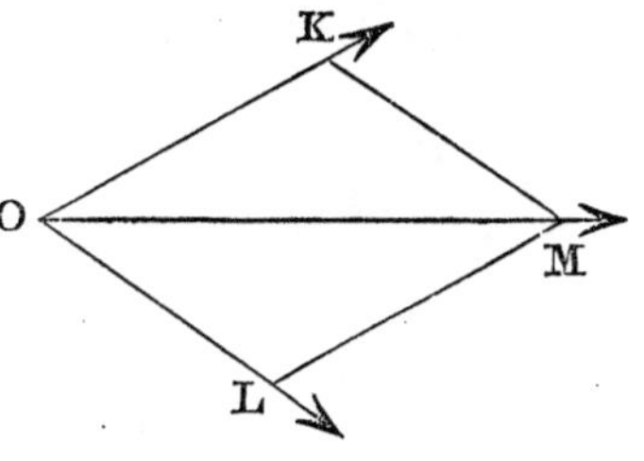

By the triangle of forces,

$$R^2=P^2+Q^2+2PQ\cos.\theta.$$

$$\therefore R.AB=AB\sqrt{P^2+Q^2+2PQ\cos.\theta},$$

$$=\sqrt{P^2.AB^2+Q^2.AB^2+2P.AB\times Q.AB\cos.\theta};$$

(since OK=P.AB, &c.),

$$=\sqrt{OK^2+OL^2+2OK.OL\cos.\theta},$$

$$=OM.$$

And OK, OL being respectively perpendicular to the planes of the component couples, OM will be perpendicular to the resultant couple.

67. Schol. The above is analogous to the parallelogram of forces, and may be called the parallelogram of couples.

If L and M represent the axes or moments of the component couples, and G the axis or moment of their resultant, then

$$G^2 = L^2 + M^2 + 2LM \cos. \theta.$$

If L, M, N were the axes of three component couples, it might be shown that G, the axis of the resultant couple, would be represented in magnitude by the diagonal of the parallelopiped formed upon them.

If the planes of the three couples are at right angles each to the other two, or if L, M, and N are at right angles each to the planes passing through the other two, the parallelopiped would be rectangular, and we should have,

$$G^2 = L^2 + M^2 + N^2.$$

CHAPTER IV.

ANALYTICAL STATICS IN TWO DIMENSIONS.

IN this chapter the points of application and the directions of the forces will be referred to co-ordinate axes at right angles to each other.

68. PROP. *To find the magnitude and direction of the resultant of any number of concurring forces in the same plane.*

Let P_1, P_2, P_3, &c. . . P_n be the n forces, and let the point

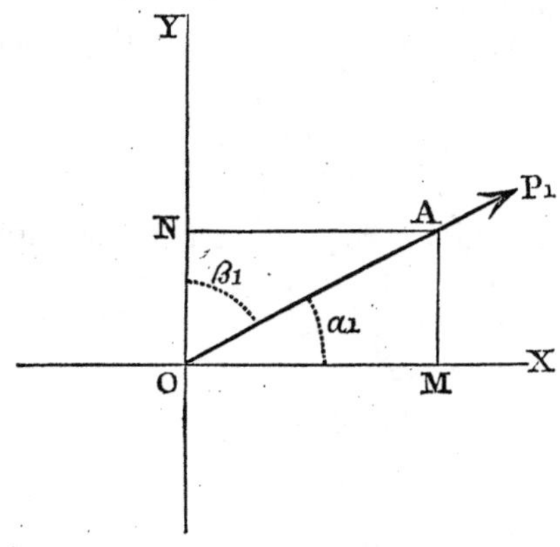

at which they act be taken for the origin of co-ordinates.

Let P_1 make the angle α_1 with OX, and β_1 with OY.

P_2	"	α_2	"	β_2	"
P_3	"	α_3	"	β_3	"
&c.,		&c.,		&c,	
P_n	"	α_n	"	β_n	"

If OA represent the force P_1, and the parallelogram OMAN be completed, OM will be the component of P_1 in OX, and ON that in OY. And

$$OM=X_1=P_1 \cos. a_1, \qquad ON=Y_1=P_1 \cos. \beta_1$$

Pursuing the same course with the forces P_2, P_3, &c., and calling the resolved parts of the forces respectively in OX,

$$X_2, X_3, X_4, \&c. \ldots X_n,$$

and those in OY $\;Y_2, Y_3, Y_4, \&c. \ldots Y_n,$

we shall obtain

$$\begin{array}{ll} X_1=P_1 \cos. a_1, & Y_1=P_1 \cos. \beta_1, \\ X_2=P_2 \cos. a_2, & Y_2=P_2 \cos. \beta_2, \\ X_3=P_3 \cos. a_3, & Y_3=P_3 \cos. \beta_3, \\ \&c., \quad \&c., & \&c., \quad \&c., \\ X_n=P_n \cos. a_n, & Y_n=P_n \cos. \beta_n. \end{array}$$

But the components in OX are equivalent to a single force,

$$=X_1+X_2+X_3+, \&c. \ldots X_n=\Sigma.X,$$

and those in OY to a single force,

$$=Y_1+Y_2+Y_3+, \&c. \ldots Y_n=\Sigma.Y,$$

the Greek letter Σ being used as the sign of summation. Hence we have

$$\Sigma.X=\Sigma.P \cos. a, \qquad (1)$$

$$\Sigma.Y=\Sigma.P \cos. \beta. \qquad (2)$$

Now if R be the resultant required, and θ the angle it makes with OX, the resolved parts of R in the axes must equal the resolved parts of the forces in the same directions.

$$\therefore\ R \cos. \theta=\Sigma.X, \qquad (3)$$

$$R \sin. \theta=\Sigma.Y. \qquad (4)$$

Hence $$\text{Tan. } \theta=\frac{\Sigma.Y}{\Sigma.X}. \qquad (5)$$

Also, $$R^2. \cos.^2 \theta+R^2. \sin.^2 \theta=R^2=(\Sigma.X)^2+(\Sigma.Y)^2;$$

or $$R=\sqrt{(\Sigma.X)^2+(\Sigma.Y)^2}. \qquad (6)$$

Equations (1) and (2) give the values of $\Sigma.X$ and $\Sigma.Y$, by which we obtain θ from (5) and R from (6); or R may be obtained, numerically, more readily from the equation

$$R \cos. \theta=\Sigma.X.$$

For $$R=\frac{\Sigma.X}{\cos. \theta}=\Sigma.X \sec. \theta. \qquad (7)$$

69. SCHOL. It is readily seen that the signs of the compo-

nents along either axis are involved in the trigonometrical expressions of their values, (1) and (2). Assume the directions OX and OY positive, and OX′, OY′ negative. The components Ox_1, Oy_1 of P_1 are positive; their values, $P_1 \cos. a_1$, $P_1 \cos. \beta_1$, will also be positive, since the cosines of angles in the first quadrant are positive. The components of P_2, which lies in the angle X′OY, are Ox_2, Oy_2, the former negative and the latter positive. The value of $Ox_2 = P_2 \cos. a_2$ is negative, since a_2, reckoned from OX around to the left, will be in the second quadrant; and that of $Oy_2 = P_2 \cos. \beta_2$ positive, since β_2, reckoned from OY around to the right, will be in the fourth quadrant; or, if reckoned from OY around to the left, will be negative, and less than 90°.

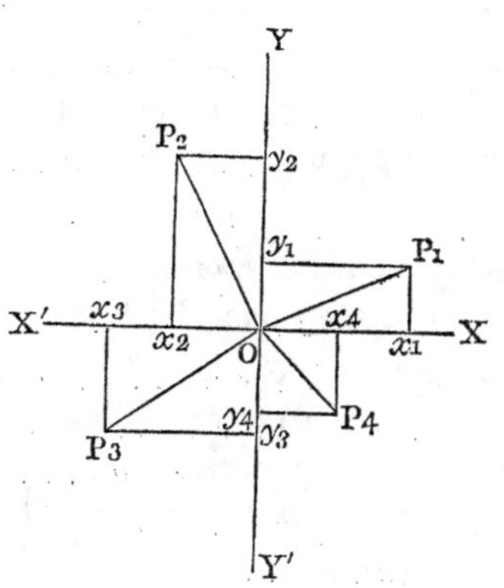

The components Ox_3, Oy_3 of P_3, in the angle X′OY′, are both negative, and their values $P_3 \cos. a_3$, $P_3 \cos. \beta_3$ are negative, because the angles a_3 and β_3 are both in the third quadrant. The components Ox_4, Oy_4 of P_4, which lies in the angle Y′OX, are, the former positive and the latter negative. The value of $Ox_4 = P_4 \cos. a_4$ is positive, because a_4 is in the fourth quadrant, or, if taken negatively, is less than 90°; that of $Oy_4 = P_4 \cos. \beta_4$ is negative, because β_4 is in the second quadrant. By the above mode of reckoning the angles, we always find $a+\beta=90°$, or one the complement of the other.

70. PROP. *Required the conditions of equilibrium of any number of concurring forces in the same plane.*

Since the forces are in equilibrium, their resultant must be equal to zero, or R=0. This gives, by (6),

$$(\Sigma.X)^2+(\Sigma.Y)^2=0.$$

But each term being a square, is essentially positive. The equation, therefore, can only be satisfied by making each term equal to zero at the same time.

Hence $\Sigma.X=0, \qquad \Sigma.Y=0,$ (8)

are *the two necessary and sufficient equations of equilibrium* of any number of concurring forces in the same plane; that is, the sums of the components of all the forces resolved in any two rectangular directions, must be separately equal to zero.

71. Prop. *Required the expressions for the resultant force and resultant couple of any number of forces acting at different points in the same plane.*

Let P_1, P_2, &c. . . . P_n be any number of forces in the same plane. Take the co-ordinate axes OX, OY in the plane of the forces, and let $\alpha_1, \alpha_2, \alpha_3 \ldots \alpha_n$ be the angles the forces make respectively with OX, and $\beta_1, \beta_2, \beta_3 \ldots \beta_n$ the angles they make with OY.

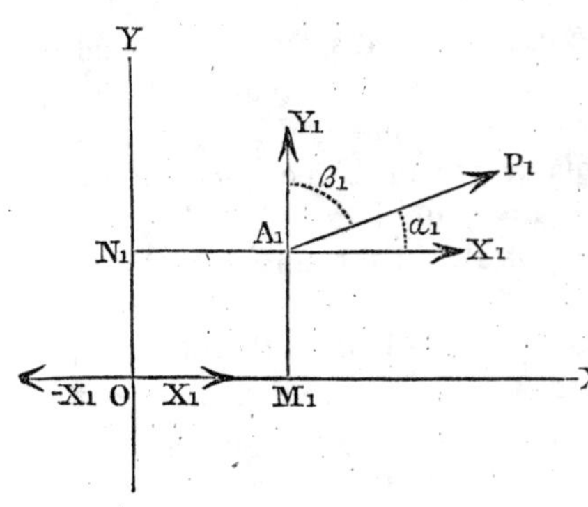

Let x_1, y_1 be the co-ordinates OM_1, M_1A_1 of A_1, the point of application of the force P_1, and x_2, y_2, &c. . . . x_n, y_n be the co-ordinates of the points of application of the others respectively. Resolving P_1 in the directions of OX and OY, we shall have for the components

$$X_1=P_1 \cos. \alpha_1, \qquad Y_1=P_1 \cos. \beta_1,$$

at the point A_1.

Now apply at O, in OX, two equal and opposite forces, each equal to X_1; this will not affect the system. Then, instead of the single force X_1 at A_1, we have X_1 at A_1, or N_1 and $-X_1$ at O, which together form a couple, and X_1 at O. That is, for the force X_1 at A_1, we may substitute

$$X_1 \text{ at O, and the couple } X_1.ON_1=X_1y_1.$$

In like manner, applying at O, in OY, two forces, opposite and each equal to Y_1, we should have Y_1 at A_1 *equivalent* to

$$Y_1 \text{ at O, and the couple } -Y.OM_1=-Y.x_1.$$

This last couple will be negative, since it tends to produce rotation in the contrary direction to that of the other.

By pursuing the same course with each of the other forces we should obviously have acting at O, in OX, a sum of forces

$$X_1+X_2+X_3+, \&c. \ldots . X_n=\Sigma.X, \qquad (9)$$

and at O, in OY,

$$Y_1+Y_2+Y_3+, \&c. \ldots . Y_n=\Sigma.Y, \qquad (10)$$

and the couples

$$X_1y_1+X_2y_2+X_3y_3 \ldots . \quad X_ny_n=\Sigma.Xy, \qquad (11)$$

$$-Y_1x_1-Y_2x_2-Y_3x_3 \ldots . -Y_nx_n=\Sigma.-Yx. \qquad (12)$$

We have now reduced the whole to a system of concurring forces and a system of couples. The couples being in the same plane, the moment of the resultant (*Art.* 63) will be equal to the algebraic sum of the moments of the components, or the resultant axis G will be equal to the algebraic sum of the component axes. Hence, taking the sum of (11) and (12),

$$G=\Sigma.Xy-\Sigma.Yx=\Sigma.(Xy-Yx). \qquad (13)$$

Now if R be the resultant force acting at O, and θ the angle it makes with OX, we have, as before,

$$R \cos. \theta=\Sigma.X,$$
$$R \sin. \theta=\Sigma.Y.$$
$$\text{Tan. } \theta=\frac{\Sigma.Y}{\Sigma.X}, \qquad (14)$$

and

$$R^2=(\Sigma.X)^2+(\Sigma.Y)^2. \qquad (15)$$

72. Schol. 1. The equations (13), (14), and (15) determine the magnitude of G and the magnitude and direction of R. *To construct these results,* draw through the origin O the line OR, making the angle θ with OX, to represent R. Then put $G=\Sigma.(Xy-Yx)=Rr$, by which the moment of the resultant couple is changed into another, whose forces are each equal to R, and arm equal to r. Let this be moved and turned round until one of its forces acts at O in an

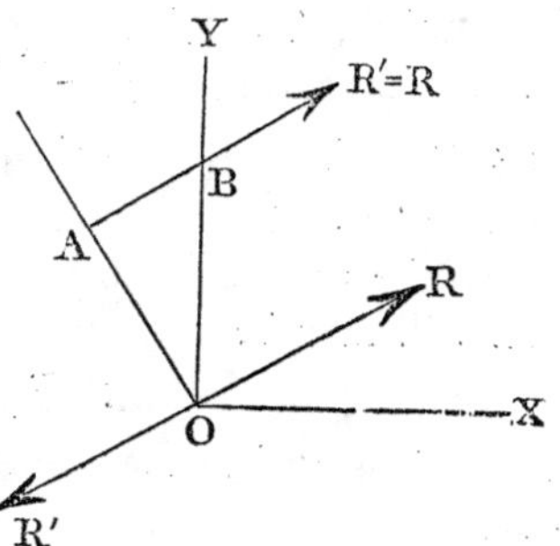

opposite direction to the resultant force; draw OA perpendicular to OR and equal to r, and AR′ parallel to OR. Then R′AOR′ represents the resultant couple. The two forces at O, being equal and opposite, may be removed, and we have the final resultant acting in AR′, which makes the angle θ with OX.

73. Schol. 2. *To find the equation of this final resultant.* It will be of the form

$$y=ax+b.$$

But
$$a=\tan.\theta=\frac{\Sigma.\text{Y}}{\Sigma.\text{X}},$$

and
$$b=\text{OB}=\frac{\text{OA}}{\cos.\theta}=\frac{r}{\cos.\theta}=\frac{\text{G}}{\text{R}\cos.\theta}=\frac{\text{G}}{\Sigma.\text{X}}.$$

Therefore, by substitution, we have for the equation

$$y=\frac{\Sigma.\text{Y}}{\Sigma.\text{X}}.x+\frac{\text{G}}{\Sigma.\text{X}} \qquad (16).$$

74. Prop. *To determine the conditions of equilibrium of any number of forces acting at different points in the same plane.*

In order that there may be no motion of translation, we must have R=0, which gives, as before,

$$\Sigma.\text{X}=0, \qquad (17)$$

and
$$\Sigma.\text{Y}=0. \qquad (18)$$

And, in order that there may be no motion of rotation, we must also have

$$\text{G}=\Sigma(\text{X}y-\text{Y}x)=0. \qquad (19)$$

Equations (17), (18), (19) are *the three necessary and sufficient conditions of equilibrium.*

75. Cor. When there is a fixed point in the system, if this point be taken for the origin, its resistance will destroy the effect of the resultant force R, and the sole condition of equilibrium will then be G=0,

or
$$\Sigma.(\text{X}y-\text{Y}x)=0;$$

or there must be no tendency to rotation around the fixed point.

EQUILIBRIUM OF A POINT ON A PLANE CURVE.

76. If a point be kept at rest on a plane curve by the action of any number of forces in the plane of the curve, the resultant must obviously be in the direction of the normal to the curve at that point, and equivalent to the pressure the curve sustains. For, if the resultant had any other direction, it might be resolved into two, one in the direction of the normal, and the other in the direction of the tangent to the curve; the former would be opposed by the reaction of the curve; the latter, being unopposed, would cause the point to move. Hence,

77. Prop. *To determine the conditions of equilibrium of a point, retained on a plane curve or line, by forces acting in its plane.*

Let N be the normal force of reaction of the curve, and a the angle made by the normal with the axis of x. Also, let $\Sigma.X$, $\Sigma.Y$ be the sums of the components of all the other forces resolved parallel to each axis respectively, and R their resultant. The resistance N may be considered a new force, which, together with the other forces, retains the point in equilibrium independently of the curve. If, therefore, N be resolved in the direction of each axis, we have (8)

$$N \cos. a+\Sigma.X=0, \qquad (a)$$

$$N \sin. a+\Sigma.Y=0; \qquad (b)$$

or, transposing, squaring, adding, and reducing,

$$N=\sqrt{(\Sigma.X)^2+(\Sigma.Y)^2}=R,$$

or the reaction is equal to the resultant of all the other forces.

Now let i be the inclination to the axis of x, of the tangent to the curve through the point, then $a=90^\circ+i$.

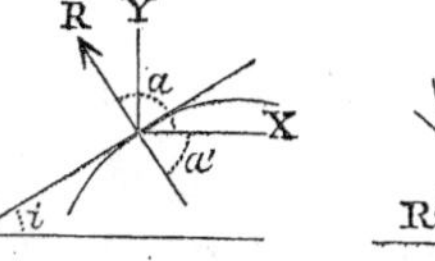

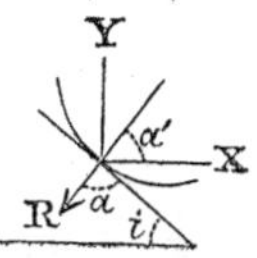

$$\therefore \cos. a=\cos. (90^\circ+i)=-\sin. i,$$

and $\qquad \sin. a=\cos. i.$

Substituting these values of cos. a and sin. a in (a) and (b), and dividing, we get

$$\text{Tan. } i = -\frac{\Sigma.X}{\Sigma.Y}, \text{ or } \Sigma.X + \Sigma.Y \tan. i = 0. \quad (20)$$

But the differential expression for the tangent of the angl which the tangent makes with the axis of x is $\frac{dy}{dx}$. Therefore, substituting and reducing,

$$\Sigma.X.dx + \Sigma.Y.dy = 0. \quad (21)$$

Whenever the line on which the point is retained is right, (20) may be used; but if the line be a curve, (21) will, in general, be necessary, and $\frac{dy}{dx}$ must be deduced from the equation of the given curve.

VIRTUAL VELOCITIES.

78. DEF. If any forces P_1 and P_2 act at the point A, and this point be displaced through an indefinitely small space Aa, and the perpendiculars ap_1, ap_2 be drawn from a on the directions of the forces, then Ap_1 and Ap_2 are called *the virtual velocities of the forces* P_1 and P_2; Ap_1, measured in the direction of the force P_1, is *positive*, and Ap_2, measured in the direction of P_2 produced, is *negative*.

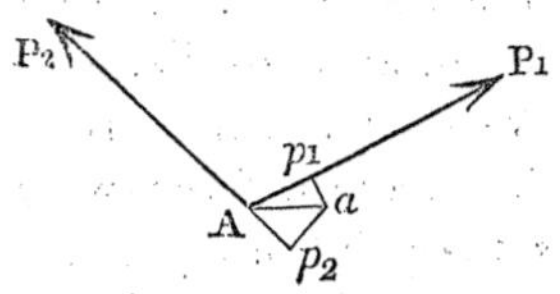

79. The principle of virtual velocities is thus enunciated:

If any number of forces be in equilibrium at one or more points of a rigid body, then if this body receive an indefinitely small displacement, the algebraic sum of the products of each force into its virtual velocity is equal to zero.

80. PROP. *To prove* the principle of virtual velocities *for concurring forces in one plane.*

Let A be the point at which the forces P_1, P_2, P_3, &c. . . . P_n act; a_1, a_2, a_3, &c. . . . a_n the angles they make respectively with OX, and $\beta_1, \beta_2, \beta_3 \ldots \beta_n$ the angles they make with OY. Let θ be the angle which the direction of the displacement Aa makes with OX

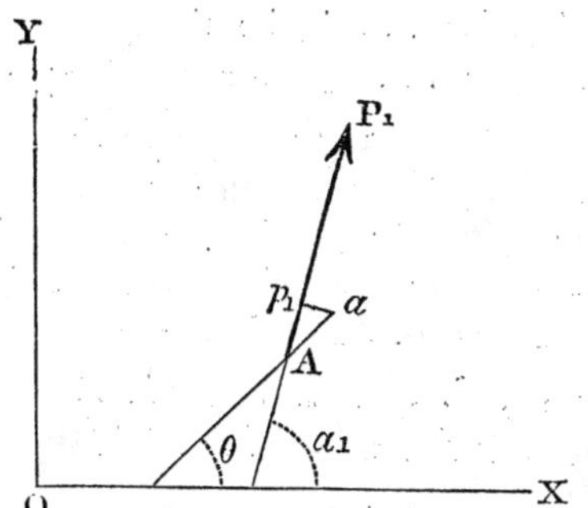

Let v_1, v_2, v_3, &c. . . . v_n be the virtual velocities of the forces respectively.

Then

$$v_1 = Ap_1 = Aa \cos. p_1 Aa = Aa. \cos. (a_1 - \theta),$$
$$= Aa(\cos. a_1 \cos. \theta + \sin. a_1 \sin. \theta),$$

and
$$P_1 v_1 = P_1.Aa(\cos. a_1 \cos. \theta + \sin. a_1 \sin. \theta),$$
$$= Aa(\cos. \theta.P_1 \cos. a_1 + \sin. \theta.P_1 \sin. a_1).$$

In like manner, for P_2 we should have

$$P_2 v_2 = Aa(\cos. \theta.P_2 \cos. a_2 + \sin. \theta.P_2 \sin. a_2),$$

and so for the other forces; and taking their sum, we should get, remembering that $\sin. a = \cos. \beta$,

$$\Sigma.Pv = P_1 v_1 + P_2 v_2 + P_3 v_3 +, \&c. \ldots . P_n v_n,$$
$$= Aa[\cos. \theta\, (P_1 \cos. a_1 + P_2 \cos. a_2 +, \&c. \ldots . P_n \cos. a_n)$$
$$+ \sin. \theta\, (P_1 \cos. \beta_1 + P_2 \cos. \beta_2 +, \&c. \ldots . P_n \cos. \beta_n)].$$

But when there is an equilibrium at a point (8),

$$P_1 \cos. a_1 + P_2 \cos. a_2 + P_3 \cos. a_3 +, \&c. \ldots P_n \cos. a_n = \Sigma.X = 0$$

and

$$P_1 \cos. \beta_1 + P_2 \cos. \beta_2 + P_3 \cos. \beta_3 +, \&c. \ldots P_n \cos. \beta_n = \Sigma.Y = 0.$$

Hence
$$\Sigma.Pv = 0,$$

or the principle is true when the forces all act at a point.

81. Prop. *To prove* the principle of virtual velocities *for forces acting at different points in the same plane.*

The points are supposed to be invariably connected by rigid lines or rods without weight, which transmit the actions and reactions of the particles or points upon each other.

Let A_1, A_2, A_3, &c. . . . A_n be the particles to which the forces P_1, P_2, P_3, &c. . . . P_n are applied, v_1, v_2, v_3 . . . v_n the virtual velocities of the forces respectively.

Let $r_{a_1 a_2}$ be the action of the particle A_1 upon the particle A_2,
$r_{a_2 a_1}$ " reaction of " A_2 " " A_1
$r_{a_1 a_3}$ " action of " A_1 " " A_3,
$r_{a_3 a_1}$ " reaction of " A_3 " " A_1.
&c., &c., &c.

Let $v_{a_1 a_2}$, $v_{a_2 a_1}$, $v_{a_1 a_3}$, $v_{a_3 a_1}$, &c., &c., be the corresponding virtual velocities.

Then $v_{a_1 a_2}=v_{a_2 a_1}$, $v_{a_1 a_3}=v_{a_3 a_1}$, &c., &c., from the nature of action and reaction.

Also, $r_{a_1 a_2}=-r_{a_2 a_1}$, $r_{a_1 a_3}=-r_{a_3 a_1}$, &c., &c. For, let A_1

and A_2 be the particles displaced to a_1 and a_2. Draw the perpendiculars $a_1 p_1$, $a_2 p_2$. Then, if the line $a_1 a_2$ is parallel to $A_1 A_2$, it is obvious that $A_1 p_1=A_2 p_2$. But $A_1 p_1$ is the virtual velocity of R_1, and $A_2 p_2$ of R_2, and they have opposite signs.

$$\therefore v_{a_1 a_2}=-v_{a_2 a_1}.$$

If $a_1 a_2$ is not parallel to $A_1 A_2$, let them meet when produced in some point C. Since the displacements are indefinitely small, the perpendiculars $a_1 p_1$, $a_2 p_2$ coincide with circular arcs whose center is C, and $Ca_1=Cp_1$, $Ca_2=Cp_2$.

But $A_1 p_1=Cp_1-CA_1=Ca_1-CA_1$,
and $A_2 p_2=Cp_2-CA_2=Ca_2-CA_2=(Ca_1+a_1 a_2)-(CA_1+A_1 A_2)$.

$$=Ca_1-CA_1.$$

$$\therefore A_1 p_1 = A_2 p_2,$$

or

$$v_{a_1 a_2}=-v_{a_2 a_1}.$$

Let the sum of the products of all the forces P_1, P_2, &c. into their virtual velocities, acting on the particle

A_1 be $\Sigma.(P_{a_1}.v_{a_1})$,

those on A_2 be $\Sigma.(P_{a_2}.v_{a_2})$,

those on A_3 be $\Sigma.(P_{a_3}.v_{a_3})$,

&c., &c.,

those on A_n be $\Sigma.(P_{a_n}.v_{a_n})$.

Since each particle is in equilibrium from the action of the external forces and the reactions of the others upon it, we have, by the last proposition,

$$0=\Sigma.(P_{a_1}.v_{a_1})+r_{a_1a_2}.v_{a_1a_2}+r_{a_1a_3}.v_{a_1a_3}+, \&c.,$$
$$0=\Sigma.(P_{a_2}.v_{a_2})+r_{a_2a_1}.v_{a_2a_1}+r_{a_2a_3}.v_{a_2a_3}+, \&c.,$$
$$0=\Sigma.(P_{a_3}.v_{a_3})+r_{a_3a_1}.v_{a_3a_1}+r_{a_3a_2}.v_{a_3a_2}+, \&c.,$$
$$\&c., \quad \&c., \quad \&c.,$$
$$0=\Sigma.(P_{a_n}.v_{a_n})+r_{a_na_1}.v_{a_na_1}+r_{a_na_2}.v_{a_na_2}+, \&c.$$

In taking the sum of the products for all the particles, the products of the reactions into their virtual velocities will disappear, being in pairs, equal in magnitude with contrary signs; therefore we have

$$\Sigma.(P_{a_1}.v_{a_1})+\Sigma.(P_{a_2}.v_{a_2})+\Sigma.(P_{a_3}.v_{a_3})+, \&c. \ldots \Sigma.(P_{a_n}.v_{a_n})=0:$$

or, generally, when there is an equilibrium,

$$\Sigma.(Pv)=0.$$

82. Prop. *Conversely.* If the sum of the products of the forces into their virtual velocities be equal to zero, or $\Sigma.(P.v.)=0$, then there will be an equilibrium.

For if the forces are not in equilibrium, they will be equivalent either to a *single force* or *a single couple* (*Art.* 74).

In the first case, let R be the single resultant force; then a force equal and opposite to R will reduce the system to equilibrium; let u be its virtual velocity for any displacement. Since, with this new force, there will be an equilibrium, we have, by the preceding proposition,

$$\Sigma.(P.v)+R.u=0.$$

But by hypothesis, $\Sigma.(P.v)=0$. $\therefore$ $Ru=0$, which, being true for all small displacements of the body, we must have $R=0$, or the body was in equilibrium from the action of the origina forces.

In the second case, if the forces were equal to a resultant couple, it would be balanced by an equal and opposite couple. Let the forces of this opposite couple be Q and Q′, and their virtual velocities for any displacement be q and q' respectively; since they will reduce the system to equilibrium, we have, by the preceding proposition,

$$\Sigma.(P.v)+Qq+Q'q'=0.$$

But $\Sigma.(P.v)=0$. $\therefore$ $Qq+Q'q'=0$, for all displacements, which is impossible, unless Q and Q′ each equal zero, since they are parallel forces, and act at different points.

CHAPTER V

THE CENTER OF GRAVITY.

83. Def. Experiment shows that *every particle* of matter is subject to a force which attracts it in a direction perpendicular to a horizontal plane or the surface of stagnant water. In reality, the directions of the forces, acting on several particles, meet nearly in the center of the earth; but as this center is very distant, compared with the distance of any particles considered together, we may, without sensible error, regard their directions as parallel.

This force is called *gravity.*

84. Experiment shows, also, that the intensity of gravity varies in different parts of the earth's surface; that it is least at the equator, and increases toward the poles in the ratio of the square of the sine of the latitude. It shows, also, that in the same latitude the intensity varies at different points in the same vertical line; that it varies inversely as the square of the distance from the center. But for any points in the same system, or any bodies nearly in the same place, this variation of intensity, as well as difference of direction, may be neglected without error. Hence,

85. Def. *A heavy body* is an assemblage of material points, or particles, acted on by equal parallel forces in the direction of the vertical to the earth's surface.

86. Def. The *weight of a body* is the *resultant* of all the efforts which gravity exerts on its component particles. This resultant (*Art.* 43) is equal to their sum, and parallel to their common direction.

87. Def. The *mass* or *the quantity of matter* of a body is the sum of all its component particles.

88. Cor. If W represent the weight of a body, M the mass, and g the ratio of the intensity of gravity at any place, to its

intensity at another place where it is assumed as unity, we shall have

$$W=gM. \quad (22)$$

For the resultant of all the parallel actions of gravity on the particles of a body is equal to their sum, or the product of its intensity into the number.

89. Def. The *density* of a body is the ratio of its mass to its volume; or, if D represent the density, M the mass, and V the volume,

$$D=\frac{M}{V}; \quad (23)$$

in which D, M, and V represent the number of units of each kind.

90. Cor. Since $W=gM$, and $M=VD$, we have

$$W=gVD. \quad (24)$$

91. Prop. *The masses of two bodies of the same density are in the direct ratio of their volumes.*

If M, V, and D be the mass, volume, and density of one body, and m, v, and d the same of the other, by (23),

$$M=VD \text{ and } m=vd; \because M:m=VD:vd; \quad (25)$$

or, since $D=d$, $M:m=V:v$.

92. Prop. *The masses being equal, the densities are inversely as the volumes.*

Since $M:m=VD:vd$; if $M=m$, $VD=vd$;

or $$D:d=v:V. \quad (26)$$

93. Prop. *The volumes being equal, the masses are directly as the densities.*

Making, in (25), $V=v$, we have $M:m=D:d$. (27)

94. Def. The *center of gravity of a body* is the center of the parallel forces of gravity on each of its component particles.

95. Cor. Hence the determination of the center of gravity involves an immediate application of the doctrine of parallel forces, and we need only refer to results already obtained for many important deductions respecting the center of gravity.

1°. The resultant of all the vertical efforts of gravity on each of the elementary particles of a body passes through its center of gravity. (*Arts.* 44 and 94.)

2°. This resultant is parallel to the forces (*Art.* 36); that is, it is vertical, and its magnitude is equal to the weight of the body.

3°. Whatever position we give to a body, this resultant will always pass through the center of gravity; since changing the position is equivalent to changing the direction of the forces, without changing their points of application or parallelism.

4°. A heavy body will be in equilibrium if its center of gravity be supported, whatever may be the situation of the body relative to the support, since, in this case, the resultant of the parallel forces of gravity will have a fixed point in its direction.

5°. When we wish to find the center of gravity of several bodies, we can suppose the mass of each concentrated at its center of gravity, since the weight of each is a force proportional to its mass, and passing vertically through its center of gravity. Hence we have only to consider a system of heavy points.

96. Def. A body is said to be *symmetrical with respect to a plane* when the lines joining its particles, two and two, are parallel, and bisected by the plane.

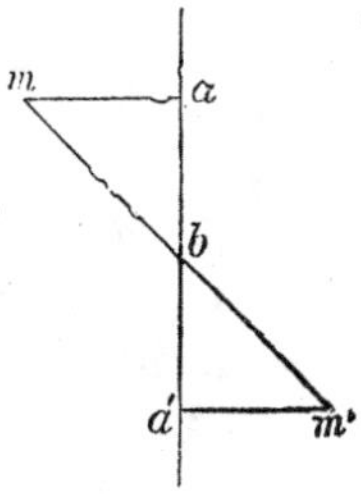

Thus, let m, m' be two symmetrical particles, so that the line mm' may be bisected in b by the plane aa'. Letting fall the perpendiculars ma, $m'a'$, the equality of the triangles mab, $m'a'b$ gives $ma=m'a'$. Hence the particles of a body, symmetrical with respect to a plane, are situated, two and two, on opposite sides of the plane, and at equal distances from it.

97. Prop. *The center of gravity of every homogeneous body of uniform density, symmetrical with respect to a plane, is situated in that plane.*

For any two particles, symmetrically placed, will be at the

same distance from the plane, and their moments will be equa and have contrary signs. But all the particles, taken two and two, are thus placed (*Art.* 96). Therefore the resultant of the system of forces will be in that plane, and, consequently, the center of gravity also.

98. Def. A body is said to be *symmetrical with respect to an axis* when it is symmetrical with respect to two planes passing through that axis.

99. Prop. *The center of gravity of a homogeneous body, symmetrical with respect to an axis, is situated in that axis.*

By *Art.* 97, it must be in each plane passing through the axis, and therefore in their common intersection, or the axis itself.

100. Cor. If a body is symmetrical with respect to two axes, its center of gravity will be at their intersection, since it must be in both axes.

101. Def. This point is also called the *center of figure.*

102. Prop. *To find the center of gravity of any number of heavy particles whose weights and positions are given.*

Let A, B, C, &c., be the particles, whose weights w_1, w_2, w_3, &c., act at their respective centers of gravity, vertically downward, and therefore constitute a system of parallel forces. w_1 and w_2 have a resultant $=w_1+w_2$ acting at some point a, such that $w_1.Aa=w_2.Ba$. The distance AB being given, the distance A a is determined by taking (*Art.* 39)

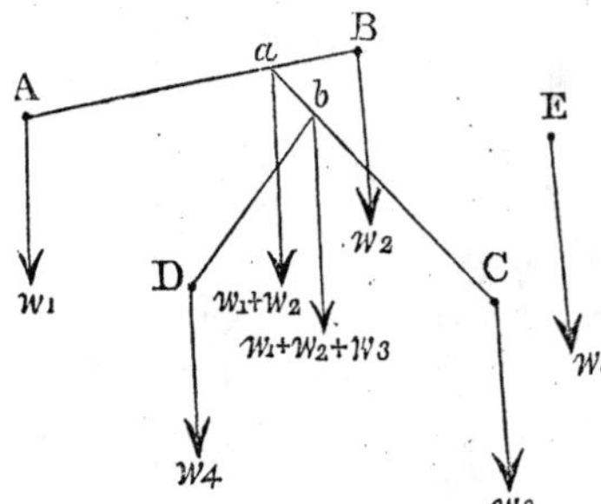

$$w_1+w_2 : w_2 = AB : Aa = AB.\frac{w_2}{w_1+w_2}.$$

Compounding the weight w_1+w_2 at a with another weight w_3 acting at C, they will have a resultant $w_1+w_2+w_3$ acting at some point b, such that

$$(w_1+w_2).ab=w_3.Cb.$$

First determining aC from the triangle aCB, in which BC Ba, and the angle aBC are supposed to be known, we can determine the distance ab by the proportion

$$w_1+w_2+w_3 : w_3 = Ca : ab = Ca.\frac{w_3}{w_1+w_2+w_3}.$$

By continuing the same process, we should determine the point at which the final resultant weight acts. The point will be the same, whatever be the order in which we compound the weights.

103. PROP. *To find the center of gravity of any number of particles in the same plane whose positions are given by their co-ordinates.*

Since the weights of the particles constitute a system of parallel forces, let P_1, P_2, P_3, &c., represent the weights of the particles which may be supposed collected in their respective centers of gravity, x_1, y_1, x_2, y_2, &c., their co-ordinates. We shall have for the co-ordinates $\overline{x}$, $\overline{y}$ of the center of parallel forces (*Art.* 44), or center of gravity of the whole body (*Art.* 94).

$$\left.\begin{aligned}\overline{x}&=\frac{P_1x_1+P_2x_2+P_3x_3+, \&c.}{P_1+P_2+P_3+, \&c.}=\frac{\Sigma.Px}{\Sigma.P}\\ \overline{y}&=\frac{P_1y_1+P_2y_2+P_3y_3+, \&c.}{P_1+P_2+P_3+, \&c.}=\frac{\Sigma.Py}{\Sigma.P}.\end{aligned}\right\}\quad(28)$$

104. COR. 1. If the particles all lie in a straight line, this line may be taken for the axis of x, and $y_1=0$, $y_2=0$, &c. $\therefore\ \overline{y}=0$, and the center of gravity will be in that line.

105. COR. 2. If the particles are homogeneous, the weights of each particle (24) will be proportional to the volumes; and if v_1, v_2, v_3, &c., denote the volumes of the particles, and V the whole volume, we have

$$\left.\begin{aligned}\overline{x}&=\frac{v_1x_1+v_2x_2+v_3x_3+, \&c.}{V}=\frac{\Sigma.vx}{\Sigma.v}\\ \overline{y}&=\frac{v_1y_1+v_2y_2+v_3y_3+, \&c.}{V}=\frac{\Sigma.vy}{\Sigma.v}.\end{aligned}\right\}\quad(29)$$

Hence *the sum of all the particles, or the whole volume, multiplied by the distance of its center of gravity from a plane, is equal to the sum of each particle into the distance of its center of gravity from the plane.*

106. Cor. 3. If the center of gravity of the whole volume be given, and the center of gravity of one of its parts, the center of gravity of the other is readily obtained.

Let V equal the whole volume, and v_1, v_2 the volumes of the two parts, $\bar{x}$, $\bar{y}$, $\bar{x}_1$, $\bar{y}_1$, $\bar{x}_2$, $\bar{y}_2$, the co-ordinates of their respective centers of gravity.

Then $$\bar{x}=\frac{v_1\bar{x}_1+v_2\bar{x}_2}{V}, \text{ and } \bar{y}=\frac{v_1\bar{y}_1+v_2\bar{y}_2}{V}.$$

But, by hypothesis, V, v_1 and $\bar{x}$, $\bar{y}$, $\bar{x}_1$, $\bar{y}_1$ are given.

Therefore, $$v_2=V-v_1,$$

and $$\bar{x}_2=\frac{V\bar{x}-v_1\bar{x}_1}{v_2}, \text{ and } \bar{y}_2=\frac{V\bar{y}-v_1\bar{y}_1}{v_2}. \quad (30)$$

107. We shall now proceed to apply the foregoing principles to specific cases.

Ex. 1. *To find the center of gravity of a uniform physical straight line.*

If AB be the uniform straight line, and C its middle point, C will be its center of gravity; for we may consider the line made up of a series of equal particles in pairs on opposite sides of C, and the weights of each pair would be equal parallel forces having their resultant at C, the middle point between them, or the resultant of all the forces will pass through C.

A C B

Or, the line AB is symmetrical with respect to a plane passing through C. The center of gravity, therefore (*Art.* 97), is in this plane, and as it is also in the line AB, it must be at their intersection C.

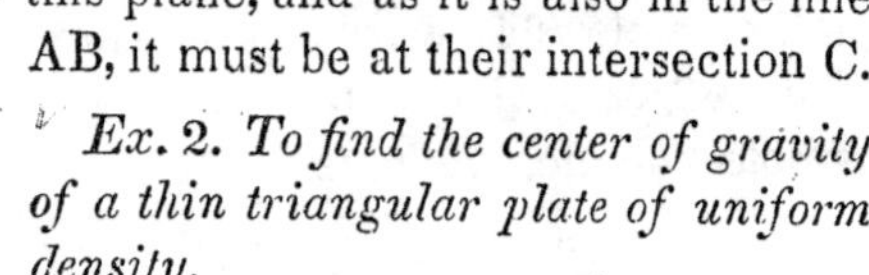

Ex. 2. *To find the center of gravity of a thin triangular plate of uniform density.*

Let ABC be the triangular plate, of which the thickness is inconsiderable. Bisect AB in D, and AC in E. Join C, D, and B, E intersecting in G. G will be the center of gravity of the plate.

Since the line CD bisects all lines drawn parallel to the base, and, consequently, divides the triangle symmetrically, the center of gravity of the triangle will be in this line. For the same reason, it will be in the line BE, and will therefore be at their intersection G.

Join D, E. DE is parallel to BC, since it divides the sides AB and AC proportionally, and DE$=\frac{1}{2}$BC.

From the similar triangles, DEG and BCG, we have

$$DE : BC = DG : GC = 1 : 2. \quad \therefore 2DG = GC.$$

Adding DG to both sides,

$$3DG = DG + GC = DC. \quad \therefore DG = \tfrac{1}{3}DC.$$

In the same way it may be shown that EG$=\frac{1}{3}$EB.

Hence the center of gravity of a triangle is one third the distance from the middle of either side to the opposite vertex.

Ex. 3. *To find the center of gravity of a parallelogram of uniform density and thickness.*

Bisect the sides AB and DC in H and K; also the sides AD and BC in E and F. The plane passing through H and K will divide the parallelogram symmetrically, since it will bisect all lines parallel to AB. The center of gravity will lie in this plane, and will therefore lie in its intersection HK with the parallelogram. For the same reason it will lie in EF, and must therefore be at G, the common point of these lines.

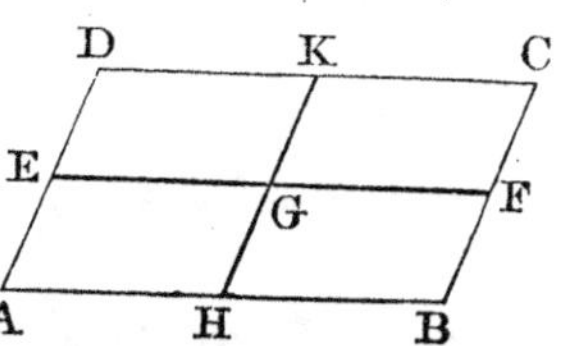

Or, since each diagonal bisects all lines drawn parallel to the other, it will be at the intersection of the diagonals.

Ex. 4. *To find the center of gravity of a thin, polygonal plate, of uniform density and thickness.*

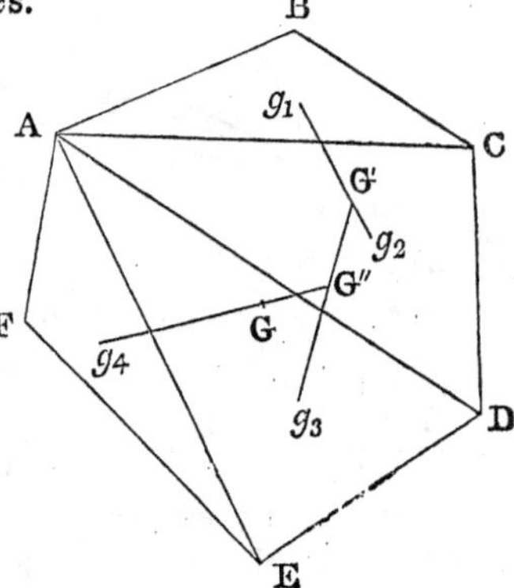

Let ABCDEF be the polygon. Draw the lines AC, AD, AE, dividing it into triangles.

When the polygon is given, these triangles will be known, and their centers of gravity, g_1, g_2, g_3, g_4, may be found by *Ex.* 2.

The mass of each triangle may be considered as a heavy particle at its center of gravity, or the weight of each, as a force acting at its center of gravity. Then (*Art.* 39),

$$g_1+g_2 : g_2=g_1g_2 : g_1G'=g_1g_2.\frac{g_2}{g_1+g_2}.$$

This determines the point G′, the center of gravity of the portion ABCD. In the same manner, we may find the point of application G″, of the resultant of g_1+g_2 acting at G′, and g_3 acting at g_3, and so on; the last point so determined will be the center of gravity G of the polygon.

Ex. 5. *To find the center of gravity of a triangular pyramid of uniform density.*

Let ABCD be the triangular pyramid. Bisect the edge BC in E, and pass a plane through E and the edge AD. This

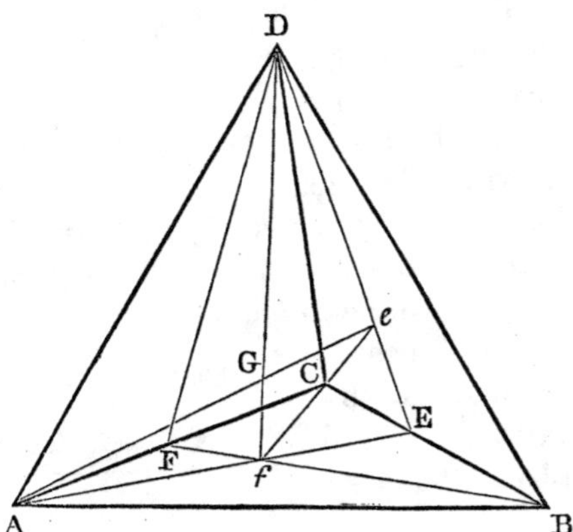

plane will bisect all lines in the pyramid parallel to BC, and will therefore divide it symmetrically. Bisect AC in F, and the plane through F and the edge BD will also divide the pyramid symmetrically; and since the center of gravity of the pyramid will be in both these planes, it must be in their intersection D*f*. But the point *f* is the center of gravity of the face ABC (*Ex* 2). Hence the center of gravity of the pyramid lies in the line drawn from a vertex to the center of gravity of the opposite face. Take $Ee=\frac{1}{3}ED$, and join A*e*. Since *e* is the center of gravity of the face BCD, the center of gravity of the pyramid will be in the line A*e*. It must therefore be at the intersection G of A*e* and D*f*.

To find *f*G, join *fe*, which will be parallel to AD, since it divides the sides ED and EA of the triangle AED proportionally. Now the similar triangles *f*G*e* and AGD, with *f*E*e* and AED, give

$$fG : GD = fe : AD = Ee : ED = 1 : 3.$$
$$\therefore\ 3fG = GD.$$

Adding *f*G to both members,

$$4fG = fG + GD = fD,$$

or

$$fG = \tfrac{1}{4} fD.$$

Hence, *the center of gravity of a triangular pyramid is one fourth the distance from the center of gravity of one face to the opposite vertex.*

Ex. 6. *To find the center of gravity of a pyramid whose base is any polygon.*

In the pyramid A BCDEF find G, the center of gravity of the polygonal base (*Ex.* 4), and join AG. Since AG passes through the center of gravity of the base, it will pass through the center of gravity of every section parallel to the base, and the center of gravity of the whole pyramid will be in AG.

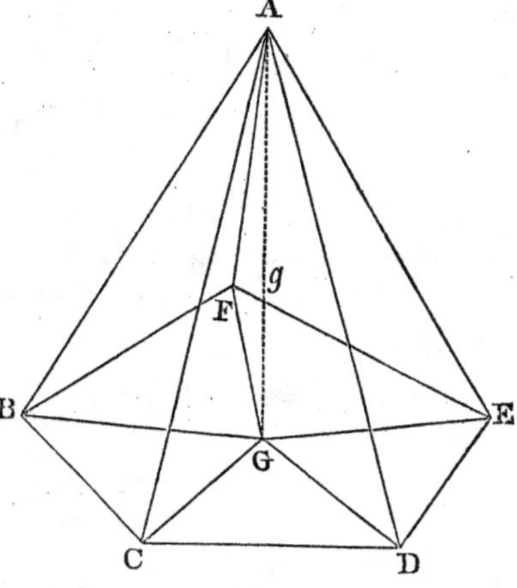

Join GB, GC, GD, GE, and GF, and conceive planes to pass through A and each of these lines, thus dividing the whole pyramid into as many triangular pyramids as the base has sides. The centers of gravity of these pyramids will be at one fourth the distances, respectively, from the centers of gravity of their triangular bases to the common vertex A. These distances being thus divided proportionally, the points of division will all lie in the same plane parallel to the base. And since the centers of gravity of all the triangular pyramids are in this plane, the center of gravity of the whole pyramid will be in it, and, being in the line AG also, will be at their intersection g. But the plane divides all lines drawn from the vertex A to the base proportionally; therefore, the center of gravity g of the whole pyramid is one fourth the distance from the center of gravity of the base to the vertex.

COR. Since the above principle is true, whatever be the number of sides of the polygon, it is true when the number becomes indefinitely great, or when the base becomes a continued closed curve, as a circle, an ellipse, &c.; or, *the center of gravity of a cone, right or oblique, and on any base, is one fourth the distance from the center of gravity of the base to the vertex.*

Ex. 7. *To find the center of gravity of a frustum of a cone or pyramid cut off by a plane parallel to the base.*

Let a be the length of the line drawn from the vertex of the cone, when complete, to the center of gravity of the base, a' that portion of it between the vertex and the smaller base of the frustum. Then (30) we have

$$\bar{x}_2 = \frac{v\bar{x} - v_1\bar{x}_1}{v_2},$$

in which $\bar{x} = \frac{3}{4}a$, $\bar{x}_1 = \frac{3}{4}a'$. Now the part of the cone or pyramid cut off is similar to the whole, and similar solids are as the cubes of their homologous dimensions, or cubes of their lines similarly situated.

Hence $v : v_1 = a^3 : a'^3$, or $v - v_1 : v = a^3 - a'^3 : a^3$.

$$\therefore\ v_1 = v\frac{a'^3}{a^3},$$

and

$$v_2 = v - v_1 = v\left(1 - \frac{a'^3}{a^3}\right).$$

$$\therefore\ \bar{x}_2=\frac{v.\frac{3}{4}a-v\left(\frac{a'^3}{a^3}\right)\frac{3}{4}a'}{v\left(1-\frac{a'^3}{a^3}\right)}=\frac{3}{4}.\frac{a^4-a'^4}{a^3-a'^3},$$

$$=\frac{3}{4}\frac{(a+a')(a^2+a'^2)}{a^2+aa'+a'^2}.$$

Subtracting this from a, we have the distance of the center of gravity of the frustum from the center of gravity of its base equal to

$$a-\frac{3}{4}\frac{(a+a')(a^2+a'^2)}{a^2+aa'+a'^2}=\frac{a}{4}-\frac{3a'^3}{4(a^2+aa'+a'^2)}.$$

Ex. 8. *To find the center of gravity of the perimeter of a triangle in terms of the co-ordinates of the angular points.*

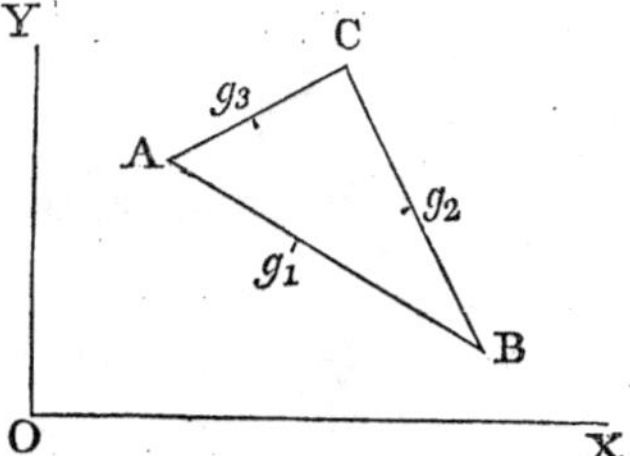

In the triangle ABC, let a, b, c represent the sides respectively opposite to the angles A, B, C. Their centers of gravity will be each at the middle point of the side; as g_1, g_2, g_3.

Let x_1y_1 be the co-ordinates of A referred to the origin O,
x_2y_2 " " " B,
x_3y_3 " " " C.

Then the co-ordinates of g_1 are $\frac{1}{2}(x_1+x_2)$, $\frac{1}{2}(y_1+y_2)$,
" " " g_2 " $\frac{1}{2}(x_2+x_3)$, $\frac{1}{2}(y_2+y_3)$,
" " " g_3 " $\frac{1}{2}(x_3+x_1)$, $\frac{1}{2}(y_3+y_1)$,

and $\bar{x}$, $\bar{y}$, being the co-ordinates required, by (28), we have

$$\bar{x}=\frac{a(x_2+x_3)+b(x_1+x_3)+c(x_1+x_2)}{2(a+b+c)},$$

$$\bar{y}=\frac{a(y_2+y_3)+b(y_1+y_3)+c(y_1+y_2)}{2(a+b+c)}.$$

Ex. 9. *To find the co-ordinates of the center of gravity of a triangle.*

Let x_1y_1, x_2y_2, x_3y_3 be the co-ordinates of the points A,

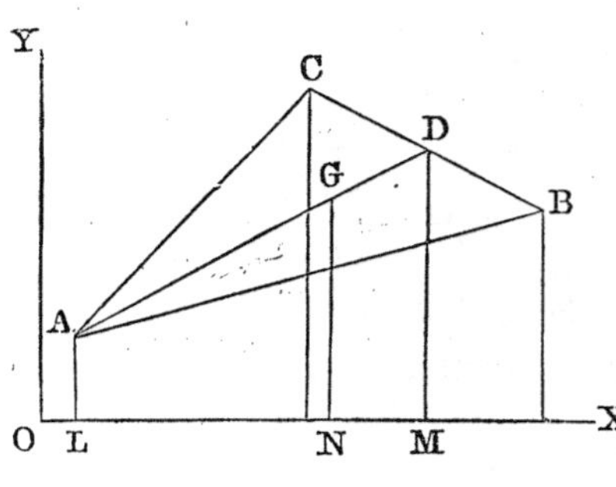

B, C respectively. Draw AD bisecting BC in D, and take $AG=\frac{2}{3}AD$: G is the center of gravity of the triangle. The co-ordinates of D are $\frac{1}{2}(x_2+x_3)$, $\frac{1}{2}(y_2+y_3)$; and if $\bar{x}=ON$, $\bar{y}=GN$, be the co-ordinates of G, we have

$$ON=OL+\tfrac{2}{3}(OM-OL),$$
$$GN=AL+\tfrac{2}{3}(DM-AL);$$

or $$\bar{x}=x_1+\tfrac{2}{3}\{\tfrac{1}{2}(x_2+x_3)-x_1\}=\tfrac{1}{3}(x_1+x_2+x_3),$$

and $$\bar{y}=y_1+\tfrac{2}{3}\{\tfrac{1}{2}(y_2+y_3)-y_1\}=\tfrac{1}{3}(y_1+y_2+y_3).$$

CONDITIONS OF EQUILIBRIUM OF BODIES FROM THE ACTION OF GRAVITY.

108. PROP. *If a body have a fixed point in it, the condition of equilibrium requires that the vertical line through the center of gravity shall pass through the fixed point.*

If the center of gravity g be in the vertical line Ag, passing through the fixed point A, the weight w of the body, being a vertical force acting at g, in the direction Ag, will be resisted by the reaction of the fixed point A. If the center of gravity be at any other point g', then drawing the vertical line through g' and the horizontal line through A, the weight w acting at g' would have an uncompensated moment w.Am, which will not vanish until the center of gravity comes into the line Ag. Or, if g' be the center of gravity, the body will be acted upon by

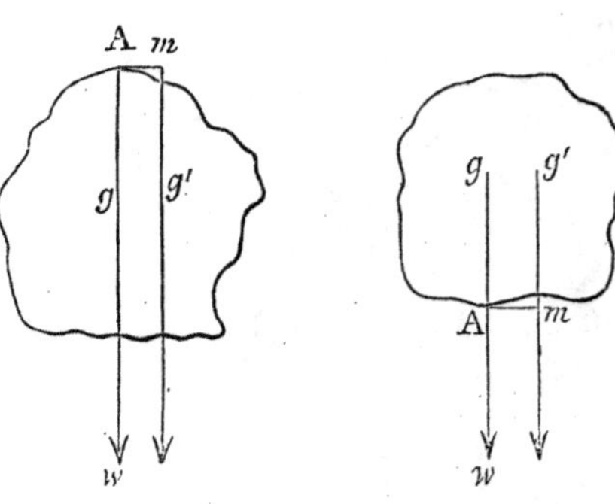

a couple of which the forces are, the weight of the body at g', and the reaction of the fixed point A_1, and the arm Am. Therefore (19), in order to equilibrium,

$$\Sigma(Xy - Yx) = w.Am = 0.$$

$\therefore Am = 0$, or the point g' must be in Ag.

109. Def. The equilibrium is said to be *stable* when the body, if slightly disturbed, tends to return to its original position. It is called *unstable* when, being disturbed, it tends to remove further from its original position; and *neutral* when, after being disturbed, it still remains in equilibrium.

110. Prop. *When the equilibrium of a body containing a fixed point is stable, the center of gravity is in the lowest position it can take; when unstable, in the highest.*

Let A be a fixed point in the bodies M and N, g their center of gravity. The center of gravity can only move on the surface of a sphere whose center is A and radius Ag. When the body M is disturbed, its center of gravity g being removed to g', will rise through the versed sine of the arc gg', and when at g', the moment $w.Am$ will obviously tend to bring it back to its original position. The equilibrium is therefore stable, and the center of gravity the lowest possible. In the body N, the center of gravity being at g is the highest possible, and being in the vertical Ag, will be in equilibrium. When removed to g', the moment $w.Am$ will obviously tend to carry it further, and the equilibrium was therefore unstable.

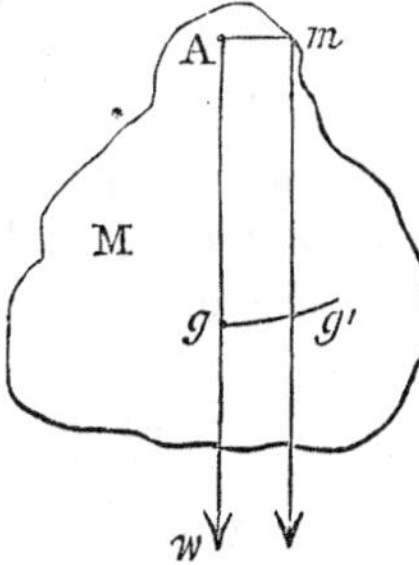

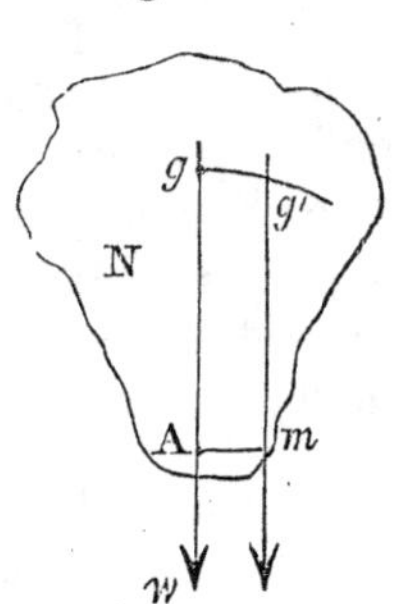

111. Cor. 1. The *pressure* on the point by which a body is suspended is clearly, in the case of equilibrium, equal to the weight of the body.

112. Cor. 2. If a body is suspended from *two points*, the position of equilibrium is that in which the center of gravity is in

the vertical plane passing through the two points of suspension, since it is then the highest or lowest possible. To determine the pressures on the fixed points A and B, let GC represent the weight of the body acting vertically at G. Resolve GC into the two forces DG and EG acting in the directions AG and BG. These will represent the pressures on A and B. Since the directions of GC, GD, and GE, and the magnitude of GC are known, the magnitudes of DG and EG may be determined.

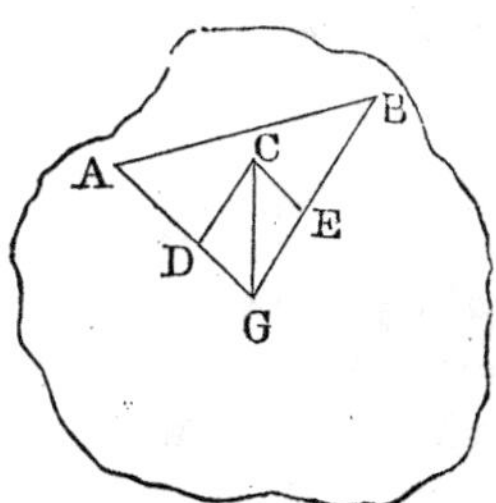

113. Cor. 3. If a body be suspended from *three fixed points* not in a right line, the body is necessarily at rest.

With regard to the *pressures* on each point, the three lines drawn from the fixed points to the center of gravity give the directions of the pressures, and the vertical is the direction of the weight, the magnitude of which is given. Hence we have, in a parallelopiped, the three sides and diagonal given in position and one given in magnitude, to determine the magnitude of the other three.

114. Cor. 4. If a body touch a horizontal plane in one point, it will be in equilibrium when the vertical through its center of gravity, and the perpendicular to the plane at the point of contact, coincide; for the weight will then be counteracted by the plane.

115. Cor. 5. If the body touch the plane in two points, it will be in equilibrium when the vertical through the center of gravity, and the perpendiculars to the plane at the points, are in the same plane. Thus, if ABC be a vertical section through the two points of support P and Q, there will not be an equilibrium unless the vertical through the center of gravity G is in the same plane, or, which is the same thing, meets

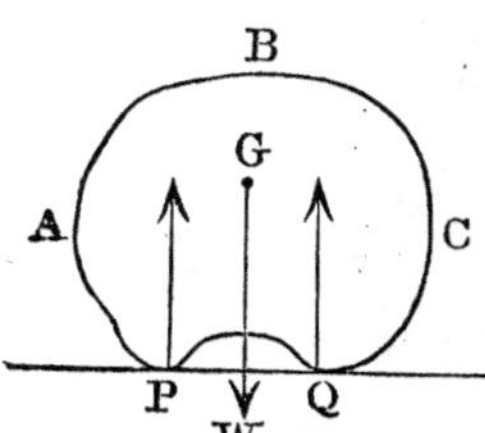

the line PQ. The pressures on P and Q may be determined by the theory of parallel forces, and

$$P : Q : W = WQ : WP : PQ,$$

P, Q, and W representing the pressures on the two points and the weight of the body respectively.

If the body touch the line PQ in more than two points, the problem of the pressures is indeterminate, as the pressures may be any how distributed.

116. Cor. 6. If the body touch the plane in three points, it will be in equilibrium when the vertical through the center of gravity falls within the triangle formed by joining these points.

To estimate the pressures in this case, let PQR be the triangle formed by joining the three points, and W the point where the vertical through the center of gravity meets it. We must resolve the weight acting at W into three others parallel to it, acting at P, Q, and R. Join PW, QW, and RW, and produce them to the opposite sides. Then, by the theory of parallel forces,

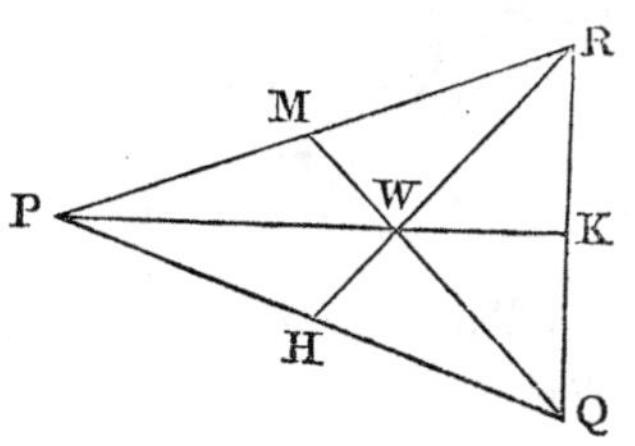

$$P : W = WK : PK = \text{triangle } WRQ : \text{triangle } PRQ,$$
$$Q : W = WM : QM = \text{"} \quad WRP : \text{"} \quad PRQ,$$
and
$$R : W = WH : RH = \text{"} \quad WPQ : \text{"} \quad PRQ.$$
$$\therefore P : Q : R : W = WRQ : WRP : WPQ : PQR.$$

If the body touches the plane in more than three points, the pressures on the points are indeterminate, but their sum is equal to the weight of the body.

117. Cor. 7. If the vertical through the center of gravity of a body on a plane meets the plane in a point within the base, the body will stand. For the resultant of the parallel forces of resistance must be within the figure formed by joining the several points of contact.

If the vertical falls without the base, we have two parallel forces in contrary, but not opposite, directions, and the body will turn over.

118. Prop. *The stability of a body is measured by the excess of the shortest line that can be drawn from the center of gravity to the perimeter of the base, above the vertical, from the center to the horizontal plane.*

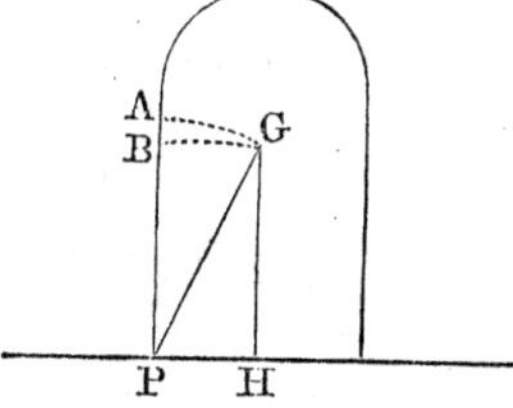

The stability will depend on the excess of GP over GH, since G must be elevated a distance equal to this difference, in order to turn the body over the edge of the base at P.

Cor. 1. The greater the base HP the greater the stability, if the height of G remain the same; and the greater HG, the less the stability if HP remain the same.

Cor. 2. The stability is measured by the versed sine of the arc, through which the center of gravity must move from rest to its highest point. For GP−GH=GP−BP=AP−BP=AB = versed sine of arc GA to radius PG.

119. Prop. *If a body be placed on an inclined plane, it will descend when there is no resistance from friction.*

For the weight of the body, represented by GA, may be resolved into two others, GB and BA, one perpendicular to the plane, and the other parallel to it, of which GB, the one perpendicular to the plane, can alone be counteracted by the plane. In this case, if the vertical GA fall within the base of the body, the body will slide; if it fall without, it will slide also.

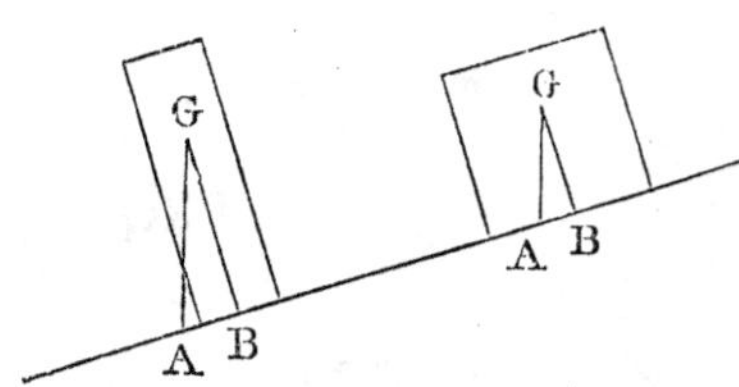

120. EXAMPLES.

1. If two right cones have the same base and their vertices in the same direction, find the distance of the center of gravity of the solid contained between their two surfaces from their common base.

Ans. $\frac{1}{4}$ sum of their altitudes

2. The center of gravity of a paraboloid being in the axis at a distance from the vertex equal to $\frac{2}{3}$ of the axis; find the center of gravity of a frustum of a paraboloid from the base, a and b being the radii of the two ends, and m the parameter to the axis (30).

$$\textit{Ans.}\ \frac{a^6-3a^2b^4+2b^6}{3m(a^4-b^4)}.$$

3. Two spheres, whose radii are a and b, touch each other internally; find the distance of the center of gravity of the solid contained between the two surfaces from the point of contact.

$$\textit{Ans.}\ \frac{a^3+a^2b+ab^2+b^3}{a^2+ab+b^2}.$$

4. The distance of the center of gravity of a hemisphere from its base being $\frac{3}{8}$ the radius, find that of a hemispherical bowl whose internal radius is a and thickness c.

$$\textit{Ans.}\ \frac{3}{8}\cdot\frac{4a^3+6a^2c+4ac^2+c^3}{3a^2+3ac+c^2}.$$

5. From the result obtained in *Ex.* 4, find the distance of the center of gravity of a hemispherical surface from the center of the base.

Ans. $\frac{1}{2}a$.

APPLICATION OF THE PRINCIPLES OF THE INTEGRAL CALCULUS TO THE DETERMINATION OF THE CENTER OF GRAVITY.

121. By the principles of the integral calculus, when the volumes v (*Art.* 105) become indefinitely small, they may be regarded as the differential elements of the body, and be represented by dv. In this case formulas (29) will take the form

$$\bar{x}=\frac{\int xdv}{\int dv},\quad \bar{y}=\frac{\int ydv}{\int dv},\qquad (31)$$

in which x and y denote the distances of the center of gravity of dv from the co-ordinate axes.

122. Prop. *Required the differential expressions for the co-ordinates of the center of gravity of a plane curve or line.*

If ds represent the differential element of the curve or line, by substituting ds for dv in (31), we have

$$\bar{x}=\frac{\int xds}{s},\quad \bar{y}=\frac{\int yds}{s}. \qquad (32)$$

If the arc is symmetrical with respect to the axis of x, the center of gravity will be in that axis (*Art.* 99), and $\bar{y}=0$.

$$\therefore\ \bar{x}=\frac{\int xds}{s}$$

is sufficient.

123. Prop. *Required the differential expressions for the co-ordinates of the center of gravity of a plane area.*

Since the differential element of a plane area is $dxdy$, $dv=dxdy$. By substitution in (31), we have

$$\bar{x}=\frac{\iint xdxdy}{\iint dxdy},\quad \bar{y}=\frac{\iint ydxdy}{\iint dxdy}.$$

Integrating in reference to y, we have

$$\bar{x}=\frac{\int xydx}{\int ydx},\quad \bar{y}=\frac{\int \frac{1}{2}y^2dx}{\int ydx}. \qquad (33)$$

If the area is symmetrical with respect to the axis of x, the center of gravity is in that line (*Art.* 99), and $\bar{y}=0$.

$$\therefore\ \bar{x}=\frac{\int xydx}{\int ydx}$$

is sufficient.

124. Prop. *Required the differential expressions for the co-ordinates of the center of gravity of a surface of revolution around the axis of* x.

The center of gravity will obviously be in the axis of x, and therefore $\bar{y}=0$; and since, for a surface of revolution, $dv=2\pi yds$, the first of equations (31) become,

$$\bar{x}=\frac{\int xyds}{\int yds}, \qquad (34)$$

and this equation is sufficient.

125. Prop. *Required the differential expressions for the co-ordinates of the center of gravity of a solid of revolution.*

In this case $dv=\pi y^2dx$. Hence, from the first of equations (31), we have

$$\bar{x}=\frac{\int y^2xdx}{\int y^2dx}, \tag{35}$$

which alone is sufficient.

By proper substitutions for dv in the fundamental equations (31), we may find expressions for the co-ordinates of the center of gravity for other forms of bodies.

126. Prop. *The surface generated by the revolution of a curve around an axis is equal to the length of the curve, multiplied by the circumference described by its center of gravity.*

From the second of equations (32), we have

$$2\pi.\bar{y}.s=2\pi\int yds.$$

Now $2\pi\bar{y}$ is the circumference of which $\bar{y}$ is the radius, and $2\pi\bar{y}.s$ is the circumference described by the center of gravity of the curve s in its revolution round the axis of x, multiplied by the length of the curve s. But this is equal to $2\pi\int yds$, which is the area of the surface generated by the revolution of the curve. Hence, &c.,

127. Prop. *The volume generated by the revolution of a plane area around an axis is equal to the product of that area by the circumference described by its center of gravity.*

For, from the second of equations (33), we have

$$2\pi\bar{y}\int ydx=\pi\int y^2dx.$$

In this equation, $\int ydx$ is the generating area, $2\pi\bar{y}$ is the circumference described by its center of gravity, and $\pi\int y^2dx$ is the volume generated. Hence the truth of the proposition.

These last two propositions comprise the theorem of *Guldin*, and their application to the determination of the surfaces and volumes of bodies constitutes the *Centrobaryc Method.* By this method, of the three quantities, viz., the generatrix, the distance of the center of gravity from the axis, and the magnitude generated, any two being given, the other may be determined.

128. EXAMPLES.

1. Required the center of gravity of a circular arc.

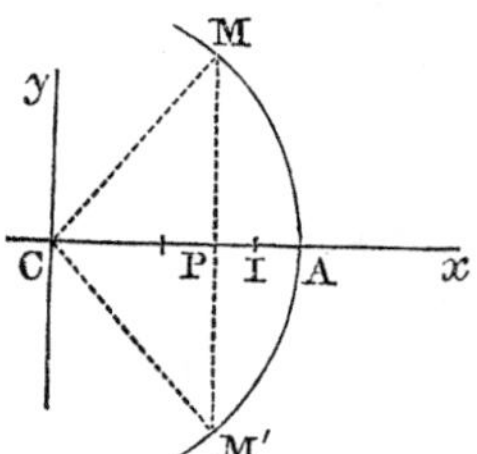

Let the axis of x bisect the arc MAM in A, the origin being at the center of the circle, and let MAM′$=2s$. From the equation of the circle $y^2=r^2-x^2$, we obtain $\frac{dy^2}{dx^2}=\frac{x^2}{r^2-x^2}$.

But $ds=dx\sqrt{1+\frac{dy^2}{dx^2}}=-\frac{rdx}{\sqrt{r^2-x^2}}$.

$$\therefore (32)\ \bar{x}=\frac{\int xds}{s}=-\frac{r}{s}\int\frac{xdx}{\sqrt{r^2-x^2}}=+\frac{r}{s}\sqrt{r^2-x^2}+C=+\frac{ry}{s}+C.$$

When $y=0$, $\bar{x}=r$, and $\frac{ry}{s}=r$. $\therefore$ C$=0$

Hence $$\bar{x}=\text{CI}=\frac{r.y}{s}=\frac{r.2y}{2s};$$

or, the distance of the center of gravity of a circular arc from the center of the circle is a fourth proportional to the arc, the radius, and the cord of the arc.

If the arc be a semicircle, $y=r$, and $s=\frac{1}{2}\pi r$

$$\therefore\ \text{CI}=\frac{2r}{\pi}=0.63662r.$$

2. Required the center of gravity of a circular segment.

Putting CP$=a$ (*Fig., Ex.* 1), and taking the center for the origin, we have $y=\sqrt{r^2-x^2}$. Hence (33),

$$\bar{x}=\frac{\int_a^r xydx}{\int_a^r ydx}=\frac{\int_a^r x(r^2-x^2)^{\frac{1}{2}}dx}{\text{MAP}}=\frac{\frac{1}{3}(r^2-a^2)^{\frac{3}{2}}}{\text{MAP}}=\frac{\frac{1}{3}\overline{\text{MP}}^3}{\text{MAP}},$$

$$\therefore\ \bar{x}=\frac{\frac{1}{24}(2\text{MP})^3}{\frac{1}{2}\text{MAM′P}}=\frac{\frac{1}{12}(\text{chord})^3}{\text{segment}}.$$

If the segment is a semicircle,

$$\bar{x}=\frac{\frac{1}{12}(2r)^3}{\frac{1}{2}\pi r^2}=\frac{4r}{3\pi}=0.42441r=\tfrac{2}{3}\text{CI}.$$

3. Required the center of gravity of the surface of a spherical segment.

Taking the origin at C (*Fig.*, *Ex.* 1), the center of the generating circle, we have $x^2+y^2=r^2$, $\frac{dy^2}{dx^2}=\frac{x^2}{y^2}$, and $yds=rdx$.

$$\therefore\ (34)\ \bar{x}=\frac{\int yxds}{\int yds}=\frac{\int xdx}{\int dx}.$$

Integrating between $x=r$ and $x=a=\text{CP}$.

$$\bar{x}=\frac{\frac{1}{2}(r^2-a^2)}{r-a}=\tfrac{1}{2}(r+a).$$

Hence the center of gravity is at the middle of PA.

4. Required the center of gravity of a spherical segment.

Taking the origin at A, the vertex of the generating circle, we have, for its equation, $y^2=2ax-x^2$.

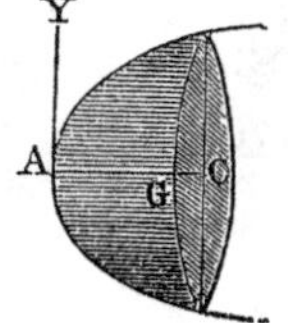

$$\therefore\ (35)\ \bar{x}=\frac{\int xy^2dx}{\int y^2dx}=\frac{\int(2ax-x^2)\,xdx}{\int(2ax-x^2)dx}.$$

Hence $\quad \text{AG}=\frac{\frac{2}{3}ax^3-\frac{1}{4}x^4}{ax^2-\frac{1}{3}x^3}=\frac{8ax-3x^2}{12a-4x}.$

If the segment is a hemisphere, $x=a$,

and $\qquad \bar{x}=\frac{5}{8}a.$

5. Required the surface of a hemisphere.

By the centrobaryc method (*Art.* 126), we have

The generatrix $=\frac{1}{2}\pi r$, the ordinate of its center of gravity $\bar{y}=\frac{2r}{\pi}$ (Art. 128, Ex. 1), and the circumference described by the center of gravity $=2\pi.\frac{2r}{\pi}=4r$. Hence the surface $=\frac{1}{2}\pi r.4r$ $=2\pi r^2$.

129. EXAMPLES ON THE PRECEDING CHAPTERS.

Ex. 1. Two beams, rigidly connected at a given angle, turn on a horizontal axis through their point of union; find the position of equilibrium by the action of their own weights.

Let AC, BC be the beams suspended from C, and making

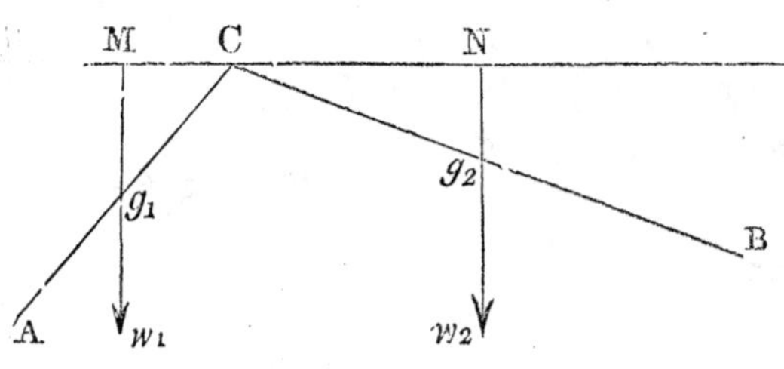

with each other the angle a. Since C is a fixed point, the only condition of equilibrium is, that the sum of the moments about C is zero (*Art.* 75).

Let g_1, g_2 be the centers of gravity of the beams, and $g_1C = a$, $g_2C = b$. Also, $w_1 =$ weight of AC, acting at g_1, and $w_2 =$ weight of BC, acting at g_2. Draw through C the line MCN horizontally, meeting the vertical directions in which w_1 and w_2 act, at M and N.

By (19), $w_1 CM - w_2 CN = 0$. Let $BCN = \theta$. The determination of θ will fix the position of the compound beam.

Since $CM = g_1C.\cos. MCA$ and $CN = g_2C.\cos. BCN$, we have

$$w_1.Cg_1 \cos. ACM - w_2.Cg_2 \cos. BCN = 0,$$

or

$$w_1.\ a.\ \cos. (180 - \overline{a + \theta}) - w_2.b.\cos.\theta = 0.$$

$$\therefore \text{Tan. } \theta = \frac{w_2 b + w_1 a.\cos. a}{w_1 a.\sin. a}.$$

Ex. 2. When a given weight W is hung from the end of one of the beams, A (*Ex.* 1), find θ in case of equilibrium.

$$\text{Tan. } \theta = \frac{w_2.b + (w_1 + 2W) a \cos. a}{(w_1 + 2W) a \sin. a}.$$

Ex. 3. Two beams, as in *Ex.* 1, are suspended from one end B; find the angle θ which the upper one makes with a horizontal line.

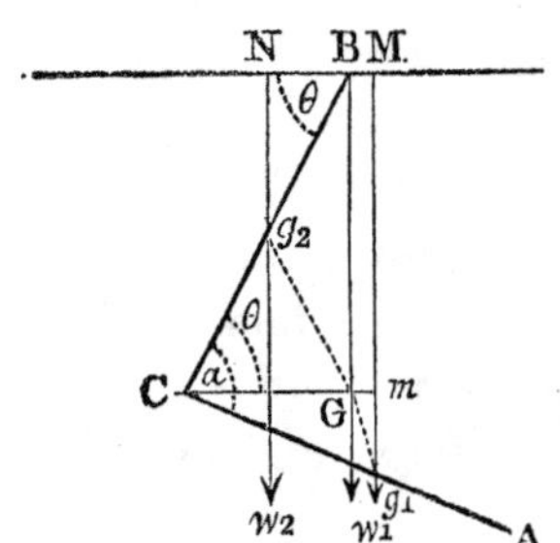

$$\text{Tan. } \theta = \frac{(2w_1 + w_2) b - w_1 a.\cos. a}{w_1 a.\sin. a}.$$

N. B. Since the common center of gravity of the two beams is in the vertical through B, $BM = Gm = Cm - CG = Cm - 2NB = a.\cos.(a - \theta) - 2b\cos.\theta$.

$\therefore w_2 b. \cos. \theta - w_1 a. \cos. (\alpha - \theta) + 2w_1 b. \cos. \theta = 0.$

Ex. 4. Two spheres of unequal radii, but of the same material, are placed in a hemispherical bowl; find the position they take when in equilibrium.

Since the reactions of the bowl upon the spheres are in the directions of the radii of the spheres through the points of contact, and since these radii produced pass through C; if C was a fixed point, and connected with A and B by a rigid rod without weight, the bowl might be removed without disturbing the equilibrium. The question, then, is reduced to finding the position of equilibrium of two weights suspended from the extremities of two rigid rods without weight, and is solved like the preceding. This position will be known when θ is known.

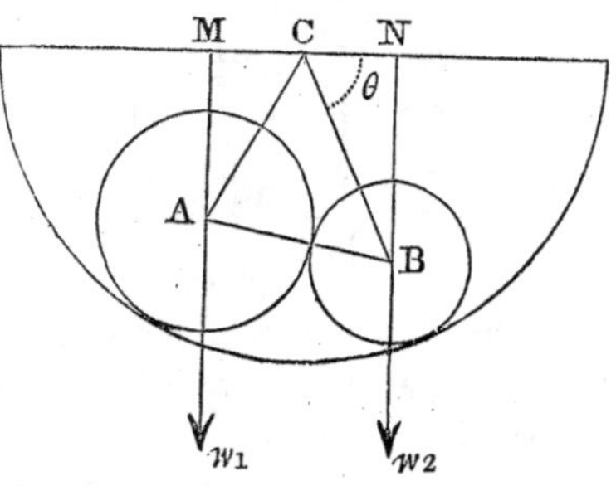

Let R be the radius of the bowl, r_1, r_2 the radii of the spheres A and B respectively, and ACB$=\alpha$.

Then $AB = r_1 + r_2$, $CA = R - r_1$, $CB = R - r_2$,

$$\text{and } \cos. \alpha = \frac{AC^2 + CB^2 - AB^2}{2AC.BC},$$

$$= \frac{(R-r_1)^2 + (R-r_2)^2 - (r_1+r_2)^2}{2(R-r_1)(R-r_2)}, \text{ which gives } \alpha.$$

Then (19) $w_1.CM - w_2 CN = 0$, or $r_1^3.CM - r_2^3.CN = 0$; since the weights of the spheres are as the cubes of their radii. Substituting the values of $CM = (R-r_1)\cos.(180-(\alpha+\theta))$ and $CN = (R-r_2)\cos.\theta$, expanding and reducing, we get

$$\text{Tan. } \theta = \frac{r_2^3(R-r_2) + r_1^3(R-r_1)\cos.\alpha}{r_1^3(R-r_1)\sin.\alpha}.$$

Ex. 5. A heavy beam rests upon a smooth peg with one end against a smooth, vertical wall; find the position of equilibrium.

Let ACB be the beam, resting at A against the wall ADE and upon the peg C.

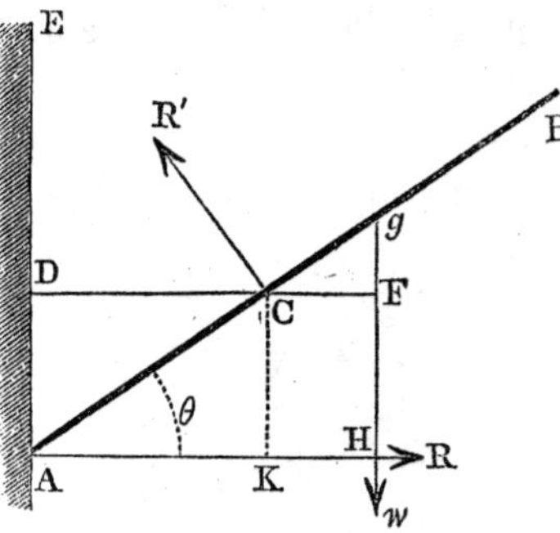

The center of gravity g, when there is an equilibrium, will evidently be at some point beyond C from A. Let $Ag=a$, $DC=b$, $w=$ the weight of the beam acting at g, $R=$ the reaction of the wall perpendicular to itself at A, and $R'=$ the reaction of the peg perpendicular to the beam at C. The angle θ, which the beam makes with the horizontal direction when in equilibrium by the action of these three forces, is required.

Employing (17), (18), and (19), and resolving the forces in vertical and horizontal directions, and about the point C, we have

w resolved in a horizontal direction $= 0$,
R " " " $= R$,
R' " " " $= -R' \sin. \theta$.

$\therefore$ (17) $\Sigma.X = R - R' \sin. \theta = 0.$ (a)

Also, w estimated vertically $= w$,
R " " $= 0$,
R' " " $= -R' \cos. \theta$

$\therefore$ (18) $\Sigma.Y = w - R' \cos. \theta = 0.$ (b)

Also, the moment of w about $C = w.CF = w.(DF - DC)$.

But $DF = AH = Ag \cos. \theta = a \cos. \theta$, and $DC = b$.

$$\therefore w.CF = w.(a. \cos. \theta - b),$$

the moment of R about $C = R.CK = R.AK. \tan. \theta = R.b. \tan. \theta$,
" R' " $=0$.

$\therefore$ (19) $\Sigma.(Xy - Yx) = w.(a. \cos. \theta - b) - R.b. \tan. \theta = 0.$ (c)

Multiplying (a) by $\cos. \theta$ and (b) by $\sin. \theta$, and subtracting, we have

$$R. \cos. \theta - w \sin. \theta = 0$$
$$\therefore R = w \tan. \theta.$$

Substituting this value of R in (*c*),

$$w(a\cos.\theta-b)-w.b.\tan.^2\theta=0,$$

or
$$a\cos.\theta=b(1+\tan.^2\theta)=b.\sec.^2\theta=b.\frac{1}{\cos.^2\theta}.$$

$$\therefore\ \cos.\theta=\sqrt[3]{\frac{b}{a}},\ \text{and } b<a,\ \text{except when } \theta=0$$

Ex. 6. Solve *Ex.* 5 by resolving the forces parallel and perpendicular to the beam, and taking the moments about either A or *g*.

The three forces resolved in the direction of the beam give

$$R\cos.\theta-w\sin.\theta=0; \qquad (a)$$

resolved perpendicularly to the beam, give

$$R'-w\cos.\theta-R\sin.\theta=0. \qquad (b)$$

The moments about A give

$$R'.b.\sec.\theta-w.a.\cos.\theta=0. \qquad (c)$$

From (*a*), we have $R=w\tan.\theta$, which, substituted in (*b*), gives

$$R'=w(\cos.\theta+\tan.\theta\sin.\theta)=w\frac{(\cos.^2\theta+\sin.^2\theta)}{\cos.\theta}=w\sec.\theta.$$

This value of R′, substituted in (*c*), gives

$$wb\sec.^2\theta-wa\cos.\theta=0.$$

$$\therefore\ \cos.\theta=\sqrt[3]{\frac{b}{a}},\ \text{as before.}$$

The moments taken about *g* give

$$R'.Cg-Ra\sin.\theta=0,$$

or $R'(a-b\sec.\theta)-R.a\sin.\theta=R'a\cos.\theta-R'b-Ra\sin.\theta\cos.\theta=0.$

Substituting the values of R and R′ above, and reducing,

$$\cos.\theta=\sqrt[3]{\frac{b}{a}}.$$

Ex. 7. A heavy beam lies partly in a smooth hemispherical bowl and partly over the edge; find the position of equilibrium.

The beam ABC will be supported by the reaction R of the bowl at A, perpendicular to the surface, or in the direction of the radius AO, by the reaction R′ of the edge of the bowl at

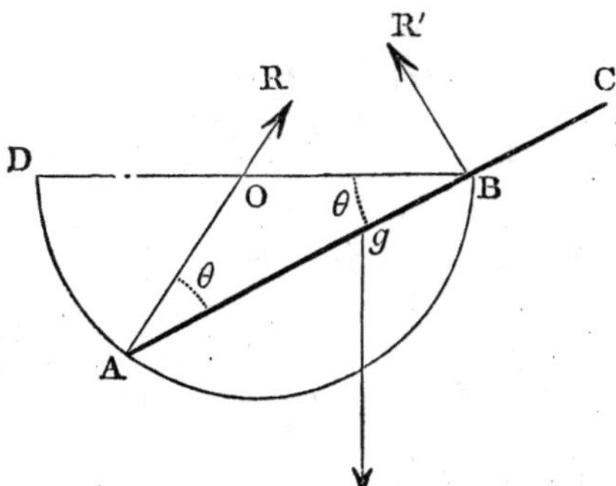

B perpendicular to the beam, and by its weight w acting at g.

Let $Ag=a$, $AO=r=$ radius of the bowl, and $\theta=ABO=BAO=$ inclination of the beam to the horizon.

If the object be to determine the angle θ solely, it will be most readily effected by resolving the forces in the direction of the beam and taking the moments about B, by which we avoid expressions involving the unknown reaction R'.

The reaction R in the direction AB $=$ R cos. θ,
the weight w " " $= -w$ sin. θ.

$$\therefore \text{R cos. } \theta - w \text{ sin. } \theta = 0, \qquad (a)$$

or

$$\text{R} = w \text{ tan. } \theta.$$

For moments about B, we have

$$\text{R.AB sin. } \theta - w.g\text{B cos. } \theta = 0. \qquad (b)$$

But $AB=2r$ cos. θ, and $gB=(2r \text{ cos. } \theta - a)$.

$$\therefore 2.\text{R}.r. \text{ sin. } \theta \text{ cos. } \theta - w(2r \text{ cos. } \theta - a) \text{ cos. } \theta = 0.$$

Substituting the value of R, and reducing,

$$2r \text{ tan. } \theta. \text{ sin. } \theta - 2r \text{ cos. } \theta + a = 0,$$

or

$$2r - 4r \text{ cos.}^2 \theta + a \text{ cos. } \theta = 0.$$

$$\therefore \text{cos. } \theta = \frac{a \pm \sqrt{32r^2 + a^2}}{8r},$$

in which the $+$ sign only is admissible.

Ex. 8. Solve the last example by following, step by step, the method of *Art.* 71, taking A for the origin of co-ordinates, and AB for the axis of x.

Ex. 9. Find the horizontal strain on the hinges of a given door, and show that the vertical pressures are indeterminate.

Let the annexed figure represent the door, of which A and B are the hinges. Let g be its center of gravity at which the weight w acts. The door is kept in equilibrium by the

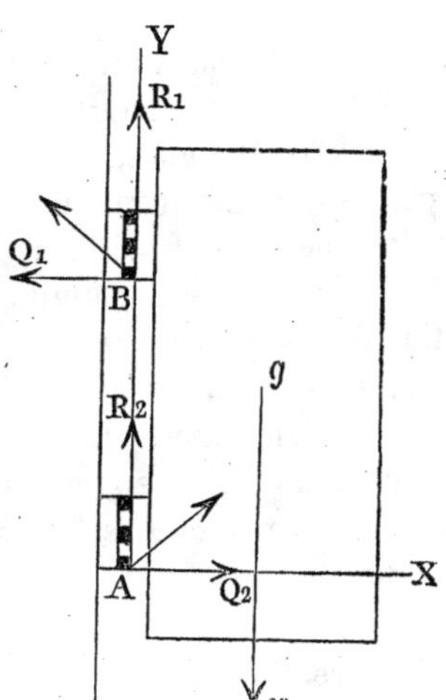

weight acting at g, and the reactions of the hinges represented by the oblique arrows at A and B.

Let A be the origin of co-ordinates, AX the axis of x, AY the axis of y; and let $x=a$, $y=b$ be the co-ordinates of g; $x=0$, $y=h$ those of the hinge B. Let the resolved parts of the reactions at B be Q_1 horizontally, and R_1 vertically, and Q_2, R_2 those at A respectively.

Then $\quad \Sigma.\ X \ = Q_2 - Q_1 = 0$, or $Q_1 = Q_2$, $\qquad (a)$

$\Sigma.\ Y \ = w - R_1 - R_2 = 0$, $\qquad (b)$

$\Sigma.(Xy - Yx) = w.a - Q_1 h = 0$. $\qquad (c)$

From (a) and (c) we have $Q_1 = \dfrac{wa}{h} = Q_2$,
which gives the horizontal strain; and it is the same in magnitude at each hinge, but opposite in direction.

Again, from (b) we have $R_1 + R_2 = w$, but we have no other relation by which we may determine the values of R_1 and R_2, which are therefore indeterminate.

Ex. 10. Two given smooth spheres rest in contact on two smooth planes, inclined at given angles to the horizon; find their position of equilibrium.

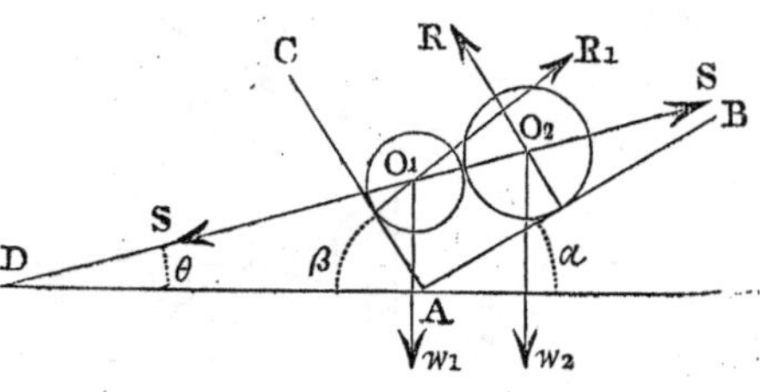

Let the planes AB, AC make the angles a and β respectively with the horizontal line through A; O_1 and O_2 be the centers of the spheres at which their weights w_1, w_2 respectively act; R_1, R_2 the reactions of the planes at the points of con tact, perpendicular to themselves, and therefore passing through the centers of the spheres to which they are tangents. Let S be equal to the mutual pressure of the spheres at their point of contact, acting in the line passing through their centers, and making the angle θ with the horizontal line AD. θ is required.

Each sphere is in equilibrium from its own weight, the reaction of the plane against which it rests, and the pressure of the other sphere.

By resolving the forces in the direction of each plane for the equilibrium of each sphere, we shall avoid equations involving the unknown reactions R_1 and R_2, and have, in the direction of AB,

$$w_2 \sin. a - S \cos. (a-\theta) = 0; \qquad (a)$$

in the direction of AC,

$$w_1 \sin. \beta - S \cos. (\beta+\theta) = 0. \qquad (b)$$

Eliminating S, we have

$$w_2 \sin. a. \cos. (\beta+\theta) = w_1 \sin. \beta. \cos. (a-\theta).$$

Expanding and reducing,

$$\tan. \theta = \frac{w_2 \sin. a. \cos. \beta. - w_1 \sin. \beta. \cos. a}{(w_1+w_2) \sin. a. \sin. \beta} = \frac{w_2 \cot. \beta - w_1 \cot. a}{w_1+w_2}.$$

Ex. 11. A sphere is sustained upon an inclined plane by the pressure of a beam movable about the lowest point of the inclined plane; given the inclination of the beam to the plane, required that of the plane to the horizon.

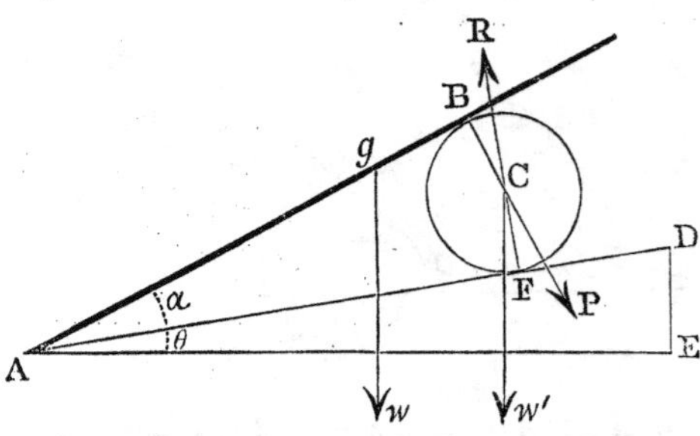

Let AgB be the beam movable about A, w= weight of beam acting at g, B the point of contact with the sphere whose center is C, w'= weight of the sphere.

The sphere is in equilibrium, from the reaction R of the plane at the point of contact F, from the pressure P of the beam at B, and from its own weight w'. These three forces all act through the center C. Ag=a, AB=b, BAD=a are given, or, instead of either of the two latter, the radius of the sphere may be given. The angle DAE$=\theta$, the inclination of the plane when in equilibrium, is required.

For the condition of equilibrium of the beam, take the moments about A.

$$P.AB=w.Ag.\cos.(a+\theta).$$

Hence $$P=\frac{wa}{b}\cos.(a+\theta).$$

For the condition of equilibrium of the sphere, resolving the forces in the direction of AD, we have

$$w'\sin.\theta-P\sin.a=0,$$

or $$w'\sin.\theta-w\frac{a}{b}\sin.a\cos.(a+\theta);$$

whence $$\tan.\theta=\frac{wa\cos.a.\sin.a}{wa\sin.^2a+w'b},$$

which gives θ the elevation of the plane, as required.

Many statical problems require, for the determination of all the unknown quantities, equations to be formed by *geometrical relations.* Take, for illustration, the following:

✓ *Ex.* 12. A heavy beam turns about a hinge, and is kept in equilibrium by a cord attached to the lower end; the cord passes over a pulley in the same horizontal line with the hinge, and sustains a given weight; find the position of equilibrium of the beam.

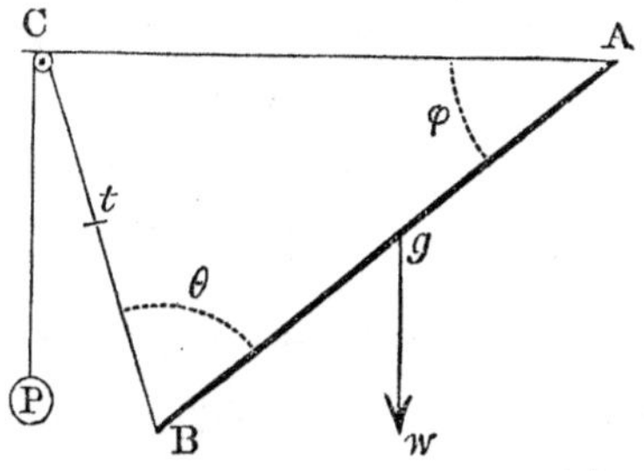

Let A be the hinge, C the pulley, AC$=c$, AB the beam $=l$ in length, g its center of gravity at which its weight w acts, A$g=a$ and P$=$ the weight hung from the cord and measuring its tension t.

Let $\theta=$ABC and $\phi=$BAC, both unknown.

Taking moments about A, we have,

$$t.AB.\sin.\theta=w.Ag\cos.\phi,$$

or $$\sin.\ \theta=\frac{w}{\mathrm{P}}.\frac{a}{l}.\ \cos.\ \phi. \qquad (a)$$

From the *geometrical* data, we have

$$\frac{\mathrm{AC}}{\mathrm{AB}}=\frac{\sin.\ \theta}{\sin.\ \mathrm{ACB}},$$

or $$\sin.\ \theta=\frac{c}{l}.\ \sin.\ (\theta+\phi). \qquad (b)$$

Equations (a) and (b) suffice to determine θ and ϕ.

Ex. 13. A uniform beam rests with its lower end in a smooth, hemispherical bowl, and its upper end against a smooth, vertical plane. Find the position of equilibrium.

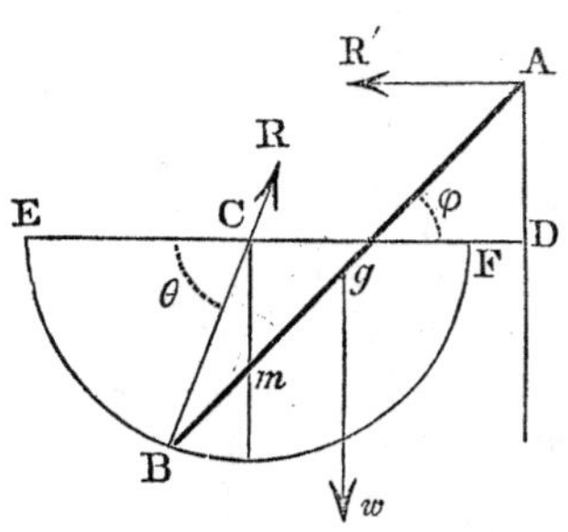

The beam AB rests against the vertical plane AD at A and upon the bowl at B, and is sustained by its weight w acting at g, the reaction R of the bowl in the radius BC, and the reaction R′ of the plane perpendicular to itself.

Let $r=$ radius of the bowl; AB $=2a$, A$g=a$, since the beam is uniform, $w=$ weight, and CD$=d$, all supposed known.

Resolving vertically, we have

$$\mathrm{R}\ \sin.\ \theta-w=0, \text{ or } \mathrm{R}=\frac{w}{\sin.\ \theta}.$$

Taking the moments about A, we have

$$\mathrm{R.AB.}\ \sin.\ (\theta-\phi)-w.\mathrm{A}g.\ \cos.\ \phi=0,$$

or $$\mathrm{R}.2a.\ \sin.\ (\theta-\phi)-w.a.\ \cos.\ \phi=0.$$

$$\therefore\ \frac{\sin.\ (\theta-\phi)}{\sin.\ \theta}-\frac{\cos.\ \phi}{2}=0,$$

or $$\cos.\ \phi-\cot.\ \theta.\ \sin.\ \phi-\tfrac{1}{2}\cos.\ \phi=0.$$

Hence $$\tan.\ \theta=2\ \tan.\ \phi. \qquad (a)$$

This equation containing two unknown quantities, a geometrical relation between them must be obtained. Cm being a vertical line meeting AB in m,

$$\frac{\cos.\theta}{\cos.\phi}=\frac{Bm}{BC}=\frac{2a-Am}{r}=\frac{2a-CD.\sec.\phi}{r}=\frac{2a-d\sec.\phi}{r}.$$

$$\therefore \cos.\theta=\frac{2a\cos.\phi-d}{r}. \qquad (b)$$

Equations (a) and (b) are sufficient to determine ϕ and θ as required.

Ex. 14. A heavy beam has one end resting against a smooth vertical wall, and the other sustained by a cord, which is fastened at a point vertically above the point where the beam rests. Find all the forces which keep the beam in equilibrium

Let CB be the beam, AB the cord, A and C the points on the vertical wall AD.

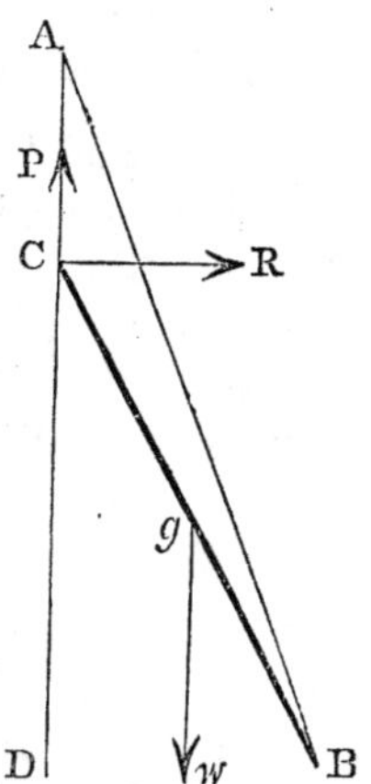

Let w= weight of the beam, g its center of gravity $Cg=a$, $CB=l$, $AB=c$, and $AC=h$, all supposed known. The angles A, B, and C will be known.

Let t= tension of the cord. The beam will press against the wall, and this pressure may be resolved in a vertical and horizontal direction; the latter, perpendicular to the wall, will be destroyed by its reaction; but, since the wall is smooth, the vertical component can be balanced only by an opposite force =P. This vertical omponent will be upward, or downward, or zero, according to the position of the point C. We shall suppose its position to be such that the component may be upward, and require a force to be applied downward to keep the end of the beam at C.

Resolve vertically and horizontally, and take the moments about C.

$$P+t\cos.A-w=0, \qquad (a)$$
$$R-t\sin.A=0, \qquad (b)$$
$$w.a.\sin.C-t.l.\sin.B=0. \qquad (c)$$

From (c), $\quad t=\frac{w.a.\sin.C}{l.\sin.B}=\frac{w.a.c}{l.h}$= tension of cord.

Substituting the value of t in (b),

$$R = w\frac{ac}{lh}.\sin.A = \text{pressure against the wall;}$$

and from (a),

$$P = w\left(1 - \frac{ac}{lh}\cos.A\right) = \text{force to be applied at C,}$$

to prevent the beam from sliding along the wall.

Ex. 15. A weight w hangs from one end of a cord of which the other end is fastened to a vertical wall; the cord is pushed from the wall by a rod tied to it, which is perpendicular to the wall. Find the pressure R of the rod on the wall when the cord makes the angle a with the wall.

Ans. $R = w.\tan.a$.

Ex. 16. A heavy beam, $AB = l$, of which the weight is w, lies with the end A against a smooth vertical wall AD, and the end B on a smooth horizontal plane DB, making with it the angle θ. The distance of its center of gravity g from B is a, and it is kept in equilibrium by a cord attached to it at B and fastened at D. Required the tension of the cord, the reaction R of the wall, and the pressure P on the plane.

Ans. $R = t = w.\frac{a}{l}.\cot.\theta$. $P = w$.

Ex. 17. A body (weight $= w$) is suspended by a cord (length $= l$) from the point A in a horizontal plane, and is thrust out of its vertical position by a rod without weight, acting from another point B in the horizontal plane, such that $AB = d$, and making the angle θ with the plane. Find the tension t of the cord.

$$t = w\frac{l}{d}\cot.\theta.$$

Ex. 18. A triangular plate of uniform thickness and density is supported horizontally by a prop at each angle. Find the pressure on each prop.

Ex. 19. A uniform beam rests on two planes inclined at angles a and β to the horizon. Find the inclination θ of the beam to the horizon.

$$\tan.\theta = \frac{\sin.(\beta - a)}{2\sin.a\sin.\beta}.$$

Ex. 20. A uniform beam AB hangs by a string BC from a fixed point C, with its lower extremity A resting on a smooth, horizontal plane. Show that, when there is an equilibrium, CB must be vertical.

Ex. 21. A uniform beam AB is placed with one end A inside a smooth hemispherical bowl, with a point P resting on the edge of the bowl. If AB= 3 times the radius R, find AP.

AP=1.838R.

Ex. 22. A beam, whose weight is w and length 6 feet, rests on a vertical prop CD (= 3 feet); the lower end A is on a horizontal plane, and is prevented from sliding by a string DA (= 4 feet). Find the tension of the string.

$$\text{Tension} = \frac{36}{125}w.$$

Ex. 23. A uniform beam, CB=$2a$, has one end C resting against a smooth, vertical wall AC, and the other sustained by a cord, whose length is c, fastened to the point A. Find AC when the beam is in equilibrium from its weight w, the tension t of the cord, and the reaction R of the wall. (*Vide Ex.* 14.)

Ex. 24. Given the inclination i of the right line AC to the horizon and the weight of a heavy body W, to determine what force or weight P_1 acting in the given direction WM will be sufficient to sustain W on the line.

The body W is kept upon the line by the action of the force P_1 in the direction WM, and its weight acting in the direction WE. The reaction of the line normal to its direction is a third force (*Art.* 76).

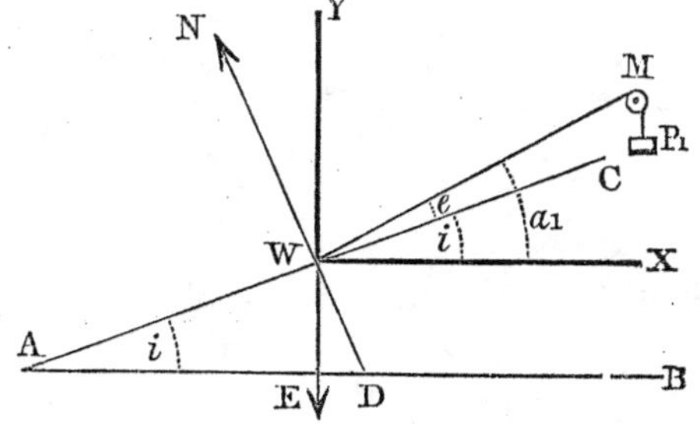

Resolving P and W horizontally and vertically, we have

$$\Sigma.X = P_1 \cos.(e+i) + W \cos. 270° = P_1 \cos.(e+i),$$
$$\Sigma.Y = P_1 \sin.(e+i) + W \cos. 180° = P_1 \sin.(e+i) - W$$

Hence (20) becomes

$$P_1 \cos.(e+i)+(P_1 \sin.(e+i)-W) \tan. i=0;$$

or, expanding and reducing,

$$P_1 \cos. e=W \sin. i,$$

or
$$P_1=\frac{W \sin. i}{\cos. e}. \qquad (a)$$

If the *reaction* of the line be required, resolve all the forces horizontally (8). This gives

$$N. \cos.(90°+i)+P_1 \cos.(e+i)=0.$$

$$\therefore N=P_1 \frac{\cos.(e+i)}{\sin. i}, \qquad (b)$$

or, by (a),
$$N=W.\frac{\cos.(e+i)}{\cos. e}. \qquad (c)$$

Ex. 25. A given weight W is kept at rest on a circular arc by a weight P attached to a cord which passes over a point M in the vertical line MX through C, the center of the circle. Required the position of W, supposing no friction at M.

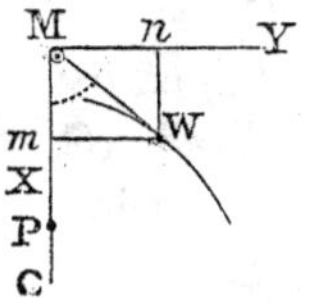

Resolving the forces P and W vertically and horizontally, we have, calling the angle PMW, e,

$$\Sigma.X=W-P \cos. e,$$
$$\Sigma.Y= \quad -P \sin. e.$$

Let the co-ordinates of the point W be x, y, and call the distance MC, a, and MW, l; then the equation referred to M as the origin, is

$$(a-x)^2+y^2=r^2,$$

or
$$y=\sqrt{r^2-(a-x)^2}.$$

Differentiating, we obtain

$$\frac{dy}{dx}=\frac{a-x}{\sqrt{r^2-(a-x)^2}}=\frac{a-x}{y}.$$

Substituting in (21), and recollecting that $\cos. e=\frac{x}{l}$ and $\sin. e=\frac{y}{l}$, we get

$$W-P.\frac{x}{l}-P.\frac{y}{l}.\frac{dy}{dx}=W-P\frac{x}{l}-P\frac{a-x}{l}=0$$

Hence $$l=\frac{aP}{W},$$

an equation which determines the distance of W from M.

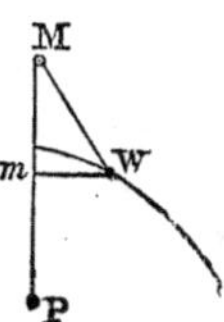

Ex. 26. Instead of a circle, as in *Ex.* 25, let the curve be a hyperbola with its transverse diameter vertical, the point M being at its center.

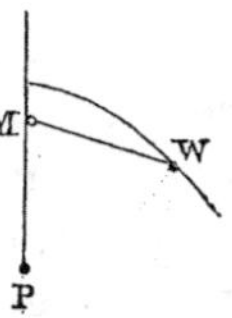

Ex. 27. Required a curve such that a given weight P, by a cord passing over a fixed point without friction, will balance another given weight W at every point of it.

Ex. 28. Required the co-ordinates of the center of gravity of a semiparabola whose equation is $y^2=px$, height $=a$, and base $=b$.

Ans. $\bar{x}=\frac{3}{5}a$, $\bar{y}=\frac{3}{8}b$.

Ex. 29. Required the center of gravity of the surface of a right cone.

Ex. 30. Find the center of gravity of a paraboloid of revolution whose altitude is a.

Ans. $\bar{x}=\frac{2}{3}a$.

Ex. 31. Find the center of gravity of a segment of a hyperboloid whose altitude is a.

CHAPTER VI.

ON THE MECHANICAL POWERS.

130. The general object of machinery is to transmit and to economize the action of certain forces at our disposal. The specific end is, sometimes to augment the action of which the power employed is capable when applied without the intervention of machinery; sometimes, merely to change the direction of the action; and sometimes to regulate the velocity of the point to which the action is transmitted.

The most simple machines are denominated *Mechanical powers*, and are reducible to three classes, viz., the *Lever*, *Cord*, and *Inclined Plane*.

The first class comprehends every machine consisting of a solid body capable of revolving on an axis, as the *Wheel and Axle*.

The second class comprehends every machine in which force is transmitted by means of flexible threads, ropes, &c., and hence includes the *Pulley*.

The third class comprehends every machine in which a hard surface inclined to the horizon is introduced, as the *Wedge* and the *Screw*.

The force which is used to sustain or overcome any opposition is called the *Power;* the opposition to be overcome is called the *Weight*. This distinction in the names of the forces employed implies none in their nature.

§ I. THE LEVER.

131. Def. A *Lever* is an inflexible rod capable of motion about a fixed point, called a *fulcrum*. The rod may be straight or any how bent.

It is generally regarded, at first, as without weight, but its weight may obviously be considered as another force applied in a vertical direction at its center of gravity.

Def. The *arms* of a lever are the portions of it intercepted between the power and fulcrum, and between the weight and fulcrum.

132. Levers are divided into three kinds, according to the relative positions of the *power*, *weight*, and *fulcrum*.

In a lever of the *first kind*, the fulcrum lies between the points at which the power and weight act.

In a lever of the *second kind*, the weight acts at a point between the fulcrum and the point of action of the power.

In a lever of the *third kind*, the point of action of the power is between that of the weight and the fulcrum.

133. Prop. *Required the condition of equilibrium and pressure on the fulcrum when two parallel forces act on a straight lever.*

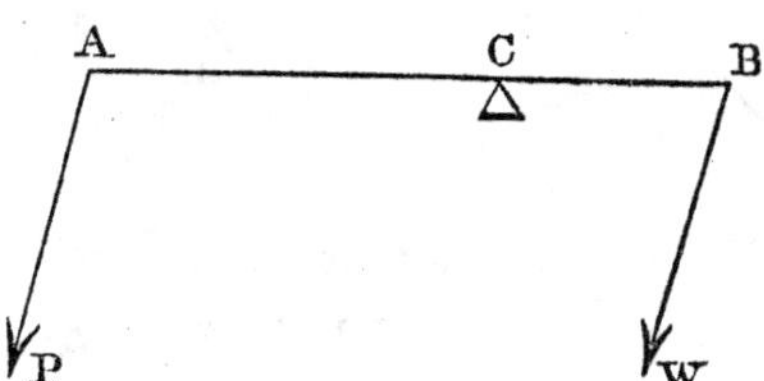

Since the fulcrum C is a fixed point, by *Art.* 75, the sum of the moments of the forces about C must be zero.

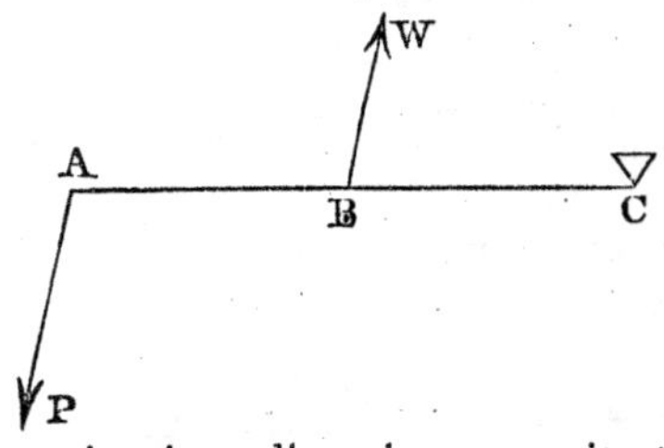

Let a be the angle made by the direction of the forces with the lever. From (19), we have

$$W.BC. \sin. a - P.AC. \sin. a = o,$$

the moment of P being negative, since it tends to produce motion in a direction opposite to that of W.

Hence $\frac{W}{P} = \frac{AC}{BC}$, or, in case of equilibrium, the weight and power are reciprocally proportional to the distances at which they act from the fulcrum.

Hence (*Art.* 36) the resultant of P and W must pass through C, and the pressure on the fulcrum is equal to the algebraic sum of P and W, and acts in the direction of the greater.

Cor. If the power equal the weight, the distances of their points of action from the fulcrum will be equal.

For $P = W$ gives $AC = BC$.

134. Prop. *Required the condition of equilibrium and the pressure on the fulcrum when any two forces in the same plane act on a straight lever.*

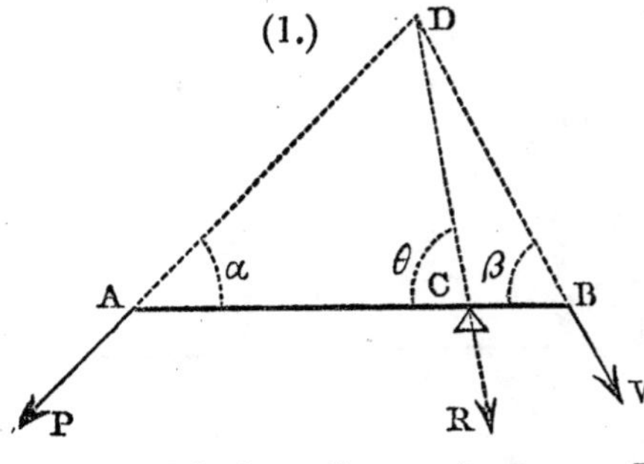

Let the forces P and W make the angles a and β respectively with the lever, and let their directions, when produced if necessary, meet in D. Since C is a fixed point in the lever, the sum of the moments of P and W about C must be zero. Hence

$$W.CB. \sin. \beta - P.AC. \sin. a = 0,$$

or $$\frac{W}{P}=\frac{AC. \sin. a}{CB. \sin. \beta}.$$

But AC. sin. $a=$ perpendicular from C on AD, and CB. sin. $\beta=$ perpendicular from C on BD. Hence the condition of equilibrium requires that *the power and weight should be inversely as the perpendiculars from the fulcrum on their respective directions.*

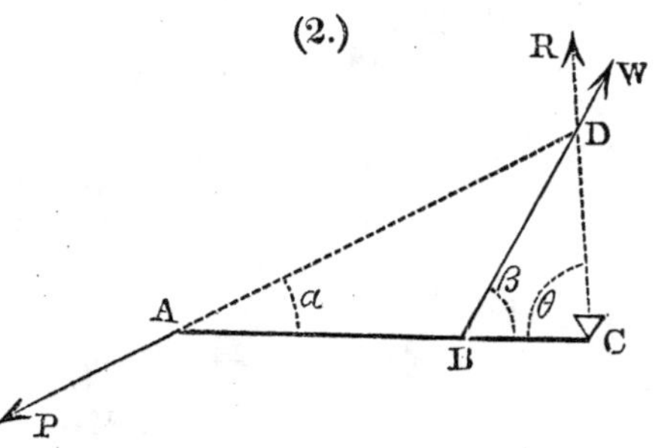

Since the lever is in equilibrium by the actions of P and W and the reaction of the fulcrum, the resultant of P and W must be equal and opposite to that reaction. It will therefore pass through C, and be equal to the pressure on the fulcrum.

To find R, we have (*Art.* 29)

$$R^2=P^2+W^2+2PW. \cos. ADB \text{ in } Fig. (1),$$

and $$R^2=P^2+W^2+2PW \cos. ADW \text{ in } Fig. (2).$$

But $ADB=180-(a+\beta)$, and $ADW=180-ADB=180-(\beta-a)$.

$$\therefore R^2=P^2+W^2-2PW \cos. (a+\beta),$$

or $$R^2=P^2+W^2-2PW \cos. (\beta-a).$$

To find the angle θ made by R with the lever, resolve parallel and perpendicular to the lever, the reaction of the fulcrum being equal and opposite to R. Hence we have, in *Fig.* (1),

$$P. \cos. a-W \cos. \beta+R \cos. \theta=0,$$
$$P \sin. a+W \sin. \beta-R \sin. \theta=0.$$

$$\therefore \text{Tan. } \theta=\frac{P. \sin. a+W \sin. \beta}{W \cos. \beta-P \cos. a};$$

in *Fig.* (2), $$P \cos. a-W \cos. \beta-R \cos. \theta=0,$$
$$P \sin. a-W \sin. \beta+R \sin. \theta=0.$$

$$\therefore \text{Tan. } \theta=\frac{W \sin. \beta-P \sin. a}{P \cos. a-W \cos. \beta}.$$

135. Cor. Whenever the lever is *bent* or *curved*, the condition of equilibrium is the same.

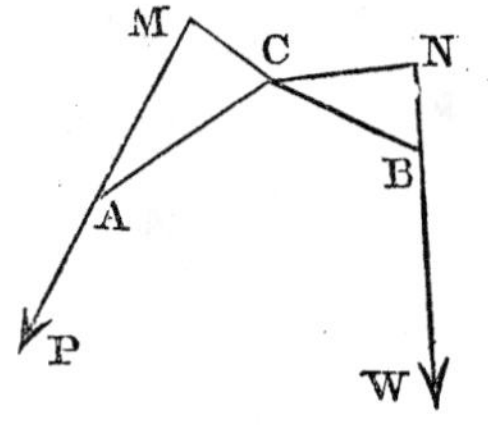

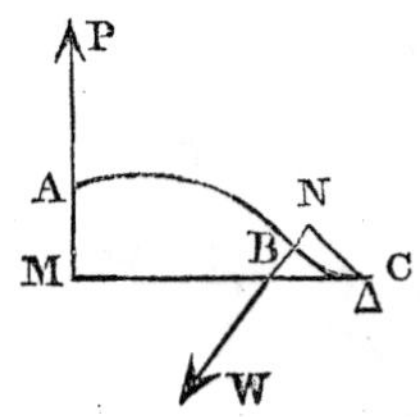

For, since the moments of the forces about the fulcrum in opposite directions must be equal, in case of equilibrium, we have

$$P.CM = W.CN,$$

or

$$\frac{W}{P} = \frac{CM}{CN}.$$

136. Prop. *Required the condition of equilibrium and the pressure on the fulcrum, when any number of forces act in any direction in one plane on a lever of any form.*

Let P_1, P_2, P_3, &c., be the forces in one plane,
p_1, p_2, p_3, " the distances of their points of application from the fulcrum,
a_1, a_2, a_3, " the angles made by the directions of P and p respectively.

Then $p_1 \sin. a_1, p_2 \sin. a_2, p_3 \sin. a_3$ are the perpendiculars from the fulcrum on the directions of the forces, and $P_1 p_1 \sin. a_1$, &c., the moments of the forces.

When there is an equilibrium, the sum of the moments of the forces about the fulcrum will be zero, or

$$P_1 p_1 \sin. a_1 + P_2 p_2 \sin. a_2 + P_3 p_3 \sin. a_3 +, \&c., = 0.$$

The signs of the moments will depend on the direction in which they tend to produce rotation.

To find the pressure on the fulcrum, we must determine the resultant of all the forces supposed to concur at the fulcrum. For each moment $P_1 p_1 \sin. a_1$ is equal to a couple of which the forces are P and the arm $p \sin. a$, and a single force P acting at the fulcrum. The magnitude of R will then be determined by the equation (6),

$$R = \sqrt{(\Sigma.X)^2 + (\Sigma.Y)^2}.$$

COR. If the lever be straight and the forces parallel,

$$a_1 = a_2 = a_3 =, \text{\&c.}$$

Hence $\qquad P_1 p_1 + P_2 p_2 + P_3 p_3 +, \text{\&c.}, = 0.$

The same result will also be obtained by following, step by step, the method of *Art.* 71.

EXAMPLES.

Ex. 1. On a straight lever AB of the first kind, without weight, 36 inches in length, a weight W=15 lbs., acting at B, is balanced by a power P=3 lbs. acting at A. Required the distance of the fulcrum C from A.

Let AC$=x$; then BC$=36-x$.

By *Art.* 133, $\qquad$ P.AC=W.BC,

or
$$3.x = 15(36-x).$$

Hence
$$x = 30 \text{ inches} = AC.$$

Ex. 2. On one arm $(=p_1)$ of a straight lever of the first kind, without weight, a body counterpoises a weight $(=a$ lbs.), on the other $(=p_2)$ a weight $(=b$ lbs.). Required the weight of the body.

A straight lever of the first kind, with unequal arms, and having the fulcrum at its center of gravity, is called a *false balance*.

Let $x=$ the unknown weight.

By *Art.* 133, $\qquad x.p_1 = a.p_2,$

and
$$x.p_2 = b.p_1.$$

By multiplying the equations member by member,

$$x^2 p_1 p_2 = ab.p_1 p_2,$$
$$x^2 = ab,$$
$$x = \sqrt{ab}.$$

Ex. 3. On a straight lever, without weight, are suspended five bodies, $P_1=4$ lbs., $P_2=10$ lbs., $P_3=2$ lbs., $P_4=3$ lbs., $P_5=7$ lbs., at the points A, B, C, D, and E, such that AB=4 feet, BC=2 feet, CD=6 feet, and DE=8 feet. Required the position of the fulcrum F, about which they will balance.

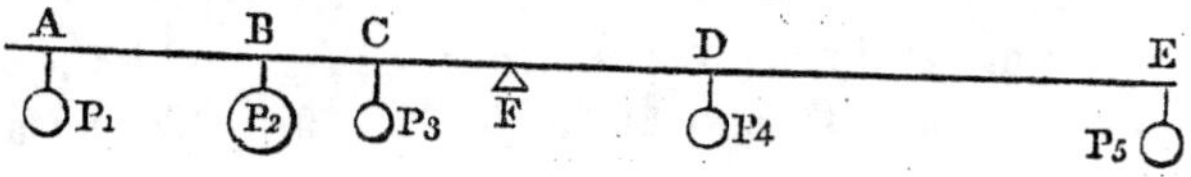

Let AF$=x$. Then, by *Art.* 136, *Cor.*, we have

$P_1 = 4$ and $p_1 = \text{AF} = x$. $\therefore P_1 p_1 = 4.x$.
$P_2 = 10$ " $p_2 = \text{BF} = \text{AF} - \text{AB} = x - 4$. $P_2 p_2 = 10(x-4)$.
$P_3 = 2$ " $p_3 = \text{CF} = \text{AF} - \text{AC} = x - 6$. $P_3 p_3 = 2(x-6)$
$P_4 = 3$ " $p_4 = \text{DF} = \text{AF} - \text{AD} = x - 12$. $P_4 p_4 = 3(x-12)$
$P_5 = 7$ " $p_5 = \text{EF} = \text{AF} - \text{AE} = x - 20$. $P_5 p_5 = 7(x-20)$.

And $\Sigma.\text{P}p = 4x + 10(x-4) + 2(x-6) + 3(x-12) + 7(x-20) = 0$ $= 26x - 228$.

Hence $x = \text{AF} = 8\frac{10}{13}$ feet.

Otherwise: Since the weights are parallel forces, their resultant R is equal to their sum. The whole system being in equilibrium, the resultant must pass through the fulcrum, and the moment of the resultant must be equal (*Art.* 45) to the sum of the moments of the components. Taking A for the origin of moments, we have

$$\text{R}.x = \text{P}_1.0 + \text{P}_2.\text{AB} + \text{P}_3.\text{AC} + \text{P}_4.\text{AD} + \text{P}_5.\text{AE},$$

or $26.x = 4 \times 0 + 1 \times 4 + 2 \times 6 + 3 \times 12 + 7 \times 20 = 228$.

$\therefore x = \text{AF} = 8\frac{10}{13}$ feet.

Ex. 4. A uniform lever AB of the first kind, 12 feet long, whose weight $w = 6$ lbs., has a weight W$=100$ lbs. suspended from the shorter arm CB$=2$ feet. Required the power P which must be applied vertically at A, to equilibrate W.

The weight of the lever has the effect of a weight equal to itself, applied at its center of gravity, which, since the lever is uniform in size and density, is at its middle point. The distance of this point from C is $\frac{\text{AC}-\text{BC}}{2}$, and its moment about C will be $w.\frac{\text{AC}-\text{BC}}{2}$.

Hence
$$\text{P.AC} + w\frac{\text{AC}-\text{BC}}{2} = \text{W.BC},$$

or
$$\text{P} = \frac{\text{W.BC} - \frac{w}{2}(\text{AC}-\text{BC})}{\text{AC}} = \frac{100 \times 2 - 3 \times 8}{10} = 17.6 \text{ lbs.}$$

Ex. 5. The arms of a bent lever are $a = 3$ feet and $b = 5$ feet, and inclined to each other at an angle $\theta = 150°$. To the arm

a a power P=7 lbs. is applied, and to the arm b a weight W =6 lbs. Required the inclination of each arm to the horizon when there is an equilibrium.

Let α be the inclination of the arm a, and β the inclination of the arm b to the horizon.

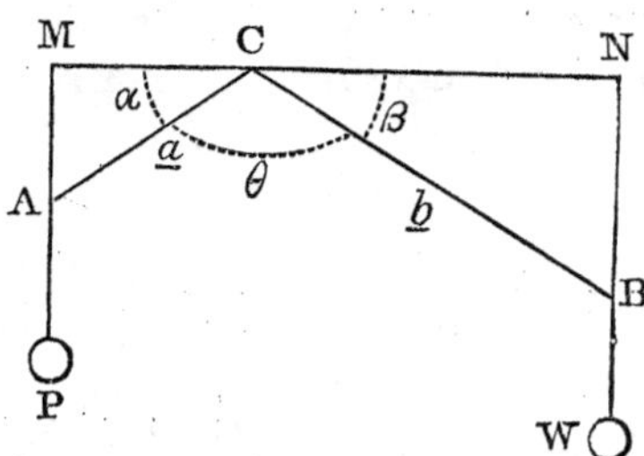

Then P.MC=W.NC,

or $P.a. \cos. \alpha = W.b. \cos. \beta = W.b. \cos. (180° - (\alpha+\theta)),$

$$= - W.b. \cos. (\alpha+\theta)$$

$$= - W.b. \cos. \alpha \cos. \theta + W.b. \sin. \alpha. \sin. \theta.$$

$$\therefore \text{Tan. } \alpha = \frac{P.a + W.b. \cos. \theta}{W.b. \sin. \theta} = \frac{21+30. \cos. 150°}{30. \sin. 150°}$$

$$= \frac{21-30 \cos. 30°}{30. \sin. 30} = \frac{21-30\times\frac{1}{2}\sqrt{3}}{30\times\frac{1}{2}} = -\frac{4{,}98}{15}$$

Log. Tan. $\alpha = 9.5211381_{n}$,

$$\alpha = -18°.22',$$

and $\beta = 180° - (\alpha+\theta) = 180° + 18°.22' - 150° = 48°.22'.$

Hence the arm AC is inclined at an angle of 18°.22′ above the horizon, and BC at an angle of 48°.22′ below the horizon.

✓ *Ex.* 6. The whole length of the beam of a false balance (*Ex.* 2) is 3 feet 9 inches. A body placed in one scale counterpoises a weight of 9 lbs., and in the other a weight of 4 lbs. Required the true weight W of the body, and the lengths a and b of the shorter and longer arms.

Ans. W=6 lbs., a=1 ft. 6 in., b=2 ft. 3 in.

Ex. 7. A false balance has one of its arms exceeding the other by $\frac{1}{m}$th part of the shorter arm. Supposing a shopkeeper, in using it, puts the weight as often in one scale as the other, does he gain or lose, and how much per cent.?

Ans. Loses $\frac{50}{m^2+m}$ per cent.

Ex. 8. The arms of a bent lever are 3 feet and 5 feet, and inclined to each other at an angle of 150°, and to the shorter arm is suspended a weight of 7 lbs. Find what the other weight must be in order, 1st, that the shorter, and 2d, that the longer arm may rest in a horizontal position.

Ans. 1st, 4.85 lbs.
2d, 3.64 lbs.

§ II. WHEEL AND AXLE.

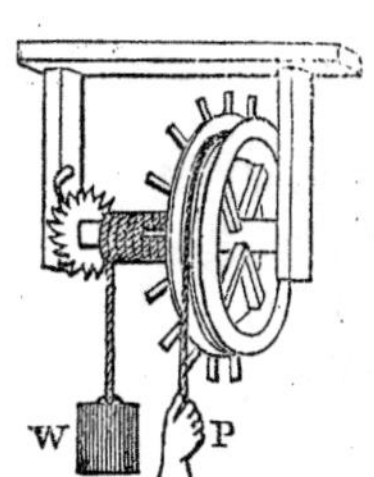

137. The *Wheel and Axle* consists of a cylinder or axle, perpendicular to which is firmly fixed a circle or wheel, whose center is in the axis of the cylinder. The whole is supposed to be perfectly rigid, and movable only round the axis of the cylinder.

In the ordinary applications of this machine, the power is applied tangentially to the surface of the wheel, and the weight, in the same manner, to the surface of the axle.

138. PROP. *Required the condition of equilibrium of the wheel and axle when two forces are applied tangentially to the circumference of the axle and of the wheel.*

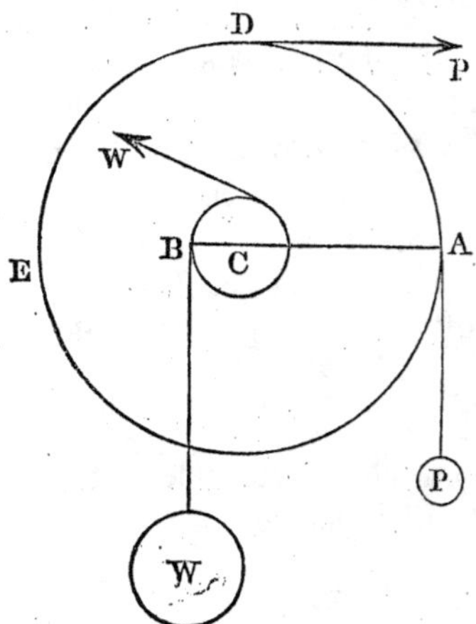

Let ADE be a section of the wheel perpendicular to the axis, and BCA a horizontal line through the center of the axis, terminated at A by the circumference of the wheel, and at B by the circumference of the axle. The power P acts in this vertical plane at A, and the weight W acts in a plane parallel to it. Since the axle and wheel are firmly connected, the action of the weight may be transferred to the plane ADE

(*Art.* 57), and supposed to act at B. Then, since the sum of the moments must be zero for equilibrium,

$$P.AC-W.BC=0.$$

$$\therefore \frac{W}{P}=\frac{AC}{BC}=\frac{\text{radius of wheel}}{\text{radius of axle}};$$

or, *the Power : Weight=radius of axle : radius of wheel.*

139. Cor. 1. The same relation will exist in all positions of the wheel, so that this machine may be called a *perpetual lever.*

140. Cor. 2. When the power and weight act vertically on opposite sides of the axis, the *pressure* on the rests or Ys equals the sum of the two; when vertically on the same side, it equals their difference; when in any other directions, it equals the diagonal of a parallelogram, whose sides represent the power and weight in magnitude and direction.

If ropes are used to transmit the action of the power and weight, we must suppose the forces applied to the axes of the ropes. Hence, if r and R represent the radii of the axle and wheel respectively, and t and T represent half the thickness of the ropes,

$$W.(r+t)=P.(R+T),$$

or

$$P : W=r+t : R+T.$$

141. Prop. *Required the condition of equilibrium of any number of forces, acting in any direction in planes perpendicular to the axis.*

Let the actions of all the forces be transferred (*Art.* 57) to the same plane; then, since the intersection of the axis with this plane is a fixed point, the condition of equilibrium is given by *Art.* 74; or, *the algebraic sum of the moments of all the forces about the axis must be zero.*

142. Def. *Toothed or cogged wheels* are those on the circumference of which are projections, called teeth or cogs. When two wheels have their cogs of such form and distance that those of one will work between those of the other, the motion of one wheel will be communicated to the other by the *pressure* of the cogs.

When the teeth are on the sides of the wheel instead of the circumference, they are called *crown wheels.*

In the preceding instance, the axes of the wheels and pinions are parallel or perpendicular to each other. When the axes of two wheels make an acute angle, the wheels take the form of frusta of cones, and are called *beveled wheels.*

Axles on which teeth are formed are called *pinions,* and the teeth *leaves.*

143. PROP. *Required the condition of equilibrium when the action of the power is transmitted to the weight by means of cogged wheels.*

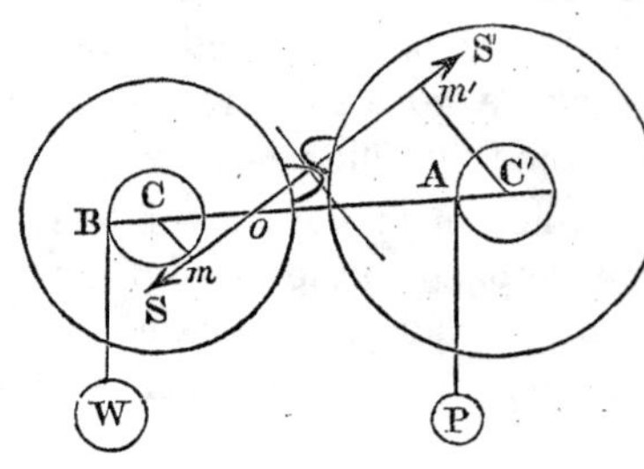

Let S be the mutual pressure of one cog upon the other. This pressure takes place in the direction of the line Sm, m'S normal to the cogs at the point of contact, Cm, C$'m'$ being perpendiculars from the centers C and C$'$, of the wheels, on that line.

Taking the moments about C$'$, when the power and weight are in equilibrium, we have

$$\text{P.C}'\text{A}=\text{S.C}'m',$$

and about C,

$$\text{W.CB}=\text{S.C}m.$$

Dividing the latter by the former, we have

$$\frac{\text{W}}{\text{P}}=\frac{\text{C}'\text{A}}{\text{CB}}\cdot\frac{\text{C}m}{\text{C}'m}.$$

Now, if the radii of the axles are equal, or C$'$A=CB, we shall have

$$\frac{\text{W}}{\text{P}}=\frac{\text{C}m}{\text{C}'m},$$

which gives the effect of the action of the cogged wheels alone or, since the triangles C$'m'o$ and Cmo are similar,

$$\frac{\text{C}m}{\text{C}'m'}=\frac{\text{C}o}{\text{C}'o}=\frac{\text{W}}{\text{P}}.$$

If the direction of the line $Sm'\,mS$ changes as the action passes from one cog to the succeeding, the point o will also change its position, and the relation of W to P become variable.

But when the cogs are of such form that the normal $Sm\,m'S$ at their point of contact shall always be tangent to both circles, the lines Cm and $C'm'$ will become radii, and their ratio constant, and the point o a fixed point, in which case

$$\frac{W}{P}=\frac{Co}{C'o}=\frac{Cm}{C'm'}=\frac{R}{R'},$$

R and R′ being the radii of the circles C and C′ respectively.

144. Cor. When the cogs are equal in breadth, the number of cogs on C will be to the number of cogs on C′ as the circumference of C to the circumference of C′, or as the radius of C to the radius of C′.

$$\therefore \frac{W}{P}=\frac{\text{number of cogs on the wheel of W}}{\text{number of cogs on the wheel of P}}.$$

Scholium. When the working sides of the cogs have the form of the involute of the circle on which they are raised, the pressure of one cog on another will always be in the direction of the common tangent to the two wheels.

Thus, let IHF, KEb be the two wheels. The acting face GCH of the cog a being formed by the extremity H of the flexible line FaH as it unwinds from the circumference, and the acting face of b by the unwinding of the thread GE, the line FCE will always be normal to the faces of the cogs a and b at their point of contact. The circles described with the radii AD and BD are called the *pitch lines* of the wheels, and will roll uniformly upon each other.

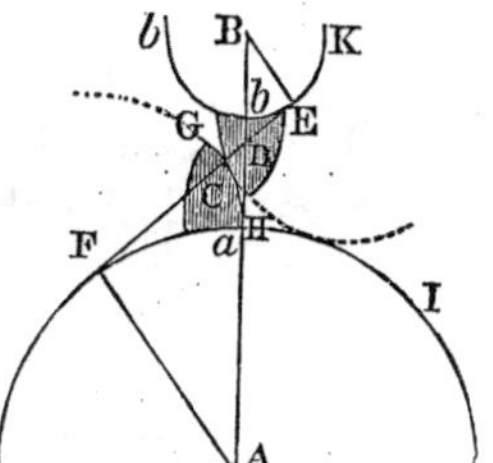

145. Prop. *Required the condition of equilibrium when the action of the power is transmitted to the weight by a system of cogged wheels and pinions.*

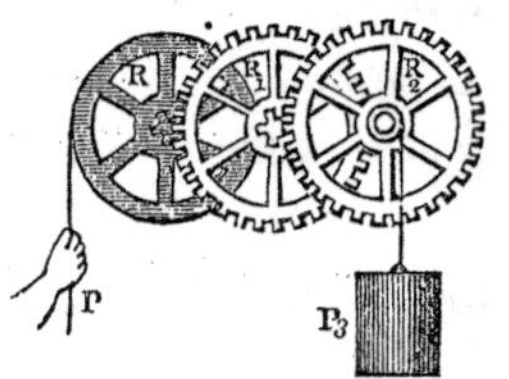

Let R, R_1, R_2, &c., be the radii of the successive wheels; r, r_1, r_2, &c., the radii of the corresponding pinions; P, P_1, P_2, &c., the powers applied to the circumferences of the successive wheels.

Taking the moments about the center of each wheel, we have

$$P.R=P_1r,\ P_1R_1=P_2r_1,\ P_2R_2=P_3r_2,\ \&c.;$$

since the power applied to the circumference of the second wheel is equal to the reaction on the first pinion.

Multiplying these equations member by member, and reducing,

$$\frac{P_n}{P}=\frac{R.R_1.R_2,\ \&c.}{r.r_1.r_2,\ \&c.};$$

or, *the power is to the weight as the product of the radii of the pinions to the product of the radii of the wheels;*

or, *as the product of the numbers expressing the leaves of each pinion to the product of the numbers expressing the cogs in each wheel.*

EXAMPLES.

Ex. 1. If a power of 10 lbs. balance a weight of 240 lbs. on a wheel whose diameter is 4 yards, required the radius of the axle, the thickness of the ropes being neglected.

Ans. r=3 inches.

Ex. 2. The radius of the wheel being 2 feet, and of the axle 5 inches, and the thickness of each rope being $\frac{3}{4}$ inch, find what power will balance a weight of 130 lbs.

Ans. P=28$\frac{2}{3}$ lbs.

Ex. 3. The radius of the wheel being 3 feet and of the axle 3 inches, find what weight will be supported by a power of 120 lbs., the thickness of the rope coiled around the axle being one inch.

Ans. W=1234$\frac{2}{7}$.

Ex. 4. There are two wheels on the same axle; the diameter of one is 5 feet, that of the other 4 feet, and the diameter of the axle is 20 inches. What weight on the axle would be supported by forces equal to 48 lbs. and 50 lbs. on the larger and smaller wheels respectively?

Ans. W=264 lbs.

§ III. THE CORD.

146. The *cord* or *rope* is employed as a means of communication of force. It is regarded as perfectly flexible and without weight, and transmits the action of a force applied at one extremity to any other point in it, unchanged in magnitude, so long as it is straight, or only passes over smooth obstacles without friction.

The force thus transmitted is called the *tension*.

Since the tension is the same throughout, from one extremity to the other, when employed alone, it affords no mechanical advantage; but when passed over or attached to certain fixed points, the resistance of these points may be employed advantageously.

147. Prop. *Required the condition of equilibrium of a cord acted upon by three forces.*

Let the forces P_1, P_2, P_3 be applied at the extremities A and B of the cord ACB, and to a knot at C. Draw any line CD in the direction of the force P_2, and DN, DM parallel to CA and CB respectively. In case of equilibrium (*Art.* 28),

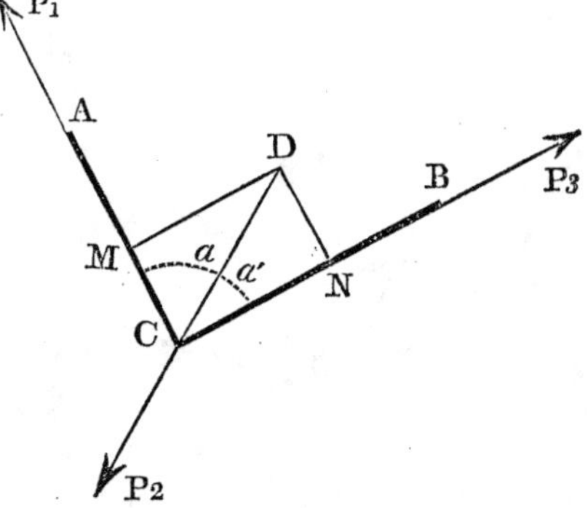

$$CM : CD : CN = P_1 : P_2 : P_3 = \sin. DCN : \sin. MCN : \sin. MCD$$
$$= \sin. a' : \sin. (a+a') : \sin. a;$$

or, *the forces are each as the sine of the angle contained between the directions of the other two.*

148. Cor. 1. If the cord be fixed at A and B, the reactions of the points A and B take the place of the forces P_1 and P_3, and are equal to the tensions of the two parts of the cord re spectively.

149. Cor. 2. If the force P_2 be applied to a running knot or ring, the points A and B being fixed, the condition of equilibrium requires that the direction of P_2 should bisect the angle ACB.

For the point C in its motion would describe an ellipse, A and B being the foci, and the force P_2 could not be in equilibrium except when normal to the curve, in which case it bisects the angle ACB. Hence, $\sin. a = \sin. a'$, and $P_1 = P_3$; and since

$$P_1 : P_2 = \sin. a' : \sin. (a+a') = \sin. a : \sin. 2a = 1 : 2 \cos. a.$$

$$\therefore P_2 = 2P_1 \cos. a.$$

Otherwise: since the tension of the cord is the same throughout, when the cord passes over an obstacle without friction. $P_1 = P_3$; $a = a'$ and $P_2 = 2P_1 \cos. a$ (*Art.* 19).

150. Prop. *Required the conditions of equilibrium when any number of forces in the same plane are applied at different points of the cord.*

Let ABCDE be a cord to which are applied the forces P_1, P_2, P_3, &c., at the points A, B, C, &c., in the directions AP_1, BP_2, CP_3, &c.

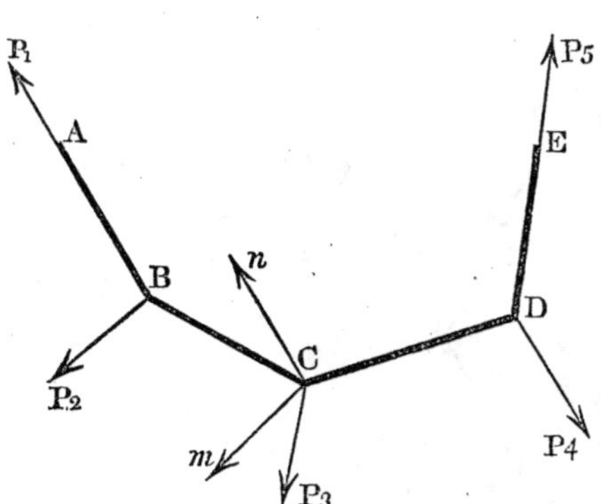

The force P_1 at A may be considered as acting at B in its direction BA; and since B, when in equilibrium, must be so from the action of P_1, P_2, and the tension of BC, the resultant of P_1 and P_2 must be in the direction of CB, and may be considered as acting at C. Suppose it thus applied, and let it be resolved into two, acting in the directions Cn and Cm parallel to the original components, and equal to them.

We have thus transferred the forces P_1 and P_2 to act at C parallel to their original directions. In the same manner, the resultant of P_1, P_2, and P_3 acting at C must be in the direction of DC, and may be applied at the point D without disturbing the equilibrium, and then replaced by P_1, P_2, and P_3 parallel to their original directions, and so on, for any number. Hence, if all the forces be supposed to act at one point parallel to their original directions, they will be in equilibrium. The conditions of equilibrium, therefore, are the same as for any number of concurring forces (*Art.* 70), or, the sum of all the forces resolved in any two rectangular directions must be zero.

The form which the cord takes under the influence of the several forces is called a *funicular polygon.*

151. Prop. *Required the relations of the forces which, acting on a cord in one plane, keep it in equilibrium.*

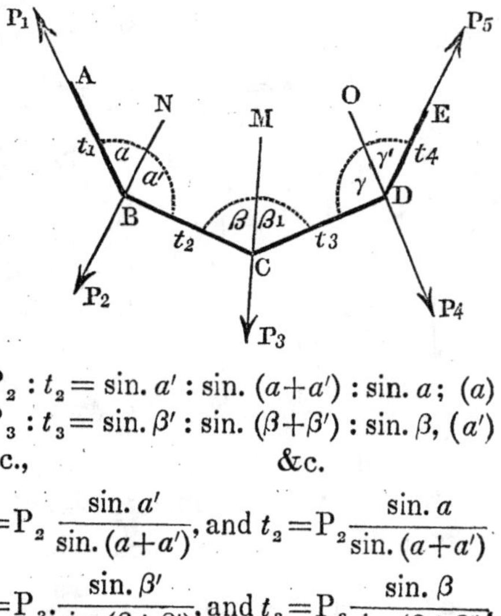

Produce P_2B to N, P_3C to M, &c., and let the

$\angle ABN = a$, $\angle NBC = a'$,
$\angle BCM = \beta$, $\angle MCD = \beta'$,
&c., &c.

Let t_1, t_2, t_3, &c., be the tensions of the several successive portions of the cord.

Then (*Art.* 147) $t_1 : P_2 : t_2 = \sin. a' : \sin. (a+a') : \sin. a;$ (a)
also, $t_2 : P_3 : t_3 = \sin. \beta' : \sin. (\beta+\beta') : \sin. \beta,$ (a')
&c., &c.

From (a) we obtain $t_1 = P_2 \dfrac{\sin. a'}{\sin. (a+a')}$, and $t_2 = P_2 \dfrac{\sin. a}{\sin. (a+a')}$

" (a') " $t_2 = P_3 . \dfrac{\sin. \beta'}{\sin. (\beta+\beta')}$, and $t_3 = P_3 \dfrac{\sin. \beta}{\sin. (\beta+\beta')}$,
&c., &c.

Equating the values of t_2, t_3, &c., we have

$$P_2 . \frac{\sin. a}{\sin. (a+a')} = P_3 . \frac{\sin. \beta'}{\sin. (\beta+\beta')}, \qquad (b)$$

$$P_3 \cdot \frac{\sin. \beta}{\sin. (\beta+\beta')} = P_4 \cdot \frac{\sin. \gamma}{\sin. (\gamma+\gamma')}, \qquad (b')$$
&c., &c.,

for the relations of the forces.

152. Cor. 1. If the cord be fixed at the points A and E, and the forces P_2, P_3, P_4, &c., be parallel, we have

$$\sin. a' = \sin. \beta, \ \sin. \beta' = \sin. \gamma, \ \&c.$$

Multiplying these equations by (b), (b') member by member in their order, we obtain

$$P_2 \cdot \frac{\sin. a \sin. a'}{\sin. (a+a')} = P_3 \frac{\sin. \beta \sin. \beta'}{\sin. (\beta+\beta')} = P_4 \frac{\sin. \gamma \sin. \gamma'}{\sin. (\gamma+\gamma')}, \ \&c.$$

But $\frac{\sin. a \sin. a'}{\sin. (a+a')} = \frac{\sin. a \sin. a'}{\sin. a \cos. a' + \sin. a' \cos. a}$

$$= \frac{1}{\frac{\cos. a}{\sin. a} + \frac{\cos. a'}{\sin. a'}} = \frac{1}{\cot. a + \cot. a'}.$$

$$\therefore \frac{P_2}{\cot. a + \cot. a'} = \frac{P_3}{\cot. \beta + \cot. \beta'} = \frac{P_4}{\cot. \gamma + \cot. \gamma'} =, \ \&c.$$

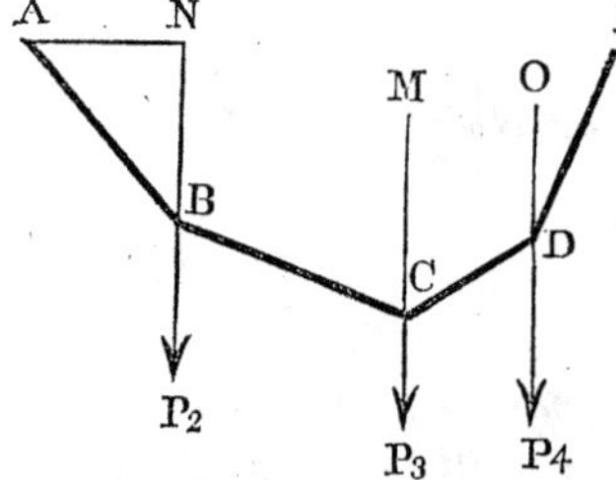

153. Cor. 2. If the cord be fixed at A and E, and the forces P_2, P_3, P_4, &c., be weights, the horizontal tension of each portion of the cord is the same.

For, by resolving the tension of each part horizontally, we have

the horizontal tension of $AB = t_1 . \sin. ABN = t_1 \sin. a$,

" " $BC = t_2 \sin. BCM = t_2 \sin. \beta$, &c.

Substituting for t_1, t_2, &c., their values found above,

horizontal tension of $AB = P_2 \frac{\sin. a \sin. a'}{\sin. (a+a')} = \frac{P_2}{\cot. a + \cot. a'}$,

" " $BC = P_3 \frac{\sin. \beta \sin. \beta'}{\sin. (\beta+\beta')} = \frac{P_3}{\cot. \beta + \cot. \beta'}$,

&c., &c.,

which, by *Cor.* 1, are all equal.

154. Cor. 3. Since the reactions of the points A and E equilibrate the resultant of all the weights, the lines AB, ED produced, must meet in some point of the vertical through the center of gravity of the system. For three forces, if in equilibrium, must meet in a point.

155. A heavy cord may be considered a funicular polygon loaded with an infinite number of small weights, and since the number of weights is infinite, the polygon will also have an infinite number of sides, or will become a curve.

The curve which a heavy cord or chain of indefinitely small links will assume, when suspended from two fixed points not in the same vertical line, is called the *catenary*.

EXAMPLES.

Ex. 1. Two equal weights balance, by a cord, over any number of fixed points without friction. Required the pressure on each.

Since the tension of the cord $P_1ABCDEP_2$ is the same throughout, each point is acted upon by three forces, viz., two equal tensions on each side of it and the reaction of the point, which last must be equal to the resultant of the other two. Hence, calling the angles at A, B, C, &c., a, b, c, &c., by *Art.* 15, and *Cor.*, *Art.* 19,

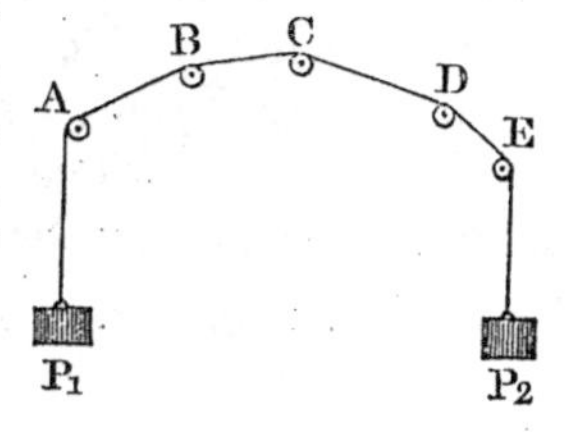

the pressure on $A=2P_1 \cos.\frac{1}{2}a$,

" " $B=2P_1 \cos.\frac{1}{2}b$,

&c., &c.

Ex. 2. A cord of given length passes over two fixed points A and B without friction, and one extremity, to which a given weight P is attached, passes through a small ring at the other extremity C. It is required to find the tension of the cord when in equilibrium, and the length of the part CP below the ring.

Since the cord passes freely over the points A and B, and through the ring C, it is of the same tension throughout, and equal to the weight P. Hence the point C is kept at rest by

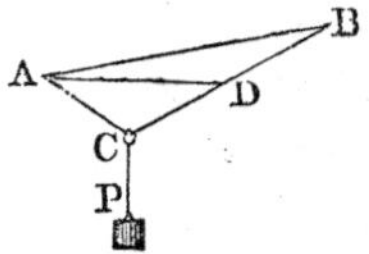

three equal forces, and, by *Art.* 18, must make angles of 120° with each other. Draw the horizontal line AD; and, since ACD=120°, ADC and CAD are each equal to 30°. The position of A and B being given, the angle BAD must be known. Hence, in the triangle ACB we have the side AB and all the angles from which AC and BC may be determined. Then the whole length of the cord, diminished by the perimeter of the triangle, will be the distance of the weight from the ring.

§ IV. THE PULLEY.

156. The *pulley* is a small grooved wheel movable about an axis, and fixed in a *block*. The cord passes over the circumference of the wheel in the groove.

The use of the pulley is to prevent the effects of friction and rigidity of the cord. The first of these it diminishes by transferring the friction from the cord and circumference of the wheel to the axle and its supports, which may be highly polished or lubricated. The effects of rigidity are diminished by turning the cord in a circular arc instead of a sharp angle.

The pulley is called *fixed* or *movable*, according as the *block* is fixed or movable.

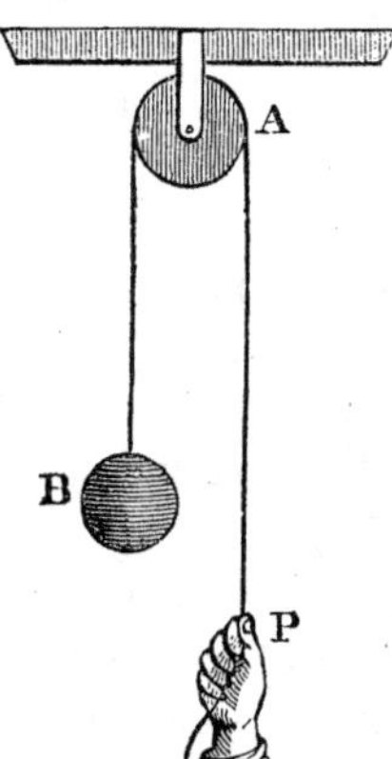

157. The *fixed pulley* serves merely to change the direction of the forces transmitted by the cord, since, neglecting the friction of the pulley, the tension of the cord is the same in every part of it. Hence the power equals the weight, and the pressure on the axis of the pulley equals their sum.

158. Prop. *Required the relation of the power to the weight in the single movable pulley.*

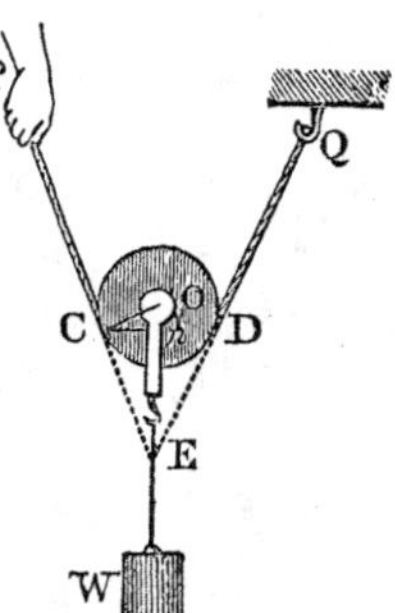

The tension t of the cord, being the same throughout, is equal to the power P; also to the pressure on the hook Q.

The resultant of the two tensions in the directions CP and DQ, being equal and opposite to the weight W, must be vertical. Let a be the angle made by the cords with this vertical, and, resolving the tensions vertically, we have

$$2t \cos. a = 2P \cos. a = W.$$

$$\therefore\ P = \frac{W}{2 \cos. a},$$

the same as obtained in *Art.* 149.

Cor. 1. If the weight w of the pulley be taken into account,

$$P = \frac{W+w}{2 \cos. a}.$$

Cor. 2. If the cords are parallel, $a=0$ and

$$P = \frac{W+w}{2}.$$

Cor. 3. If $a=90°$, $2a=180°$, or the cord becomes straight and horizontal. In this case

$$P = \frac{W+w}{0} = \infty,$$

or the power must be infinite. In other words, no power can reduce the cord to a horizontal straight line while the weight is finite.

159. Of the various combinations of pulleys there are three, which we shall distinguish by the *first, second,* and *third systems* of pulleys.

160. Prop. *Required the relation of the power to the weight in the first system of pulleys.*

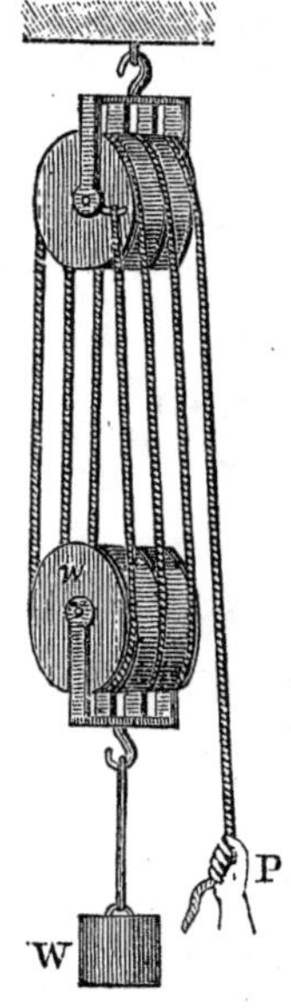

The annexed figure represents this system. Neglecting friction, the tension t of the cord is the same throughout, and equal to P. The weight W and the weight of the lower block w are sustained by the tensions of the several cords at the lower block. Hence, if n be the number of cords at this block,

$$nt = n\mathrm{P} = \mathrm{W} + w$$

$$\therefore\ \mathrm{P} = \frac{\mathrm{W} + w}{n}.$$

If the weight of the lower block be neglected,

$$\mathrm{P} = \frac{\mathrm{W}}{n}.$$

Of this system of pulleys there are various modifications. The annexed form is the one in most common use.

161. Prop. *Required the relation of the power to the weight in the second system of pulleys.*

The annexed figure represents this system with three movable pulleys, each pulley having its own rope.

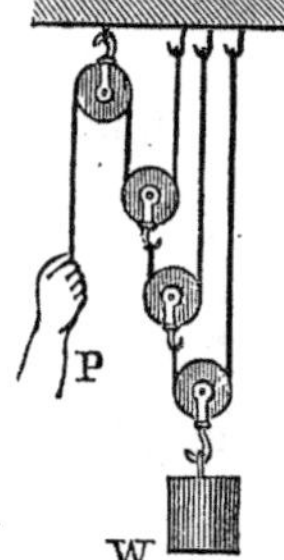

Designate by a_1, a_2, a_3, &c., the pulleys respectively in their order from the weight W, by w_1, w_2, w_3, &c., their weights, and by t_1, t_2, t_3, &c., the tensions of their respective cords.

Then, for the equilibrium of a_1, we have

$$2t_1 = \mathrm{W} + w_1, \text{ or } t_1 = \frac{\mathrm{W} + w_1}{2}.$$

For the equilibrium of a_2, $2t_2 = t_1 + w_2 = \dfrac{\mathrm{W} + w_1 + 2w_2}{2}$,

or

$$t_2 = \frac{\mathrm{W} + w_1 + 2w_2}{2^2}.$$

For the equilibrium of a_3, $2t_3 = t_2 + w_3 = \dfrac{\mathrm{W} + w_1 + 2w_2 + 2^2 w_3}{2^2}$, or $$t_3=\frac{W+w_1+2w_2+2^2w_3}{2^3}.$$

. . . .

For the equilibrium of a_n,

$$P=t_n=\frac{W+w_1+2w_2+2^2w_3+2^3w_4+\ldots 2^{n-1}w_n}{2^n}.$$

Cor. 1. If the weight of each pulley is the same, and equal w_1

$$P=\frac{W+w_1(1+2+2^2+2^3+\ldots . 2^{n-1})}{2^n}$$

$$=\frac{W}{2^n}+\frac{w_1(2^n-1)}{2^n}$$

$$=\frac{W}{2^n}+w_1\left(1-\frac{1}{2^n}\right).$$

Cor. 2. If the weights of the pulleys be neglected,

$$P=\frac{W}{2^n}, \text{ or } W=2^n.P.$$

162. Prop. *Required the relation of the power to the weight in the third system of pulleys.*

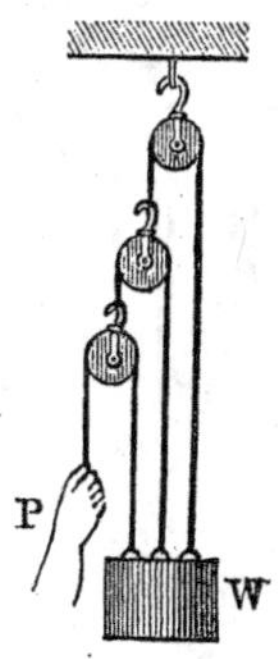

In this system each cord is attached to the weight, and the number of *movable* pulleys is one less than the number of cords. Designating the pulleys in their order from the weight by a_1, a_2, a_3, &c., their weights respectively by w_1, w_2, w_3, &c., and the tensions of the successive cords by t_1, t_2, t_3, &c., we have

$$t_1=P,$$
$$t_2=2t_1+w_1=2P+w_1,$$
$$t_3=2t_2+w_2=2^2P+2w_1+w_2,$$
$$t_4=2t_3+w_3=2^3P+2^2w_1+2w_2+w_3.$$

. . . .

And if there be n cords,

$$t_n=2^{n-1}P+2^{n-2}w_1+2^{n-3}w_2+\&c. \ldots . 2w_{n-2}+w_{n-1}.$$

But $W=t_1+t_2+t_3+,\&c. \ldots . t_n$

$$=P(1+2+2^2+2^3+,\&c. \ldots 2^{n-1})+w_1(1+2+2^2+,\&c. \ldots 2^{n-2})+w_2(1+2+2^2+,\&c. \ldots 2^{n-3})+,\&c.$$

$$=P(2^n-1)+w_1(2^{n-1}-1)+w_2(2^{n-2}-1)+,\&c.\ldots w_{n-1}(2-1).$$

If the $(n-1)$ pulleys are of the same weight w_1,

$$W=P(2^n-1)+w_1(2^{n-1}+2^{n-2}+2^{n-3}+,\&c.\ldots.2^2+2+1-n)$$
$$=P(2^n-1)+w_1(2^n-1)-w_1n$$
$$=(P+w_1)(2^n-1)-w_1n.$$

If the weights of the pulleys be neglected,

$$W=P(2^n-1).$$

EXAMPLES.

Ex. 1. At what angle must the cords of a single movable pulley be inclined in order that P may equal W?

Ex. 2. In the *first* system of pulleys, if there be 10 cords at the lower block, what power will support a weight of 1000 lbs.?

Ex. 3. In the *second* system of pulleys, if 1 lb. support a weight of 128 lbs., required the number of pulleys supposed without weight.

Ex. 4. In the *third* system of 6 pulleys, each weighing 1 lb., find what weight will be supported by a power of 12 lbs

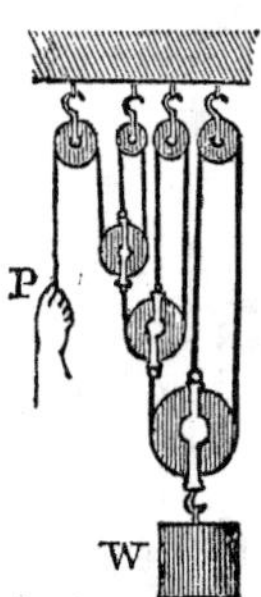

Ex. 5. Find the ratio of the power to the weight in the annexed modification of the second system of pulleys.

§ V. THE INCLINED PLANE.

163. The *Inclined plane,* as a mechanical power, is supposed perfectly hard and smooth, unless friction be considered. It assists in sustaining a heavy body by its reaction. This reaction, however, being normal to the plane, can not entirely counteract the weight of the body, which acts vertically down-

ward. Some other force must therefore be made to act upon the body, in order that it may be sustained.

164. Prop. *Required the conditions of equilibrium of a body sustained by any force on an inclined plane.*

Let AB be a section of an inclined plane, of which AB is the length, BC the height, and AC the base. Let i be the inclination of the plane to the horizon, ε the angle made by the direction of the power P with the plane AB, W = the weight of the body a, and R = the reaction of the plane. The body is kept at rest by the action of P, W, and R. Resolving the forces parallel and perpendicular to the plane, we have

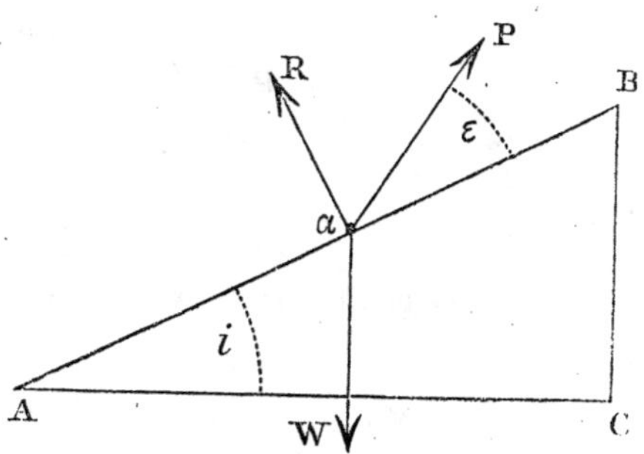

$$P \cos. \varepsilon - W \sin. i = 0, \qquad (a)$$

$$R + P \sin. \varepsilon - W \cos. i = 0. \qquad (b)$$

From (a) we obtain

$$\frac{W}{P} = \frac{\cos. \varepsilon}{\sin. i}. \qquad (c)$$

From (b), $R = W \cos. i - P \sin. \varepsilon = W \cos. i - \dfrac{W \sin. i \sin. \varepsilon}{\cos. \varepsilon}$

$$= W . \frac{\cos. (i+\varepsilon)}{\cos. \varepsilon},$$

or

$$\frac{W}{R} = \frac{\cos. \varepsilon}{\cos. (i+\varepsilon)}, \qquad (d)$$

the same relations as obtained in *Ex.* 24, *Art.* 129.

165. Cor. 1. If the force P act parallel to the plane, $\varepsilon = 0$ and (c) becomes

$$\frac{W}{P} = \frac{1}{\sin. i} = \frac{AB}{BC},$$

or, *the power is to the weight as the height of the plane to its length.*

From (d) we get $\frac{W}{R}=\frac{1}{\cos. i}=\frac{AB}{AC}$,

or, *the reaction of the plane is to the weight as the base to the length.*

166. Cor. 2. If the power act parallel to the base of the plane, $\epsilon=-i$, and (c) becomes

$$\frac{W}{P}=\frac{\cos. i}{\sin. i}=\frac{AC}{BC},$$

or, *the power is to the weight as the height to the base.*

From (d),
$$\frac{W}{R}=\frac{\cos. i}{1}=\frac{AC}{AB},$$

or, *the reaction of the plane is to the weight as the length to the base.*

167. Prop. *Required the conditions of equilibrium of two bodies resting on two inclined planes having a common summit, the bodies being connected by a cord passing over a pulley at the summit.*

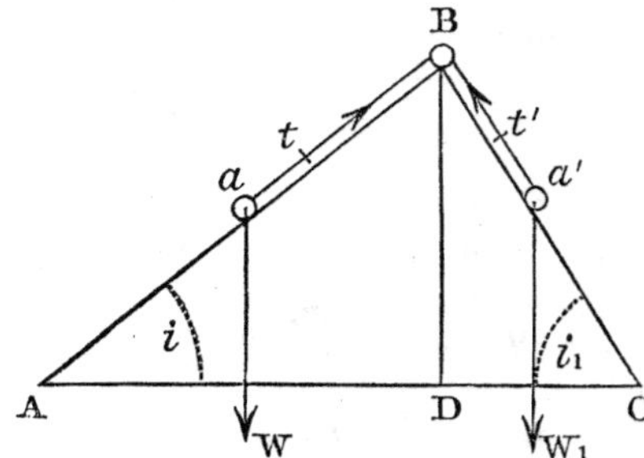

Let W and W_1 be the weights of the bodies, and i, i_1 the inclinations of the planes.

If t be the tension of the cord, we have for equilibrium on the plane AB (*Art.* 165),

$$t=W. \sin. i,$$

on the plane BC
$$t=W_1 \sin. i_1.$$

$$\therefore W \sin. i=W_1 \sin. i_1,$$

or
$$W.\frac{BD}{AB}=W_1\frac{BD}{BC},$$

or
$$\frac{W}{W_1}=\frac{AB}{BC},$$

or, *the weights are proportional to the lengths of the planes on which they rest respectively.*

EXAMPLES.

Ex. 1. What force acting parallel to the base of the plane is necessary to support a weight of 50 lbs. on a plane inclined at an angle of 15° to the horizon?

Ex. 2. If the weight, power, and reaction of the plane are respectively as the numbers 25, 16, and 10, find the inclination of the plane, and the inclination of the power's direction to the plane.

§ VI. THE WEDGE.

168. The *wedge* is a triangular prism whose perpendicular section is an isosceles triangle. The dihedral angle formed by the two equal rectangular faces, is called the angle of the wedge. The other rectangular face is called the *back*. It is used to separate the parts of bodies, by introducing the angle of the wedge between them by a power applied perpendicularly to the back. The equal rectangular faces are regarded as perfectly smooth, in which case the only effective part of the resistance must be perpendicular to these faces.

169. PROP. *To determine the conditions of equilibrium in the wedge.*

Let ABC be a section of the wedge perpendicular to the *angle* or edge A. Draw AD bisecting the angle, and let BAD=CAD=a. Let 2P be the power applied to the back BC of the wedge, which must be in equilibrium with the pressures R on the two faces AB and AC. If an equilibrium exist, the forces 2P, R, and R must meet at some point in AD (*Art.* 74).

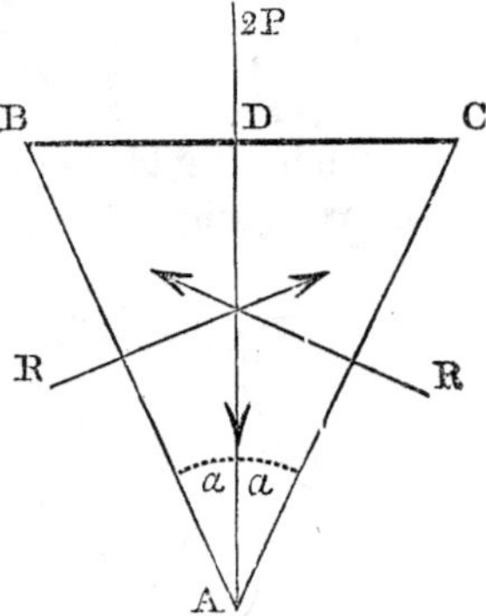

Resolving the forces vertically, or in the direction AD, we have

$$2R. \sin. BAD - 2P = 0,$$

or

$$P = R. \sin. a,$$

or

$$\frac{P}{R} = \sin. a = \frac{BD}{BA} = \frac{BD.l}{BA.l} = \frac{\frac{1}{2} \text{ the back of the wedge}}{\text{face of the wedge}},$$

where l= the length of the edge or breadth of the face.

170. In the foregoing investigation of the theory of equilibrium in the wedge, we have omitted the consideration of the *friction*, and have supposed the power to be a pressure; whereas, in practice, the wedge is kept at rest by *friction* alone, and the power arises from *percussion*. The following problem will serve to elucidate the theoretical view here taken of the wedge.

171. Prob. A heavy beam is attached, by a hinge at one end, to a smooth, horizontal plane, while the other rests on the smooth face of a semi-wedge. Required the horizontal force necessary to keep the wedge from moving.

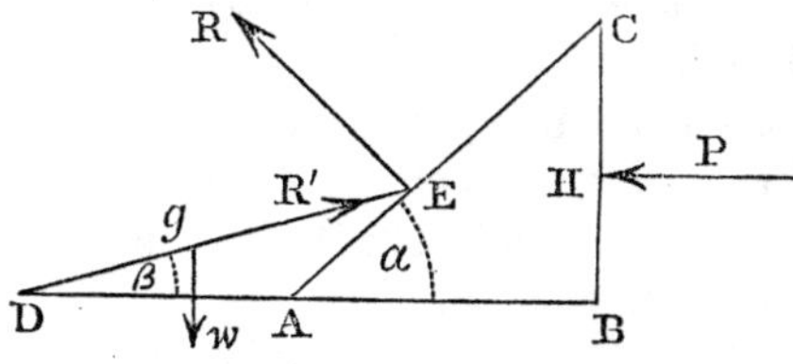

Let DE be the beam and BAC the wedge, $\angle BAC = a$, $\angle ADE = \beta$, l= the length of the beam, g the center of gravity, and $Dg = a$.

The wedge is kept in equilibrium by the pressure of the beam upon it at E, and the horizontal force P acting upon it at some point H. The beam is kept in equilibrium by its weight w acting at g and the reaction R of the face of the wedge at E.

Taking the moments about D for the equilibrium of the beam, we avoid expressions involving the unknown thrust R′ and have

$$w.Dg. \cos. ADE - R.DE. \sin. DER = 0,$$

or

$$w.a. \cos. \beta - R.l. \cos. (a-\beta) = 0.$$

$$\therefore \ R = w.\frac{a}{l}.\frac{\cos. \beta}{\cos. (a-\beta)}.$$

But the principles of the wedge give

$$P = R. \sin. a.$$

$$\therefore P = w.\frac{a}{l}.\frac{\sin. a \cos. \beta}{\cos. (a-\beta)}.$$

By an examination of this value of P, it will be seen that the power necessary to keep the wedge from moving will diminish as the wedge advances beneath the beam.

§ VII. THE SCREW.

172. If we divide the rectangle ABCD into equal parts by the lines *mn*, *m'n'*, &c., parallel to AB, and draw the parallel diagonals of the rectangles thus formed, and if we suppose the whole rectangle to be wrapped round the surface of a cylinder, the perimeter of whose base is equal to AB and altitude to BC, the diagonals of the rectangles will trace on the surface of the cylinder a *continuous* curve, which is called the helix.

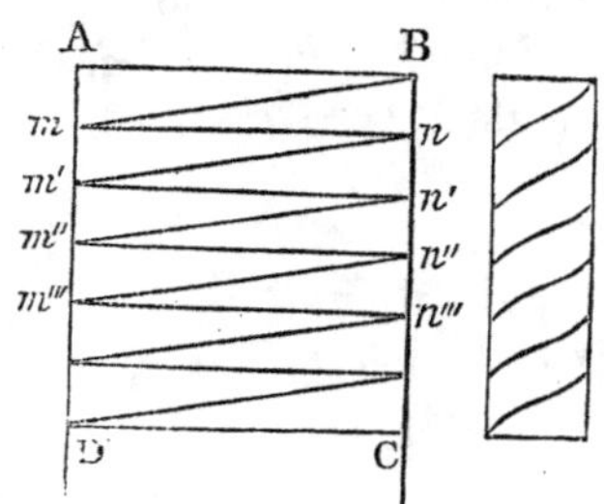

If a projecting *thread* or *rib* be attached to the cylinder upon this curve, we have the *screw*, sometimes called the *external screw*. Similarly, if we take a hollow cylinder of exactly the same radius as the solid one, and generate a groove in the same curve, we have the *internal screw* or *nut*.

The screw works in the nut, either of which may be fixed and the other movable.

173. Prop. *To determine the conditions of equilibrium in the screw.*

From the construction of the screw, it appears that the thread of it is an inclined plane, of which the base is the circumference of the cylinder and the height the distance between the threads. The force is generally applied perpendicularly to the end of a lever inserted into the

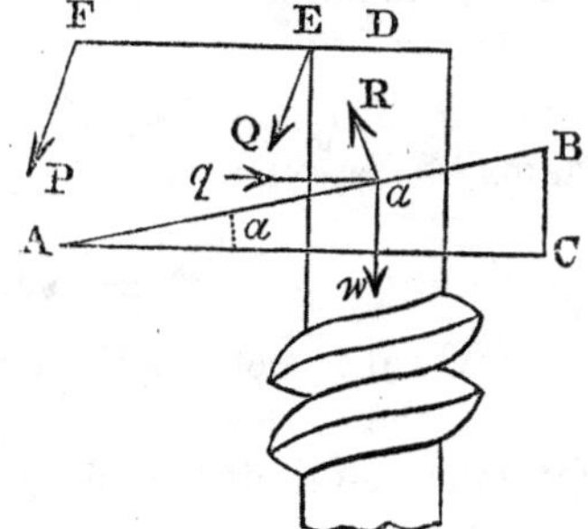

cylinder, and in the plane perpendicular to the axis of the cylinder. The power P thus applied, in turning the screw round, produces a pressure on the threads of the screw in the direction of the axis of the cylinder. In case of equilibrium, let the counterpoise of this pressure be W. Let ABC be the inclined plane formed by unwrapping one revolution of a thread, and let $w=\frac{1}{n}$th part of W, be supported at a, q being the same part of Q, the force applied at the circumference of the cylinder. Put FD$=a$, ED the radius of the cylinder $=r$, and angle BAC$=\alpha$. Then, *Art.* 133,

$$Q=P.\frac{FD}{ED}=P.\frac{a}{r}=P.\frac{2\pi a}{2\pi r}. \qquad (a)$$

By *Art.* 166, $$q=w.\frac{BC}{AC},$$

or $$\frac{Q}{n}=\frac{W}{n}.\frac{\text{distance of two threads}}{\text{circumference of the cylinder}},$$

and the same holds for each of the other portions of W at the other points of the plane. Therefore, we have

$$Q=W.\frac{\text{distance of two threads}}{\text{circumference of the cylinder}}.$$

By (a), $$Q=P.\frac{2\pi a}{2\pi r}=P.\frac{\text{circumference described by P}}{\text{circumference of cylinder}}.$$

$$\therefore \frac{W}{P}=\frac{\text{circumference described by P}}{\text{distance of two threads}},$$

or, the power is to the weight as the distance of two threads is to the circumference described by the power in one revolution of the screw.

It will be seen that the ratio of P to W is independent of the radius of the cylinder.

EXAMPLES.

Ex. 1. What force must be exerted to sustain a ton weight on a screw, the thread of which makes 150 turns in the height of 1 foot, the length of the arm being 6 feet?

Ex. 2. Find the weight that can be sustained by a power

of 1 lb. acting at the distance of 3 yards from the axis of the screw, the distance between the threads being 1 inch.

Ex. 3. What must be the length of a lever at whose extremity a force of 1 lb. will support a weight of 1000 lbs. on a screw, whose threads are $\frac{3}{4}$ inch apart?

§ VIII. BALANCES AND COMBINATIONS OF THE MECHANICAL POWERS.

174. The *common balance*, in its best form, is a bent lever, in which the weight of the lever must be taken into consideration.

In the annexed figure the points A and B, from which the scale-pans and weights are suspended, are called the points of suspension; C is the fulcrum, being the lower edge of a prismatic rod of steel projecting on each side of the beam; when the balance is in use, these edges on each side of the beam, as at C, rest on hard surfaces, so that the beam turns freely about C as a fulcrum.

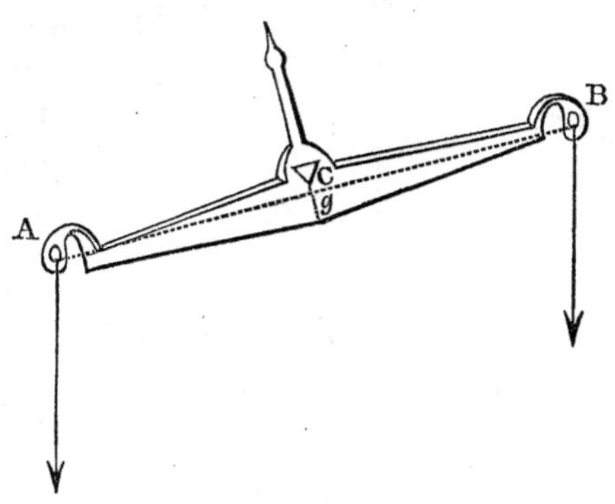

175. The requisites of a good balance are,

1°. That the beam rest in a horizontal position when loaded with equal weights.

2°. That the balance possess great sensibility.

3°. That it possess great stability.

176. PROP. *To determine the conditions that the beam rest in a horizontal position when loaded with equal weights.*

Supposing the beam horizontal, in case of equilibrium, if we neglect the weight of the beam, the moments of the weights must be equal, and, therefore, the *arms must be equal* (*Art.* 133).

But taking into consideration the weight of the beam, its center of gravity must be in the vertical through C, the center of motion (*Art.* 110), and, in order to this, the beam must be *symmetrical on opposite sides of the fulcrum* (*Art.* 97). The line AB, joining the points of suspension, is obviously bisected by the vertical through C and the center of gravity of the beam, and the point of intersection, for reasons which will appear, should be below C.

177. Prop. *To determine the conditions that the balance may possess great sensibility.*

Let C, A, and B be the fulcrum and points of suspension, as in the preceding figure, and join Cg, the centers of motion and of gravity. Cg is perpendicular to AB and bisects it, if the beam is constructed in accordance with the first requisite. Let M C E N be a horizontal line through C, meeting AB in E making with it an angle equal to θ.

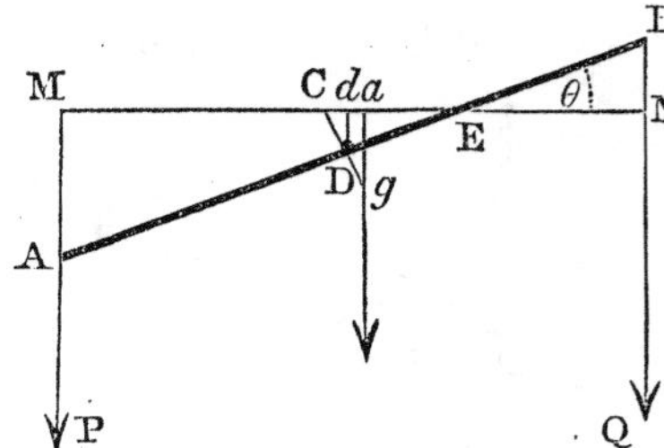

The sensibility is measured by the amount of deflection θ of the line AB from a horizontal position by a given small difference P—Q of the weights.

Draw the vertical lines Dd and ga, and put $AD=BD=a$, $CD=d$, $Cg=h$, and weight of the beam $=w$. Now MN is bisected in d, and $Md=a \cos. \theta$, $Cd=d. \sin. \theta$, and $Ca=h. \sin. \theta$.

If the system is in equilibrium, the moments about C give

$$P.CM-Q.CN-w.Ca=0,$$

or $$P.(Md-Cd)-Q(Nd+Cd)-w.Ca=0,$$

or $$(P-Q)a. \cos. \theta.-(P+Q)d \sin. \theta-w.h. \sin. \theta=0,$$

or $$(P-Q)a-\{(P+Q).d+w.h\} \tan. \theta=0.$$

$$\therefore \tan. \theta=\frac{(P-Q).a}{(P+Q)d+w.h}.$$

Hence the angle θ, and, therefore, the sensibility, is increased for given values of P and Q, *by increasing the lengths of the arms* (a), *by diminishing the weight of the beam* (w), *or by diminishing the distances of the fulcrum from the center of gravity of the beam* (h) *and from the line joining the points of suspension* (d).

178. PROP. *To determine the conditions that the balance may possess great stability.*

If the balance be loaded with equal weights and disturbed from its position of equilibrium, the *rapidity* with which it returns to that position is a measure of its *stability*. But this rapidity of return to a horizontal position will depend upon the moment which urges it back.

But this moment is, since $P=Q$,

$$P.CN-P.CM+w.Ca,$$

or

$$P.2.Cd+w.Ca,$$

or

$$(2P.d+w.h)\sin.\theta.$$

Hence, for given values of P and θ the stability is greater, as d, h, and w are increased.

179. COR. Hence, by increasing the *stability*, we diminish the sensibility, but the sensibility may be increased by increasing the length of the arms, without affecting the stability.

For commercial purposes, when expedition is required, and the material weighed is not of great value, sensibility is sacrificed to stability; but for philosophical purposes great sensibility is required, and stability is of little comparative importance.

THE STEELYARD BALANCE.

180. The *steelyard balance*, or *Roman steelyard*, is a lever of the first kind, with unequal arms. The body W to be weighed is hung at the shorter arm A, and a given constant weight P is moved along the other arm till it balances W; then the weight of W is known from the place of the counterpoise P.

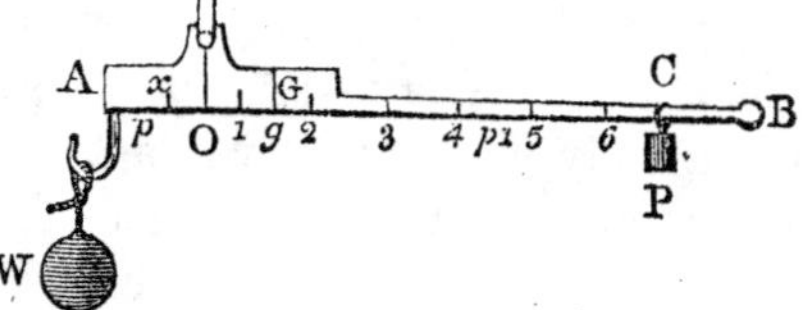

181. Prop. *To determine the law according to which the longer arm of the steelyard must be graduated.*

Let G be the center of gravity of the beam AB and w its weight. Put $OA=p$, $OC=p_1$, $OG=g$, and taking moments about the fulcrum O, in case of equilibrium we have

$$Wp-wg-Pp_1=0.$$

$$\therefore p_1+\frac{wg}{P}=\frac{Wp}{P}.$$

Taking Ox, a fourth proportional to P, w, and g, we have

$$p_1+\frac{wg}{P}=CO+Ox=Cx=\frac{Wp}{P}.$$

Hence the distance from x to the counterpoise P varies as the weight; and if the weights be in arithmetical progression, the distances $x1$, $x2$, $x3$, &c., will also be in arithmetical progression.

THE BENT LEVER BALANCE.

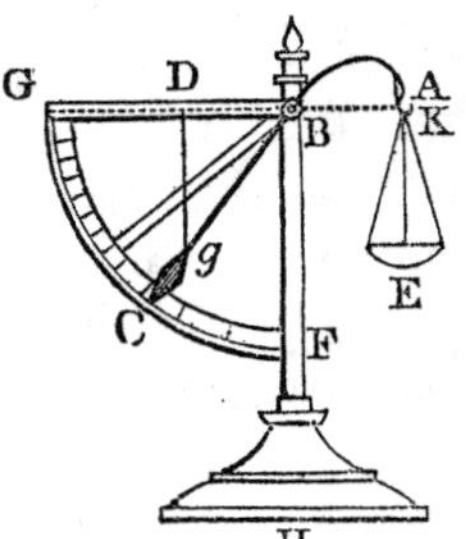

182. This balance is represented by the annexed figure, where ABC is the bent lever turning about a pivot at B. A scale (E) hangs from A, and at C an index points to some division on the graduated arc GCF.

183. Prop. *To determine the principle of graduation in the bent lever balance.*

Let g be the center of gravity of the beam at which its weight w acts. The weight of the scale (E) and the weight (W) of a body placed in it will act vertically through A. Let the horizontal line through B meet the vertical lines through g and A respectively in D and K. Then the moments about B, in equilibrium, give

$$w.DB-(E+W)BK=0. \qquad (a)$$

As greater weights are put into the scale E, the point A approaches more nearly the vertical through B from the bent form of the beam, or BK diminishes, while BD increases.

Suppose the point A to be at K when the scale is unloaded, and let, in this position of the beam, the angle $DBg=\theta$. When a weight is put into the scale the point A will descend through some angle ϕ, and the arm Bg will rise through the same angle. In this new position the angle DBg will become $\theta-\phi$. Let $Bg=a$ and $BA=b$; then

$$DB=a\cos.(\theta-\phi),\text{ and } BK=b\cos.\phi.$$

These substituted in (a), give

$$w.a.\cos.(\theta-\phi)-E.b.\cos.\phi-W.b.\cos.\phi=0,$$

$$\text{or } w.a.\cos.\theta\cos.\phi+w.a\sin.\theta\sin.\phi-E.b\cos.\phi-W.b.\cos.\phi=0,$$

$$\text{or } w.a.\cos.\theta+w.a\sin.\theta\tan.\phi-E.b-W.b=0.$$

$$\therefore\ \tan.\phi=\frac{W.b}{w.a.\sin.\theta}+\frac{E.b-w.a\cos.\theta}{w.a.\sin.\theta}.$$

Hence, tan. ϕ varies as W, and the limb GF must be divided into arcs whose tangents are in arithmetical progression.

Practically, the limb may be graduated from the positions of the index at C for a succession of weights put into E. This instrument possesses great *stability*.

ROBERVAL'S BALANCE.

184. This instrument is of greater interest from its paradoxical appearance than from its use as a machine for weighing bodies. Its discussion affords an interesting application of the doctrine of couples.

It consists of an upright stem upon a heavy base A, with equal cross-beams turning about pivots at a and b. These cross-beams are connected by pivots at c, d, e, and f, with two other equal pieces in the form

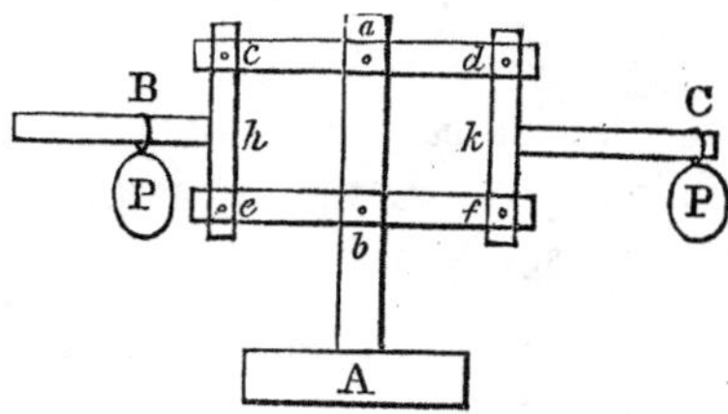

of a T. The weights are suspended from the horizontal arms of the latter pieces.

185. Prop. *In Roberval's balance, equal weights balance at all distances from the upright stem.*

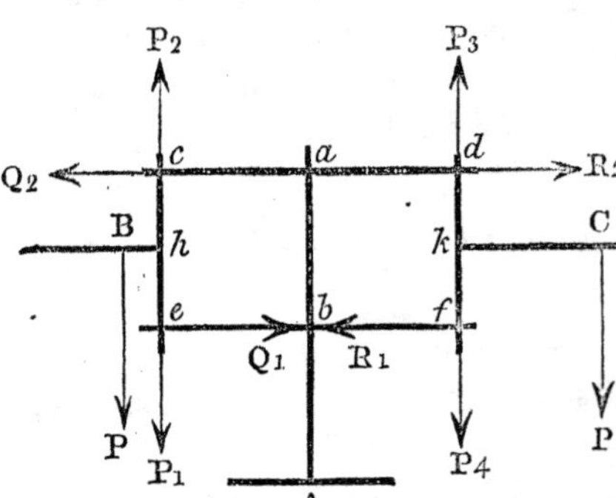

Let the letters in the annexed figure indicate the same parts as in the former.

Let equal and opposite forces P_1 and P_2 each equal to P act in *ec*, and, similarly, let P_3 and P_4 act in *df*. These forces P_1, P_2, P_3, and P_4 do not disturb the equilibrium. Now P at B, as in *Art.* 71, is equivalent to P_1 at *e* and the couple P, B*h*, P_2, and, similarly, P at C is equivalent to P_4 at *f* and the couple P, C*k*, P_3; and since P_1 at *e* balances P_4 at *f*, we have only the two couples to dispose of.

Now for the couple P, B*h*, P_2, *Arts.* 58 and 59, we may substitute a couple Q_1, *ec*, Q_2 in its own plane and of equal moment, in which the forces Q_1 and Q_2, acting in the directions of the cross-beams *cd* and *ef* (which always remain parallel to each other as they turn on the pivots *a* and *b*), are destroyed by the resistance of the pivots *a* and *b*. Similarly, the couple P, C*k*, P_3 may be replaced by the equivalent couple R_1, *fd*, R_2, in which R_1 and R_2 are destroyed by the resistance of *b* and *a*. These new couples, therefore, do not disturb the equilibrium, and the original forces P at B and P at C must be in equilibrium.

If the beams *cad* and *ebf* be moved round the pivots into any oblique position, the same reasoning would apply, and the equilibrium still subsist.

Cor. Unequal weights can not balance from whatever points suspended

186. PROP. *To determine the ratio of the power to the weight in a combination of levers.*

Let the power P act at A, and the weight W at B‴. The first three levers are of the second kind, and the last one of the first kind, the fulcrums being at C, C′, C″, and C‴. Let BA′, B′A″, B″A‴ be rigid rods connecting the levers, and let the action of the first lever on the rod BA′ be W′, which becomes the power acting on the second, W″ and W‴ the weights to the second and third respectively, and powers to the third and fourth.

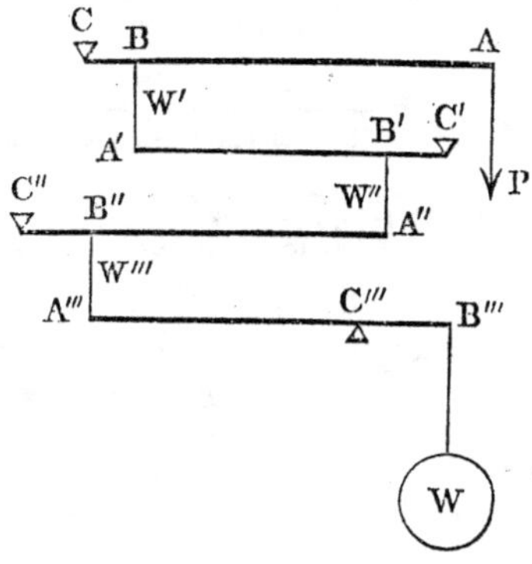

Then $\frac{P}{W'}=\frac{CB}{CA}$, $\frac{W'}{W''}=\frac{C'B'}{C'A'}$, $\frac{W''}{W'''}=\frac{C''B''}{C''A''}$, and $\frac{W'''}{W}=\frac{C'''B''}{C'''A''}$.

Taking the continued product of these equations member by member, we have

$$\frac{P}{W}=\frac{CB\times C'B'\times C''B''\times C'''B'''}{CA\times C'A'\times C''A''\times C'''A'''},$$

or, *the ratio of the power to the weight in the combination is equal to the product of the ratios of the power to the weight in each lever.*

187 PROP. *To determine the ratio of the power to the weight in the endless screw.*

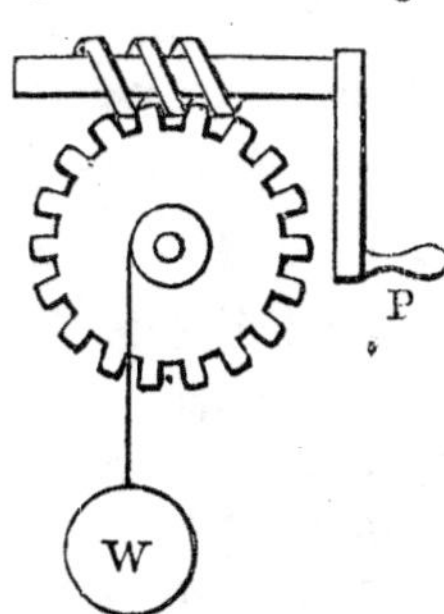

This machine is a combination of the screw and wheel and axle.

Let P be the power applied to the handle of the winch, W′ the pressure of the screw on the teeth of the wheel, and W the weight suspended from the axle of the wheel. Then

$$\frac{P}{W'}=\frac{\text{distance between the threads of the screw}}{\text{circumference described by the power}},$$

and $$\frac{W'}{W}=\frac{\text{radius of axle}}{\text{radius of wheel}}.$$

$$\therefore \frac{P}{W}=\frac{\text{distance between two threads of screw}}{\text{circumference described by the power}}$$

$$\times\frac{\text{radius of axle}}{\text{radius of wheel}};$$

or, *the ratio of the power to the weight in the endless screw is equal to the product of the ratios of the power to the weight in the screw and in the wheel and axle.*

188. Prop. *To determine the ratio of the power to the weight in any combination of the mechanical powers.*

Let P= the power for the whole combination,
W^n= " weight " " "
W' = " " to the 1st in the series and power to the 2d,
W''= " " " 2d " " " 3d,
&c., &c., &c.

Let a_1= the ratio of the power to the weight in the 1st,
a_2= " " " " " 2d,
a_3= " " " " " 3d, &c.

Then $\frac{P}{W'}=a_1$, $\frac{W'}{W''}=a_2$, $\frac{W''}{W'''}=a_3$ . . . . . $\frac{W^{n-1}}{W^n}=a_n$;

and, taking the product,

$$\frac{P}{W^n}=a_1.a_2.a_3 \,.\,.\,.\,.\,.\, a_n.$$

Hence the ratio of the power to the weight in any combination of elementary machines *is equal to the product of the ratios of the power to the weight in each of the simple machines.*

189. Prop. *To determine the ratio of the power to the pressure in the combination of levers called* the knee.

This combination of levers is used with advantage where very great pressure is required to act through only a very small space, as in coining money, in punching holes through thick plates of iron, in the printing-press, &c The lever AB turns about a firmly fixed pivot at A, and is connected by another pivot at B to the rod BC, whose extremity C produces the pressure on the obstacle at E.

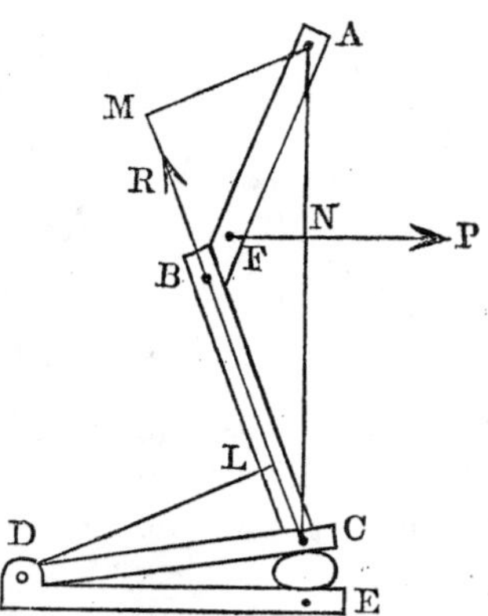

Let the power P act horizontally at some point F in the lever AB, ANC be a vertical line meeting the direction of P in N, and DE a horizontal plane, on which, at E, is the substance subject to pressure. Let R= the reaction of the rod BC in the direction of its length, AM, DL perpendiculars upon its direction from A and D, and W the vertical resistance of the substance at E.

Taking the moments about A and D in equilibrium, we have

$$P.AN=R.AM, \text{ and } W.DE=R.DL,$$

or

$$\frac{P}{W}=\frac{AM\times DE}{AN\times DL}.$$

When BC becomes nearly vertical, DL becomes nearly equal to DE, and AN to AF, while AM becomes very small.

In this situation, $\frac{P}{W}=\frac{AM}{AF}$ nearly, so that P is a very small fraction of W.

CHAPTER VII.

APPLICATION OF THE PRINCIPLE OF VIRTUAL VELOCITIES TO THE MECHANICAL POWERS.

190. In *Arts.* 80 and 81, it is shown that the principle of virtual velocities obtains for all cases of equilibrium of a free body under the actions of any number of external forces in the same plane.

In the mechanical powers, the parts by which the actions of the forces are transmitted being rigid or inextensible, the forces may be considered as acting in the same plane, and the internal reactions and tensions will not enter the fundamental equation $\Sigma.P.v=0$. Also, the virtual velocities of the supporting parts will in general be zero for the possible displacements of the system.

In some of the mechanical powers, the principle applies to *all* possible displacements, however great, since they must be in the direction of the forces. This is true in the wheel and axle, toothed wheels, pulleys with parallel cords, the inclined plane, the wedge, and screw. In the lever, and pulleys with inclined cords, the displacements must be taken *indefinitely small.*

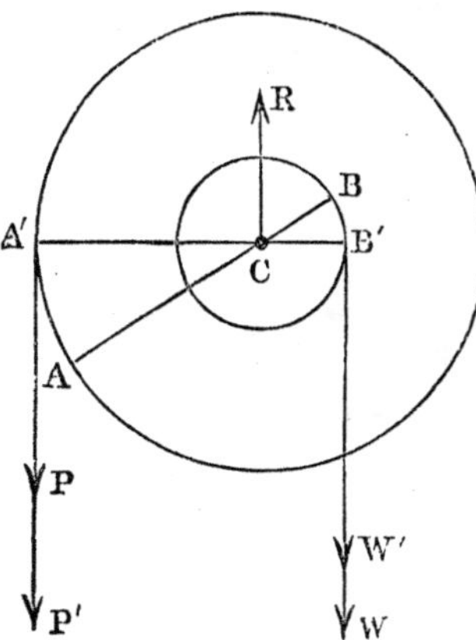

191. Prop. *The principle of virtual velocities obtains in the wheel and axle in equilibrium.*

The forces which act on the wheel and axle are the power P, the weight W, and the reaction R of the steps which support each end, C, of the pivot about which it turns, and, in consequence of the rigidity of the system, they may be considered as acting in the same plane. Also, the wheel and axle receiving a displacement turning about C, the virtual velocity of R equals 0.

Let A and B be the points at which the cords left the wheel and the axle respectively before displacement; A′, B′ afterward. Then W ascends through the space WW′= arc BB′, and P descends through PP′=arc AA′. PP′ is the virtual velocity (*Art.* 78) of P, and positive; WW′ is the virtual velocity of W, and negative. Hence (*Art.* 81),

$$P.PP'-W.WW'=0,$$

or

$$P.\text{ arc } AA'-W.\text{ arc } BB'=0,$$

or

$$P.AC.\angle A'CA-W.BC.\angle BCB'=0.$$

$$P.AC-W.BC=0, \text{ or } \frac{W}{P}=\frac{AC}{BC},$$

the condition of equilibrium found in *Art.* 138.

192. Prop. *The principle of virtual velocities obtains in a pair of toothed wheels.*

Let the circles in the annexed figure represent the pitch-lines of the wheels (*Art.* 144), and D_1, D_2 the points which were in contact in the line CC′ before displacement. Since the pitch-lines roll on each other without slipping, arc DD_1 = arc DD_2, and

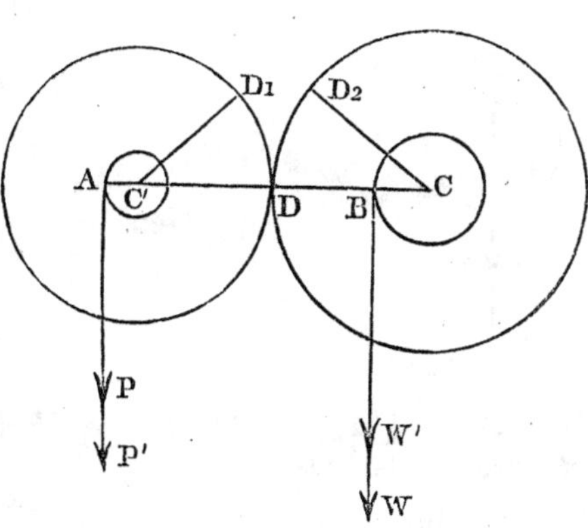

$$\text{P's displacement } = PP'=AC'.\angle D_1C'D=AC'.\frac{\text{arc } D_1D}{C'D}.$$

$$\text{W's displacement } = WW = CB.\angle D_2CD=CB.\frac{\text{arc } D_2D}{CD}.$$

By the principle of virtual velocities,

$$P.PP'-W.WW=0,$$

or $$P.AC'.\frac{\text{arc } D_1D}{C'D} - W.CB.\frac{\text{arc } D_2D}{CD} = 0,$$

or $$P.\frac{AC'}{DC'} - W.\frac{CB}{CD} = 0.$$

$$\therefore \frac{W}{P} = \frac{AC'}{BC}.\frac{CD}{C'D}; \text{ and when } AC' = BC, \frac{W}{P} = \frac{CD}{C'D},$$

as found in *Art.* 143.

193. PROP. *The principle of virtual velocities obtains in the single movable pulley with parallel cords.*

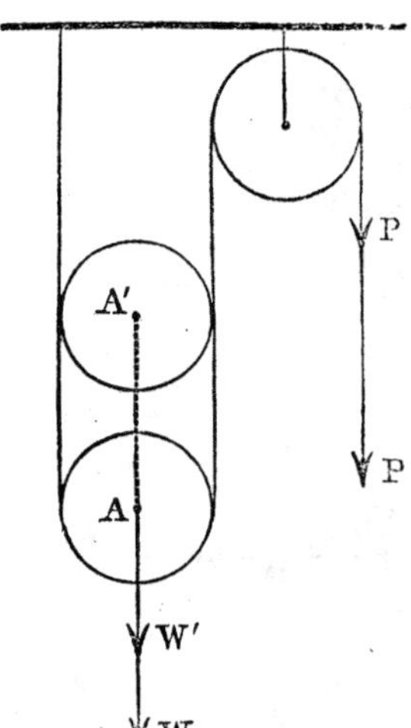

If the pulley A be raised to A we shall have $AA' = WW' = \frac{1}{2}PP'$, since each of the cords passing round the pulley A must be shortened by a length $= WW'$. WW', the virtual velocity of W, is negative.

$$\therefore P.PP' - W.WW' = 0,$$

or $$P.PP' - W.\tfrac{1}{2}PP' = 0.$$

Hence $\frac{W}{P} = 2$, as found in *Art.* 158.

194. PROP. *The principle of virtual velocities obtains in the first system of pulleys.*

In the figure of *Art.* 160 we see that, if W be raised through a space WW', each of the n cords at the lower block will be shortened by the same quantity, or that P will descend through a space $n.WW'$. Hence the equation of virtual velocities $P.PP' - W.WW' = 0$ becomes

$$P.n.WW' - W.WW' = 0 \text{ or } \frac{W}{P} = n,$$

as in *Art.* 160.

195. Prop. *The principle of virtual velocities obtains in the second system of pulleys.*

Referring to the figure in *Art.* 161, we see that, if P descends through the space PP′,

the pulley a_n	would rise through a space			$=\frac{PP'}{2}$,
" a_{n-1}	"	"	"	$\frac{PP'}{2^2}$,
&c.,				&c.,
" a_3	"	"	"	$\frac{PP'}{2^{n-2}}$,
" a_2	"	"	"	$\frac{PP'}{2^{n-1}}$,
" a_1 or the weight W			"	$\frac{PP'}{2^n}$,

And the equation of virtual velocities $P.PP'-W.WW'=0$ becomes

$$P.PP'-W.\frac{PP'}{2^n}=0,$$

or

$$\frac{W}{P}=2^n,$$

the same as in *Art.* 161, *Cor.* 2.

196. Prop. *The principle of virtual velocities obtains in the third system of pulleys.*

Referring to figure of *Art.* 162, and designating the pulleys as in that article, we see that, if W be raised a space $=WW'$, each cord will be shortened by a space equal to WW'. The highest movable pulley a_{n-1} will descend a distance $=WW'$. The next pulley a_{n-2} will descend a distance $=2.WW'$ by the descent of a_{n-1}, and a distance WW' by the elevation of W, or will descend on the whole $(2+1)WW'$.

Similarly, the pulley a_{n-3} will descend through

$$\{2(2+1)+1\}W.W'=(2^2+2+1)WW'.$$

Proceeding in the same way, we find that pulley $a_{n-(n-3)}$, or a_3, will descend through the space

$$(2^{n-4}+2^{n-5}+, \&c. \dots 2+1)WW',$$

and a_2 through the space $(2^{n-3}+2^{n-4}+, \&c. \dots 2+1)WW$,
and a_1 " " " $(2^{n-2}+2^{n-3}+, \&c. \dots 2+1)WW'$,
and P will descend through twice the last found space by the descent of the pulley a_1, and through the space WW′ by the elevation of the weight;

or $$PP'=WW'.\{2(2^{n-2}+2^{n-3}+, \&c. \dots 2+1)+1\}$$
$$=WW'.(2^{n-1}+2^{n-2}+, \&c. \dots 2^2+2+1)$$
$$=WW'(2^n-1).$$

The equation of virtual velocities is

$$P.PP'-W.WW'=0,$$

which becomes

$$P.WW'(2^n-1)-W.WW'=0,$$

or $$\frac{W}{P}=2^n-1$$

as in *Art.* 162.

197. Prop. *The principle of virtual velocities obtains in the inclined plane.*

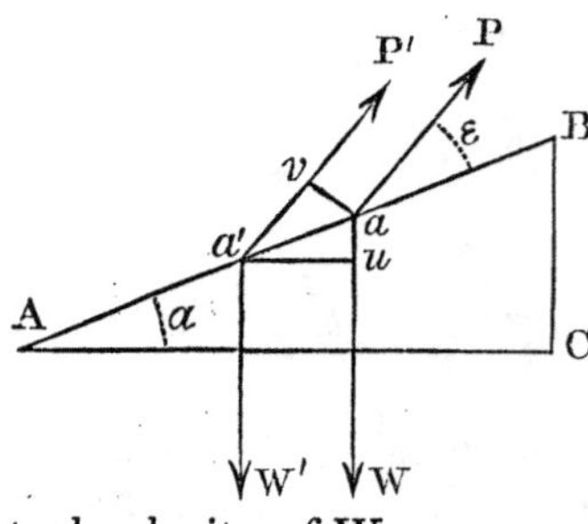

Let the force P make any angle ε with the plane, $a=\angle BAC$, a the first position of the body whose weight is W, a' the position of it after displacement.

Drawing the perpendiculars av, $a'u$, we have $-a'v=-aa'\cos.\varepsilon=$ the virtual velocity of P, and $au=aa'.\sin.a=$ the virtual velocity of W.

By the equation of virtual velocities,

$$P.a'v-W.au=0,$$

or $$P.aa'.\cos.\varepsilon-W.aa'.\sin.a=0.$$

Hence $$\frac{W}{P}=\frac{\sin.\varepsilon}{\sin.a},$$

as found in *Art.* 164.

198. Prop *The principle of virtual velocities obtains in the wedge.*

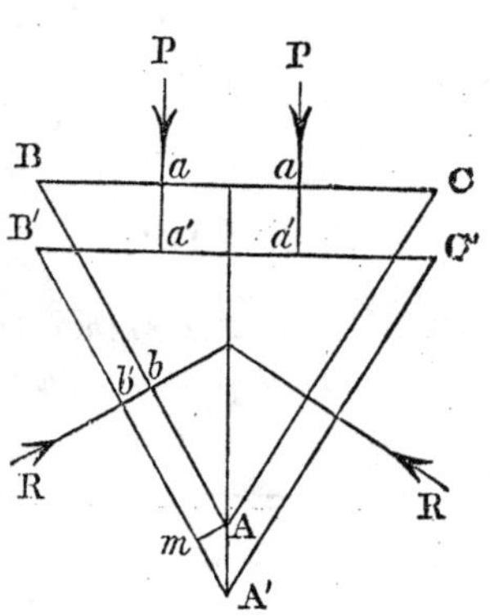

Let 2P be the whole power, R and R the pressures perpendicular to the faces of the wedge ABC, which produce equilibrium. Let the wedge be displaced to the position A'B'C'. The displacement of the point of application of P is $aa'=$ AA'; that of b, the point of application of R, is $bb'=Am$, a perpendicular from A on A'B', and $Am=AA'$ $\sin.\frac{BAC}{2}$.

The equation of virtual velocities is

$$P.aa'-R.bb'=0,$$

or

$$P.AA'-R.AA'.\sin.\frac{BAC}{2}=0.$$

$$\therefore R=\frac{P}{\sin.\frac{BAC}{2}},$$

as found in *Art.* 169.

199. Prop. *The principle of virtual velocities obtains in a lever of any form.*

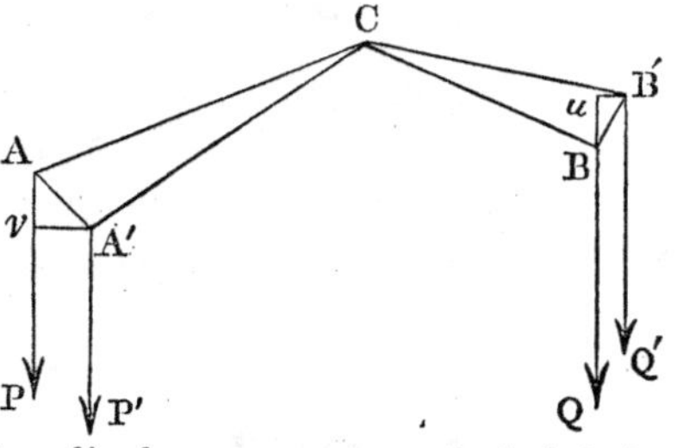

Let ACB be the lever before displacement, A'C'B' its position afterward. From A' draw A'v perpendicular to AP, and from B', B'u perpendicular to BQ produced. Av is the virtual velocity of P, and Bu that of Q. Now when the displacements are indefinitely small, the circular arcs AA', BB' become straight lines, and

$$Av=AA'\cos.A'Av=AC.\angle ACA'.\cos.(PAC-90°)$$
$$=AC.\sin.PAC.\angle ACA';$$
$$Bu=BC.\sin.QBC.\angle BCB',$$

and

$$\angle ACA'=\angle BCB'.$$

The equation of virtual velocities is

$$P.Av - Q.Bu = 0,$$

or
$$P.AC. \sin PAC - Q.BC. \sin. QBC = 0.$$

$$\therefore \frac{P}{Q} = \frac{BC. \sin. QBC}{AC. \sin. PAC},$$

as found in *Art.* 134.

200. Prop. *The principle of virtual velocities obtains in the single movable pulley with cords inclined.*

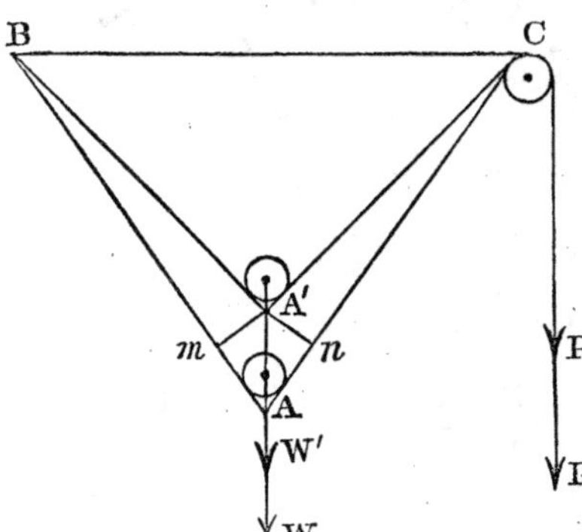

Let A be the point where the cords produced would meet at the first position of the pulley, when P and W are the positions of the power and weight.

Let P be displaced to P′, when the weight is raised to W′, or the point of meeting of the cord is raised to A′. Draw the circular arcs A′*m*, A′*n* with centers B and C. When the displacement is indefinitely small, the arcs A′*m*, A′*n* become straight lines, and

$$Am = AA' \cos. BAA' = An,$$

$$PP' = Am + An = 2AA' \cos. \frac{BAC}{2}, \quad WW' = AA'.$$

The equation of virtual velocities is

$$P.PP' - W.WW' = 0,$$

which becomes

$$P.2AA'. \cos. \frac{BAC}{2} - W.AA' = 0,$$

or
$$\frac{W}{P} = 2 \cos. \frac{BAC}{2},$$

as found in *Art.* 158.

Scholium. In the preceding propositions, the expression

$$P.PP' = W.WW',$$

or $$\frac{WW'}{PP'}=\frac{P}{W},$$

explains the principle that, "in using any machine, what we gain in power we lose in time." For, in order that W may be moved through any given space, we must have the space moved through by the power P, increased in the same ratio tnat P is diminished.

CHAPTER VIII.

FRICTION.

201. The surfaces on which bodies pressed have hitherto been regarded as perfectly smooth, so that they offered no resistance to motion parallel to themselves, their only reaction being perpendicular.

When rough surfaces are in contact, the motion, or tendency to motion, parallel to the surfaces, is affected by the roughness, and the effect is called *Friction.*

Friction may be divided into two kinds: *sliding* friction, when one rough surface slides on another, and *rolling* friction, when one rolls on the other. The former only will be considered here, under the term *Statical Friction.*

202. The laws of friction are determined by experiment.

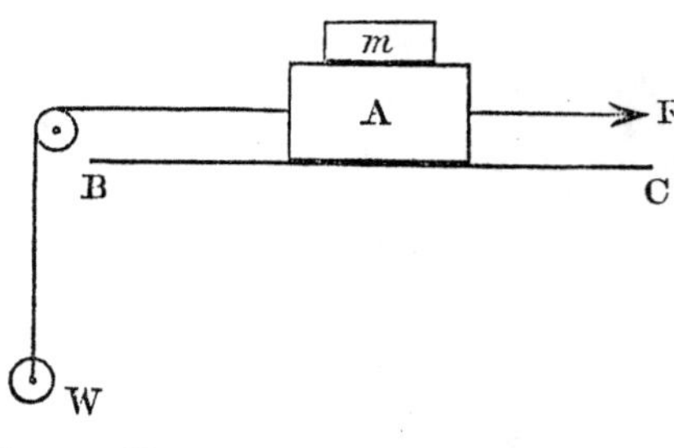

If the body A rest upon the perfectly smooth plane BC, the smallest possible force applied to it will cause it to move. But if the body or the plane, or both, be rough, a force within certain limits of magnitude may be applied to it without causing motion. The greatest force which can be so applied to the body in the direction of the plane will measure the friction.

Let W be this force, acting by a cord over a pulley on the body A, F being the opposing force of friction. Then $F=W$.

203. The following laws of friction are deduced by this or some similar process.

1°. *The friction of the same body, or a body of the same material, when the weight is the same, is independent of the extent of the surfaces in contact,* except in extreme cases, where the weight is very great compared with the surfaces in contact. Thus the friction of the body A will be the same whichever side rest on the plane, or whatever be the form within the excepted limits.

2°. *The friction is proportional to the pressure on the plane, or the reaction of the plane, within moderate limits.* If other weights, as m, be placed on A, W or F will vary as the whole reaction R of the plane.

COR. The friction is therefore some function of this pressure, and we may represent it by μR. μ is called the *coefficient* of friction, and is equal to $\frac{F}{R}$, or the ratio of the friction to the reaction of the plane.

204. PROP. *The coefficient of friction between two given substances is equal to the tangent of the inclination of the plane formed of one of the substances, when the body formed of the other is about to slide down it.*

Let the inclination a of the plane AC be increased till the body a is just on the point of sliding down it. The body a will then be in equilibrium from the normal reaction of the plane R, the friction μR acting up the plane, and its weight W acting vertically downward.

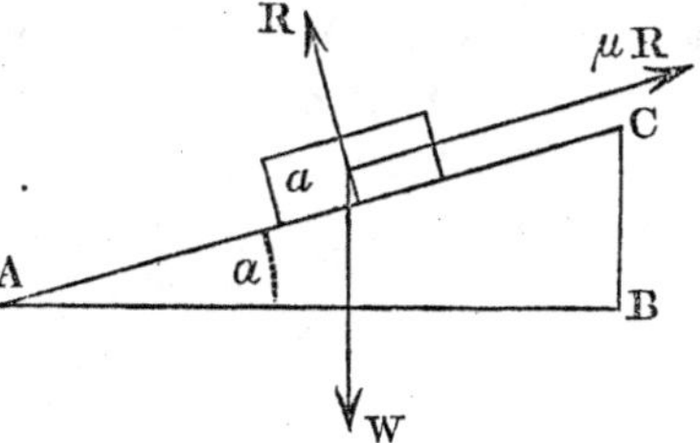

Resolving parallel and perpendicular to the plane, we have

$$\mu R - W.\sin. a = 0,$$
$$R - W \cos. a = 0.$$

Eliminating R, we have

$$\mu \cos. a - \sin. a = 0,$$

or

$$\mu = \tan. a.$$

205. PROP. *To determine the limits of the ratio of* P *to* W *on an inclined plane, when friction acts up or down the plane.*

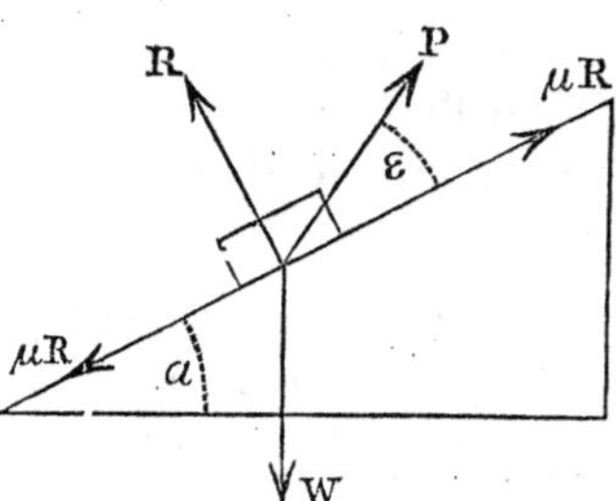

Let the power P make an angle ε with the plane whose inclination is a, and W the weight of the body.

1°. To determine the greatest value P can have without causing the body to move up the plane. In this case, the friction μR, opposing the motion up the plane, will act down it, and, resolving parallel and perpendicular to the plane, we have

$$\text{P}.\cos.\varepsilon-\mu\text{R}-\text{W}\sin.a=0, \qquad (a)$$
$$\text{P}.\sin.\varepsilon+\text{R}-\text{W}\cos.a=0. \qquad (b)$$

Multiplying (b) by μ and adding to (a), we have

$$\text{P}(\cos.\varepsilon+\mu\sin.\varepsilon)-\text{W}.(\sin.a+\mu\cos.a)=0,$$

or

$$\text{P}=\frac{\text{W}.(\sin.a+\mu\cos.a)}{\cos.\varepsilon+\mu\sin.\varepsilon}.$$

2°. To determine the least value P can have without causing the body to move down the plane. In this case the friction will oppose the descent, and will therefore act up the plane. Hence equations (a) and (b) become

$$\text{P}.\cos.\varepsilon+\mu\text{R}-\text{W}\sin.a=0, \qquad (a')$$
$$\text{P}.\sin.\varepsilon+\text{R}-\text{W}\cos.a=0. \qquad (b')$$

Multiplying (b') by μ and subtracting, we find

$$\text{P}=\frac{\text{W}.(\sin.a-\mu\cos.a)}{\cos.\varepsilon-\mu\sin.\varepsilon}.$$

No motion will take place while the value of P is between these two, which are its limits.

These two values of P may be combined so as to take the form

$$\frac{\text{W}}{\text{P}}=\frac{\cos.\varepsilon\pm\mu\sin.\varepsilon}{\sin.a\pm\mu\cos.a},$$

in which the upper sign is to be taken when friction acts down the plane, and the lower when the friction acts up the plane.

206. Prop. *To determine the limits of the ratio of* P *to* W *in the screw, when friction acts assisting the power or the weight.*

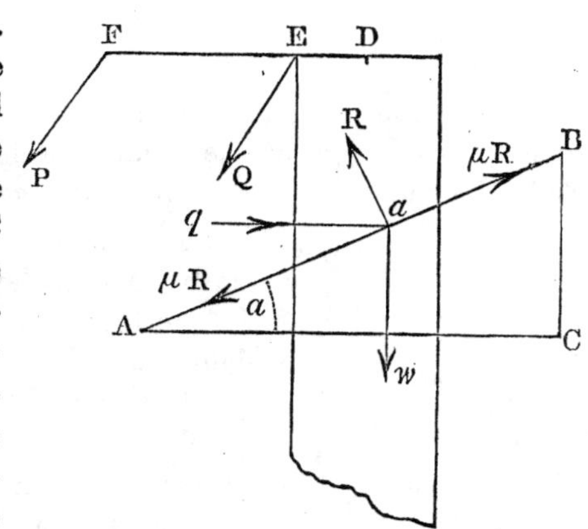

Proceeding as in *Art.* 173, let ABC be the inclined plane formed by unwrapping one revolution of the thread, the angle BAC $=a$; let W= the whole weight sustained by the screw, w= that part of it supported at a, Q= the whole force acting at the circumference of the cylinder, r=ED the radius of the cylinder, a=FD the lever at which the power acts, and q that part of Q which supports w at a. Then $Q=P\frac{a}{r}$.

The forces which are in equilibrium at a are the weight w, the reaction R, the horizontal force q, and the friction μR acting *up* or *down* the plane.

Resolving parallel and perpendicular to the plane, we have

$$q. \cos. a \pm \mu R - w. \sin. a = 0, \qquad (a)$$

$$q \sin. a - R + w \cos. a = 0. \qquad (b)$$

Multiplying (b) by μ, and adding and subtracting, we have

$$q(\cos. a \pm \mu \sin. a) - w(\sin. a \mp \mu \cos. a) = 0,$$

or

$$\frac{q}{w} = \frac{Q}{W} = \frac{\sin. a \mp \mu \cos. a}{\cos. a \pm \mu \sin. a}.$$

$$\therefore \frac{P}{W} = \frac{r}{a} \cdot \frac{\sin. a \mp \mu \cos. a}{\cos. a \pm \mu \sin. a}.$$

The two values of this expression give the limits required, and $\frac{P}{W}$ may have any intermediate value.

207. Cor. From the two preceding propositions it will appear that, when we have obtained one of the limits, the other may be had by simply changing the sign of μ in the former.

EXAMPLES.

Ex. 1. A uniform straight beam rests on a rough cylinder of given radius; required the greatest weight that can be suspended from one end of the beam without causing it to slide off.

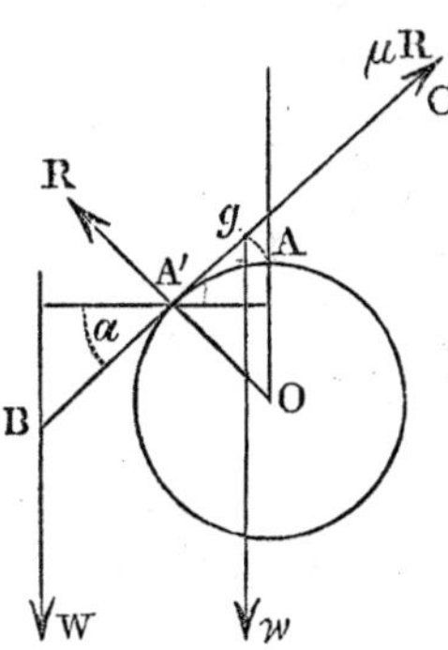

Let g be the center of gravity of the beam BC, whose length is $2d$, and $\mathrm{B}g=d$ since the beam is uniform, $w=$ its weight, and W the weight suspended from B. Before the weight W was suspended from the beam the point g must have been at A. Let A′ be the point of contact with the cylinder when the beam is on the point of sliding off, a the angle it then makes with the horizon, and $r=$ the radius of the cylinder. Resolving parallel and perpendicular to the beam, we have

$$\mu \mathrm{R}-w.\sin. a-\mathrm{W}\sin. a=0, \qquad (a)$$

$$\mathrm{R}-w\cos. a-\mathrm{W}\cos. a=0; \qquad (b)$$

whence $\mu=\tan. a$.

Taking the moments about A′, we have

$$\mathrm{W.BA'}\cos. a-w.\mathrm{A'}g.\cos. a=0,$$

or

$$\mathrm{W.(B}g-\mathrm{A'}g)-w.\mathrm{A'}g=0.$$

But $\mathrm{A'}g=$ arc A′A$=$ radius $\times$ angle AOA′$=r.a$.

$$\therefore\ \mathrm{W}.(d-ra)-w.ra=0,$$

or

$$\mathrm{W}=\frac{w.ra}{d-ra}=\frac{w.r\tan.^{-1}\mu}{d-r\tan.^{-1}\mu}, \text{ the weight required.}$$

Ex. 2. A ladder rests with one end on a rough horizontal plane, and the other on a rough vertical wall; given $l=$ its length, $d=$ the distance of its center of gravity from its lower end, μ and $\mu'=$ the coefficients of friction on the horizontal and vertical planes respectively; required its inclination θ when on the point of sliding down.

Let AB be the ladder and the forces acting upon it, as represented in the figure.

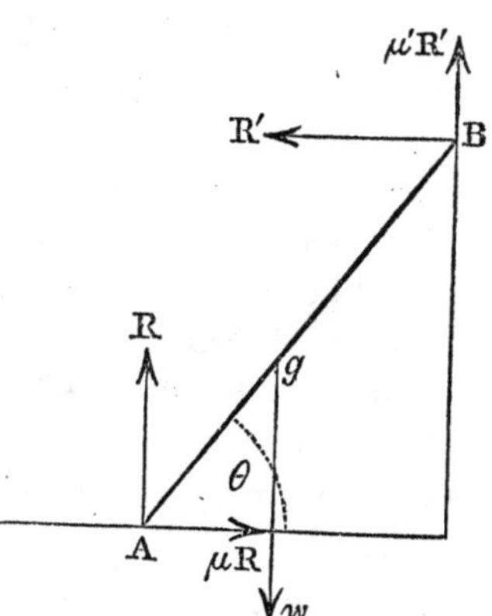

Resolving horizontally and vertically, we have

$$R'-\mu R=0, \qquad (a)$$
$$R+\mu' R'-w=0. \qquad (b)$$

Multiplying (*a*) by μ', and subtracting from (*b*), we find

$$w = R(1+\mu\mu'),$$

and from (*a*), $R'=\mu R$.

Taking moments about A, we get

$$w.d. \cos. \theta - R'.l. \sin. \theta - \mu' R'.l. \cos. \theta = 0,$$

or
$$\tan. \theta = \frac{w.d-\mu' R' l}{R' l}.$$

Substituting the values of w and R',

$$\tan. \theta = \frac{d.(1+\mu\mu')-l\mu\mu'}{\mu l}.$$

If the center of gravity of the ladder be at the middle point, $l=2d$, and

$$\tan. \theta = \frac{1-\mu\mu'}{2\mu}.$$

208. EXAMPLES ON CHAPTERS VI., VII., AND VIII.

Ex. 1. A beam 30 feet long balances on a prop $\frac{1}{3}$ of its length from the thicker end; but when a weight of 10 lbs. is suspended from the other end, the prop must be moved 2 feet toward it to maintain the equilibrium. Required the weight of the beam.

Ex. 2. The forces P and Q act at arms a and b respectively of a straight lever, which rests on a fixed point to which it is not attached. When P and Q make angles α and β with the lever, required the conditions of equilibrium.

Ex. 3. A uniform beam is sustained by three persons, one at one end, and the other two by a hand-spike placed at some point beneath it. At what point must the hand-spike be placed that each person may sustain one third of the weight?

Ex. 4. A Roman steelyard, whose weight is 10 lbs., has its center of gravity 2 inches from the fulcrum, and the weight to be determined is supported by a pan placed at a distance of 3 inches on the other side. Find the respective distances from the fulcrum at which the constant weight of 5 lbs. must be placed, in order to balance 10, 20, 30, &c., lbs. placed successively in the pan.

Ex. 5. Find the ratio of the power to the pressure in the *common vice.*

Ex. 6. Find the ratio of the power to the pressure in the screw by the principle of virtual velocities.

Ex. 7. An isosceles triangle, whose base is to one of its equal sides as $1 : \sqrt{7}$, is placed with its base on an inclined plane; and it is found that, when the body begins to slide, it also begins to roll over. Find the coefficient of friction.

Ex. 8. A ladder rests against a vertical wall, to which it is inclined at an angle of 45°; the coefficients of friction of the wall and of the horizontal plane being respectively $\frac{1}{3}$ and $\frac{1}{2}$, and the center of gravity of the ladder being at its middle round. A man whose weight is half the weight of the ladder ascends it. Find to what height he will go before the ladder begins to slide.

Ex. 9. In a uniform lever of the second kind, which weighs 2 ounces per inch, required the length of the lever, in order that the power may be the least possible when in equilibrium with a weight of 48 ounces placed at a distance of three inches from the fulcrum.

Ex. 10. Two given weights, P and Q, are suspended from two given points in the circumference of the wheel, a being the angle made by the radii drawn to the points of suspension. Required the angle θ which the lower radius makes with the vertical when the weights cause the greatest pressure on the axle.

DYNAMICS.

INTRODUCTION.—DEFINITIONS.

209. In Statics we have investigated the relations of the intensities and directions of forces necessary to produce equilibrium, this result being entirely independent of the *time* during which the forces act

In Dynamics, forces are regarded as producing motion or change of motion in bodies, and these effects must obviously depend on the duration of the action of the forces. In dynamics, therefore, *time* becomes an element in our investigations.

210. *Motion* is the transit of a material point or body from one position to another in space.

211. The *absolute motion* of a body is its transit from one fixed point in space to another.

212. *Relative motion* is a change of distance from a point which is itself in motion.

All motions are relative in any practical view which we can take of them, since we have no means of determining the absolute *rest* of any point in space.

213. The *velocity* of a body is its *rate* of motion. It may be uniform or variable.

The velocity of a body is *uniform* when it passes over equal spaces in equal times, and is measured by *the space passed over in the unit of time.*

Let v be the velocity, or space passed over in one second, then the space described in two seconds will be $2v$, in three seconds $3v$, and so on; and if s be the space described from the commencement of motion, and t the number of seconds also reckoned from the commencement of motion, then

$$s = vt. \qquad [I.]$$

The units of time and space are arbitrary. It is usual to take *one second* for the unit of time, and *one foot* for the unit of space. When no mention is made of different units, these will be understood.

214. *Variable velocity* is that which continually increases or decreases, so as to be the same at no two successive instants. To find a measure of the velocity of a point or body so moving, let us assume, at first, that the velocity which the body has at the end of t_1 seconds is uniform from t_1 seconds to t_2 seconds, a very small interval, the space passed over at the end of t_1 seconds being s_1, and the space passed over at the end of t_2 seconds being s_2. Then, using the symbol Δ to signify finite difference, by [I.] we have

$$v = \frac{s_2 - s_1}{t_2 - t_1} = \frac{\Delta s}{\Delta t}.$$

If, however, there is no time, however small, during which the velocity is uniform, then the smaller we take Δt, and consequently Δs, the more nearly

$$v = \frac{\Delta s}{\Delta t}. \qquad [II.]$$

That is, the velocity, when variable, is measured by the *limit* of the ratio of the space described to the time of describing it.

215. *Relative velocity* is the velocity with which two bodies approach or recede from each other.

216. Matter at any given moment must be in one of the two states, motion or rest. The *inertia of matter* is the entire absence of power in itself to change this state. It implies equally a disability, when in motion, to change its rate or its direction. Hence

A body, when not acted on by any external forces, if at rest, will remain so, or, if in motion, will continue to move in a straight line and with a uniform velocity.

This is called the first law of motion.

217. It is a consequence of the inertia of matter, that when a force is applied to a body to move it, each of its particles opposes a resistance to motion in directions parallel but oppo-

site to the direction of the applied force. The center of these parallel forces (*Art.* 44) of resistance is called the *center of inertia* of the body. It is the same point which, in statics, was called the center of gravity of the body in reference to the force which was there supposed to act on the body.

The sum of these parallel resistances, or their resultant (*Art.* 43), is obviously proportional to the number of particles in the body or to the whole mass. Hence the inertia of a body is a surer test of the quantity of matter or mass of a body than its weight is; for the latter (*Arts.* 84 and 88) varies by a change of position on the earth, while the former is always the same.

218. The *path* of a body is the line, straight or curved, which its center of inertia describes when it passes from one point to another in space.

219. A body is said to be *free* or move *freely* when its path depends on the action of the impressed forces only. Its motion is said to be *constrained* when its path is limited to a given line, straight or curved, or limited to a given surface.

220. An *impulsive force* is one which acts instantaneously and without sensible duration.

221. An *incessant force* is one which acts without intermission. If a material point move from rest by the action of an incessant force, its rate of motion or velocity must continually increase. The amount of this increase, or the *increment* of the velocity in the unit of time, will obviously be greater or less as the intensity of the force is greater or less. This increment of the velocity in one second is therefore the measure of the intensity of the force.

222. A *constant force* is an incessant force whose intensity is at all times the same. If a material point move by the action of a constant force, the increments of the velocity in each successive unit of time must all be equal, and each increment will be a measure of the force. If, therefore, we put ϕ for the increment of velocity, or the velocity generated by the force in one second, ϕ will represent the force. The increments being all equal, the velocity generated in two seconds will be

2ϕ, in three seconds 3ϕ, and so on. Hence, if v be the velocity generated in t seconds,

$$v=\phi t. \qquad [III.]$$

223. A *variable force* is an incessant force whose intensity either increases or decreases, so as to be the same at no two successive instants. To find an expression for a variable force, let us assume it to be constant from the end of the time t_1, when the velocity is v_1, to the end of the time t_2, when the velocity is v_2. Then, by [III.], we have

$$\phi=\frac{v_2-v_1}{t_2-t_1}=\frac{\Delta v}{\Delta t}.$$

If, however, there be no interval during which the force is constant, then the smaller we take Δt, and consequently Δv the more nearly will

$$\phi=\frac{\Delta v}{\Delta t}. \qquad [IV.]$$

That is, a variable force is measured by the *limit* of the ratio of the velocity caused by it to the time of causing it.

224. The *momentum of a body* is its quantity of motion, and is measured by *the product of the mass of the body by its velocity*. For the motion of a single particle is its velocity, and the motion of any number of particles, having the same velocity, is obviously as much greater as the number of particles is greater. Hence the whole motion is equal to the whole number of particles in the body, or its mass, multiplied by their common velocity. Or, if Q be the quantity of motion of a body, M its mass, and V its velocity,

$$Q=MV. \qquad [V.]$$

The mass, multiplied by the square of the velocity, is called the *vis viva*, or the *living force* of a body.

225. In estimating the effects of incessant forces, we have considered only the acceleration or velocity which each force will produce when acting on a free material point or a unit of mass. When so measured, they are called *accelerating* forces. If the mass moved differs from that which we have called the unit of mass, and it is taken into consideration in estimating

the effects of the forces, they are then called *moving forces*. Let ϕ be the acceleration, or velocity generated in a unit of time, M the number of units of mass; then Φ, the moving force, will be measured by the quantity of motion generated in the unit of time, or

$$\Phi=\phi M. \qquad [VI.]$$

Hence $\phi=\frac{\Phi}{M}$, or the accelerating force, is equal to the moving force divided by the mass.

226. If a body already in motion be acted on by a force in the direction in which the body moves, the *superadded motion* is just the same as that which would have been produced in the body if at rest when the force began to act.

If the force act in a direction different from that in which the body moves, the new motion produced by the force, *estimated in the direction of the force*, will be the same as if the body had been at rest.

If the force act in a direction opposite to that in which the body moves, the motion destroyed in the body is equal to that which the force would produce in the body if at rest when the force began to act.

These facts are consequences of the inertia of matter, and will receive additional illustration in the sequel. They are embodied in the following enunciation, called the second law of motion:

All motion or change of motion in a body is proportional to the force impressed and in the direction of that force.

227. When one body impinges on another at rest or in motion, the quantity of motion, or momentum of the two bodies after impact, is the same as before impact. For matter being incapable of originating motion, can not add to the motion of other matter, or take from it except by imparting its own motion. Hence whatever motion the second body receives in the direction of the striking body, just so much must be lost by the striking body.

This fact is usually enunciated as follows, and is called the third law of motion:

Action and reaction are equal and in opposite directions.

CHAPTER I.

UNIFORM MOTION.

228. In considering the effect of an impulsive force, we shall suppose the force applied, at the center of inertia (*Art.* 217). When so applied the parallel forces of resistance of all the particles situated on opposite sides of the direction of the force will balance each other, and the body will not rotate, but all its particles will describe parallel lines with a common velocity.

229. Prop. *To find the general equation of uniform motion.*

A′ O A B

Let OB be the path of the body (*Art.* 218), v its velocity (*Art.* 213), and t the time of its motion in seconds. Let O be the origin, or point from which we estimate the successive positions of the body, s the distance of the body from O at the end of the time t, and $OA=s_1$ its distance from the origin O at the commencement of the time.

When the body is at B, we have [I.]

$$AB=vt.$$

But

$$OB=OA+AB,$$

or

$$s=s_1+vt, \tag{36}$$

which is the general equation of uniform motion.

If the body, instead of receding from the origin O, approach it, then v will be negative; or if the body begin to move from A′, then s_1 will be negative.

Cor. 1. From (36) we obtain

$$v=\frac{s-s_1}{t}=\frac{AB}{t},$$

or *the velocity of a body is equal to the constant ratio of the space described to the time of describing it.*

Cor. 2. If we estimate the position of the body from the

point where the body is when $t=0$, or suppose the space and time to commence together, then $s_1=0$, and

$$s=vt. \qquad (37)$$

230. Prop. *If two bodies move during the same time, their velocities will be proportional to the spaces described by them respectively.*

Taking the points of departure for the origin of spaces, we have from (37)

$$s=vt \text{ and } s'=v't'.$$

$$\therefore\ s : s'=vt : v't'.$$

And, since $t=t'$, $\qquad s : s'=v : v'.$

231. Prop. *If the velocities of two bodies are equal, the spaces described are proportional to the times.*

As before, $\qquad s : s'=vt : v't'$;

and if $v=v'$, $\qquad s : s'= t : t'$.

232. Prop. *If the spaces described by two bodies are equal. their velocities are reciprocally proportional to the times.*

For, since $\qquad s : s'=vt : v't'$,

if $s=s'$, $\qquad vt=v't'$.

$$\therefore\ v : v'=t' : t.$$

233. Prop. *An impulsive force is measured by the momentum it can produce in any mass.*

If v be the velocity produced by the force F_1 in a body containing M units of mass, by [V.] the momentum will be Mv. Now if the mass M move from rest by the action of the force, all the motion it receives is the effect of the force. And, admitting the principle that effects are proportional to their causes, if λ be the constant ratio of the force to the momentum, then

$$F_1=\lambda.Mv.$$

But the unit of force being arbitrary, we may assume it to be λ. Hence, putting F for the number of units of force, or the ratio of F_1 to λ, we have

$$F=\frac{F_1}{\lambda}=Mv. \qquad (38)$$

Cor. If $M=1$, $F_1=\lambda.v$, in which λ is constant. Hence the force *is proportional to the velocity produced in the unit of mass.*

SCHOL. Since we know nothing of the nature of forces, we can not determine their effects *à priori*. The foregoing proposition ought, therefore, to be regarded as depending ultimately on observation and experiment. The fact that on the earth, which is subject to the double motion of rotation and translation, forces are found to produce precisely the same effects as if the earth were at rest, is a confirmation of its truth; also that a pendulum performs its vibrations in the same time, whatever be the direction of these vibrations in reference to an east and west line.

234. *If any number of forces act upon a body in the same direction, the velocity imparted to the body will be equal to the sum of the velocities imparted by each.*

Let F, F′, F″ . . . be any number of forces acting upon a body, v, v', v'' . . . the velocities imparted by each respectively, ϕ the resultant force, and u the velocity due to ϕ.

Then $F=\lambda v$, $F'=\lambda v'$, $F''=\lambda v''$, &c. Since the forces are conspiring, their resultant will be equal to their sum.

$$\therefore\ F+F'+F''+\ \&c., =\phi=\lambda(v+v'+v''+\ \&c.).$$

But $\phi=\lambda u$.

Hence $$u=v+v'+v''+\ \&c.$$

235. PROP. *If the adjacent sides of a parallelogram represent in magnitude and direction the velocities which two forces, by their separate action, would respectively produce in a body, the diagonal of the parallelogram will represent the actual velocity produced by their joint action.*

Let AB and AC represent the velocities which the forces P and Q respectively impart to a body placed at A, or, in other words, the spaces over which the body would pass in a unit of time by the separate action of the forces, then AD will represent the velocity produced by their resultant, or the space over

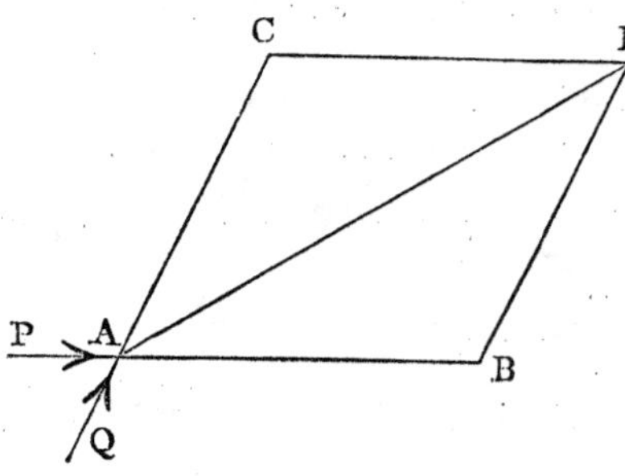

which the body will pass in the unit of time by the joint action of the forces.

We may assume AB to represent the force P; then, by *Art.* 233, *Cor.*,

$$P : Q = AB : AC,$$

and AC will represent the force Q. Also, *Art.* 21, AD will be the resultant R of P and Q. Let x be the velocity due to the resultant R, then (*Art.* 233, *Cor.*)

$$P : R = AB : x.$$

But $$P : R = AB : AD.$$

$$\therefore x = AD,$$

and AD being the direction of the resultant, the velocity due to it will have the same direction.

The two preceding propositions illustrate the second law of motion (*Art.* 226).

236. PROP. *The velocity of a body being given, to find the component velocities in any directions at right angles to each other; and two component velocities at right angles to each other being given, to find the resultant velocity.*

Let AD represent the velocity v of the body in direction and magnitude, AX and AY the rectangular directions in which the components are required, and θ the angle which AD makes with AX. Completing the rectangle, AB will represent the velocity a in AX, and AC the velocity b in AY.

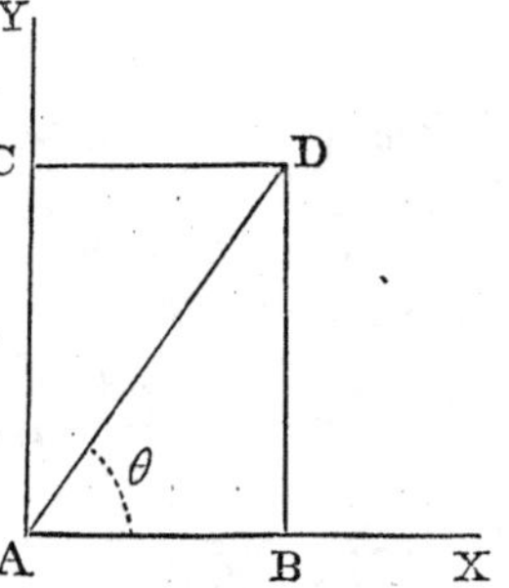

Then $$AB = AD. \cos. \theta, \; AC = AD. \sin. \theta,$$

or $$a = v. \cos. \theta, \quad b = v. \sin. \theta. \qquad (39)$$

Again, if a and b represent the given component velocities in AX and AY, to find the resultant velocity v we have

$$v = \sqrt{a^2 + b^2}, \qquad (40)$$

and $$\tan. \theta = \frac{b}{a}, \qquad (41)$$

as required.

Cor. Hence velocities may be compounded and resolved like forces in statics.

237. Prop. *Two bodies,* A *and* B, *describe the same path with the velocities* v *and* v', *and at the commencement of their motion are at a distance* a *from each other: to find the time* t *when they will be at a distance* b *from each other and the position of each at the end of that time.*

Taking the position of A when $t=0$ for the origin of spaces, the equations of their motions are, by (37) and (36),

$$s=vt, \tag{a}$$

and

$$s'=a+v't'. \tag{b}$$

By the conditions of the question, $t=t'$.

Also $s-s'=b$ or $s'-s=b$, $\therefore\ s-s'=\pm b$.

Subtracting (b) from (a), and putting $t=t'$, we have

$$s-s'=vt-a-v't=\pm b.$$

$$\therefore\ t=\frac{a\pm b}{v-v'}. \tag{42}$$

Also, from (a)

$$s=vt=v.\frac{a\pm b}{v-v'}, \tag{43}$$

and from (b)

$$s'=a+v't=a+v'.\frac{a\pm b}{v-v'}=\frac{va\pm v'b}{v-v'}. \tag{44}$$

Cor. 1. If the bodies move in opposite directions, v' will be negative, and

$$t=\frac{a\pm b}{v+v'},\ s=v.\frac{a\pm b}{v+v'},\ s'=\frac{va\mp v'b}{v+v'}.$$

Cor. 2. If the time when the bodies are together be required, then $b=0$, and

$$t=\frac{a}{v-v'},\ s=\frac{va}{v-v'}=s',\ s'-a=\frac{v'a}{v-v'}.$$

238. Prop. *When two bodies* A *and* B *move in the circumference of a circle with uniform velocities, to determine the circumstances of their motion.*

Let v and v' be their velocities, c the circumference of the circle, and a the distance apart at the commencement of the time. Then, putting $b=0$ in (42), we have

$$t_1=\frac{a}{v-v'} \quad = \text{the time of their first meeting.}$$

$$t_2=\frac{a+c}{v-v'} \quad = \text{" \quad " \quad second \quad "}$$

$$t_3=\frac{a+2c}{v-v'} \quad = \text{" \quad " \quad third \quad "}$$

$$\cdot \quad \cdots \qquad\qquad \cdots$$

$$t_n=\frac{a+(n-1)c}{v-v'}= \text{" \quad " \quad } n\text{th \quad "}$$

also, $$s=v\,\frac{a+(n-1)c}{v-v'} = \text{space described by A,}$$

and $$s-a=\frac{v'a+v(n-1)c}{v-v'}= \text{" \quad " \quad B.}$$

COR. The interval between two successive conjunctions is

$$t_2-t_1=t'=\frac{c}{v-v'}. \tag{45}$$

239. EXAMPLES.

Ex. 1. If an iron rod have one end against the sun and the other resting on the earth, the distance of the sun from the earth being 95,125,000 miles, in what time will a blow applied to the end on the earth be felt by the sun, the velocity of an impulse in iron being 11,865 feet per second?

Ans. 490 days.

Ex. 2. When the earth is in that part of its orbit nearest to Jupiter, an eclipse of one of Jupiter's satellites is seen 16 minutes 36 seconds sooner than it would be if the earth were in that part of its orbit most remote from Jupiter. The radius of the earth's orbit being 95,125,000 miles, what is the velocity of light?

Ex. 3. The star 61 Cygni is ascertained to be 56,319,996,-600,000 miles distant from us; were its light suddenly extinguished, in what time would the intelligence reach us, the velocity of light being 191,000 miles?

Ex. 4. Suppose 964 tons of ice to be floating directly to the east at sunrise on the 21st of March, with a velocity of 12 feet per minute, how many grains of light from the sun would be sufficient to stop it?

Ex. 5. A train of cars moving with a velocity of 20 miles an hour, had been gone three hours, when a locomotive was dispatched in pursuit, with a velocity of 25 miles an hour; in what time did the latter overtake the former?

Ex. 6. Had the trains in *Ex.* 5 started together and moved in opposite directions around the earth, 24,840 miles, in what time would they meet?

Ex. 7. Suppose it to be 12 minutes past noon by a clock, in how long a time will the hour and minute hands of the clock be together?

Ex. 8. The daily motion of Mercury in his orbit is 4°.09239; that of Venus 1°.60216; that of the earth 0°.98563: what are the intervals between the epochs at which Mercury and Venus respectively will be in the same direction from the sun as the earth?

Ex. 9. A man being caught in a shower in which the rain fell vertically, ran with a velocity of 12 feet per second. He found that the drops struck him in the face, and estimated that the apparent direction of the drops made an angle of 10° with the vertical line. What was the velocity of the drops?

Ans. 68 feet.

Ex. 10. When the path of the earth in its orbit is perpendicular to a line drawn from a star to the earth, the path of the light from the star *appears* to make an angle of 20″,445 with the perpendicular to the path of the earth. The velocity of the earth being 68,180 miles per hour, what is the velocity of light?

CHAPTER II.

IMPACT OF BODIES.

240. Def. When two bodies in motion impinge, if their centers of inertia move in the same straight line perpendicular to a plane tangent to the bodies at their point of contact, the impact is said to be *direct* and *central.*

If the straight line described by the center of inertia of one of the bodies is not perpendicular to the tangent plane, the impact is said to be *oblique.*

In the cases discussed the bodies will be supposed, spherical and of uniform density.

241. Def. When the bodies impinge, they exert a mutual but varying pressure during the interval between contact and separation, an interval of time which is generally very short, and we *suppose* them to suffer a degree of compression, by which, during a portion of this interval, their centers will approach each other, and during the remaining portion will recede by the action of an internal force tending to restore them to their original form. The force urging the approach of their centers is called the *force of compression;* the opposing force causing them to separate again is called the *force of restitution* or *elasticity.* The *ratio* of the force of restitution to that of compression is called the *modulus of elasticity.*

When this ratio is unity, or the force of restitution is equal to that of compression, the bodies are *perfectly elastic;* when it is zero, or the force of restitution is nothing, they are *inelastic.* If the value of the ratio is intermediate between zero and unity the bodies are *imperfectly elastic.*

242. Def. If the bodies suffer no compression, they are called *hard;* if, when compressed, they exert no force to recover their original form, they are called *soft.*

There are no known bodies either perfectly elastic or perfectly inelastic, but these states may be considered as limits to the various degrees of elasticity presented in nature.

243. PROP. *To determine the velocity of two inelastic bodies after direct impact.*

Since the bodies are inelastic, the force of restitution is zero, and the bodies will move on together with a common velocity.

1°. Let m_1 and m_2 be the masses or bodies moving in the *same* direction with the velocities v_1 and v_2 $(v_1>v_2)$, and v their common velocity after impact.

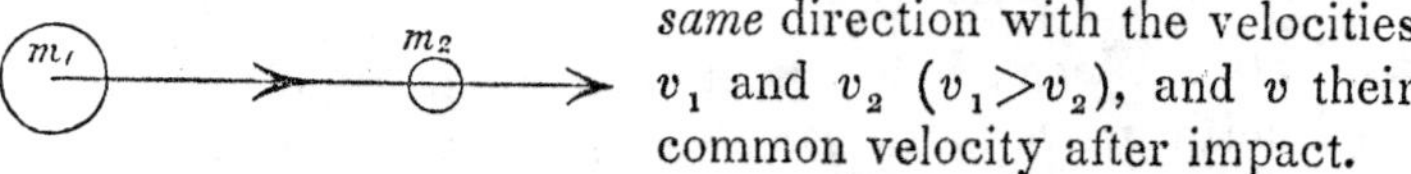

If F and F′ are the forces which impress on the bodies their respective velocities, then (38)

$$F=m_1v_1, \text{ and } F'=m_2v_2,$$

and their resultant, $F+F'=m_1v_1+m_2v_2$.

After impact the bodies move on together as one mass, and its momentum $(m_1+m_2)v$ must be a measure of the force $F+F'$.

$$\therefore\ (m_1+m_2)v=m_1v_1+m_2v_2,$$

or

$$v=\frac{m_1v_1+m_2v_2}{m_1+m_2}. \tag{46}$$

2°. If the bodies move in opposite directions, the resultant of the forces will be $F-F'$.

$$\therefore\ F-F'=m_1v_1-m_2v_2=(m_1+m_2)v,$$

or

$$v=\frac{m_1v_1-m_2v_2}{m_1+m_2}. \tag{47}$$

The same result will be obtained by changing the sign of v_2 in (46).

Hence the velocity of two inelastic bodies after impact is *equal to the algebraic sum of their momenta before impact, divided by the sum of their masses.*

COR. 1. *If two bodies move in opposite directions, with velocities reciprocally proportional to their masses, they will rest after impact.*

For $$m_1 : m_2=v_2 : v_1$$
gives $$m\ v_1=m_2v_2,$$
which, substituted in (47), gives $v=0$.

COR 2. If m_2 is at rest before impact, $v_2=0$, and

$$v=\frac{m_1 v_1}{m_1+m_2};$$

and if, at the same time, the masses are equal,

$$v=\frac{v_1}{2}.$$

COR. 3. If the masses are equal, and move in the same direction,

$$v=\frac{v_1+v_2}{2};$$

if in opposite directions,

$$v=\frac{v_1-v_2}{2}.$$

244. PROP. *In the impact of inelastic bodies there is a loss of living force, and this loss is equal to the sum of the living forces due to the velocities lost and gained by the bodies respectively.*

For the living force before impact $=m_1v_1^2+m_2v_2^2$ (*Art.* 224),
and " " " after " $=(m_1+m_2)v^2$.

∴ the loss of living force by impact is

$$m_1v_1^2+m_2v_2^2-(m_1+m_2)v^2=m_1v_1^2+m_2v_2^2-2(m_1+m_2)v^2+(m_1+m_2)v^2$$

(46),

$$=m_1v_1^2+m_2v_2^2-2(m_1v_1+m_2v_2)v+(m_1+m_2)v^2$$

$$=m_1(v_1-v)^2+m_2(v_2-v)^2,$$

which is necessarily positive, and in which v_1-v is the velocity lost by m_1, and $v-v_2$ the velocity gained by m_2.

From this proposition it appears that in machinery made of inelastic materials all abrupt changes of motion are attended with a loss of living force, by which loss the efficiency of the machinery is impaired.

245. PROP. *To find the velocities of two imperfectly elastic bodies after direct impact.*

Let m_1 and m_2 be the masses, v_1 and v_2 their velocities before impact, and v'_1, v'_2 their velocities after impact.

The bodies being elastic, will suffer compression. Let v be their common velocity at the instant of greatest compression or when the distance between their centers is least. Then the velocity lost by m_1 at this instant wil be v_1-v.

Let ε be the modulus of elasticity, or the ratio of the force of restitution to that of compression. Since these forces are proportional to the velocities they generate or destroy in the same mass, the velocity destroyed in m_1 by the force of restitution will be $\varepsilon(v_1-v)$.

Hence the whole velocity lost by m_1 will be

$$v_1-v+\varepsilon(v_1-v)=(1+\varepsilon)(v_1-v).$$

This, subtracted from the velocity of m_1 before impact, will give its velocity after impact, or

$$v'_1=v_1-(1+\varepsilon)(v_1-v)=v-\varepsilon(v_1-v). \qquad (a)$$

In like manner, the velocity gained by m_2 during compression will be $v-v_2$, and the velocity gained by the force of restitution $\varepsilon(v-v_2)$.

Hence the whole velocity gained by m_2 will be $(1+\varepsilon)(v-v_2)$. This, added to the velocity before impact, will give the velocity after impact, or

$$v'_2=v_2+(1+\varepsilon)(v-v_2)=v+\varepsilon(v-v_2). \qquad (b)$$

Substituting in (a) and (b) the value of v (46), and reducing,

$$v'_1=\frac{m_1v_1+m_2v_2}{m_1+m_2}-\frac{m_2\varepsilon(v_1-v_2)}{m_1+m_2}. \qquad (48)$$

$$v'_2=\frac{m_1v_1+m_2v_2}{m_1+m_2}+\frac{m_1\varepsilon(v_1-v_2)}{m_1+m_2}. \qquad (49)$$

As in *Art.* 243, if the bodies move in opposite directions, we must change the sign of v_2, or, if one of them be at rest before impact, make $v_2=0$. Also, if we put $\varepsilon=1$, the results will be those for perfectly elastic bodies, or make $\varepsilon=0$, the result will be that for inelastic bodies.

Cor. 1. If the bodies are perfectly elastic, their relative velocities before and after impact are the same. For, making $\varepsilon=1$ in (a) and (b), and subtracting the latter from the former, we have

$$v'_1-v'_2=v_2-v_1. \qquad (50)$$

Cor. 2. In the impact of bodies no motion is lost.

For, multiplying (48) by m_1, and (49) by m_2, adding and reducing, we have

$$m_1v'_1+m_2v'_2=m_1v_1+m_2v_2, \qquad (51)$$

in which the first member is the sum of their momenta after impact, and the second member the sum of their momenta before impact.

COR. 3. If the bodies are perfectly elastic and equal, they will interchange velocities by impact.

For, making $m_1=m_2$ and $\varepsilon=1$ in (48) and (49), we have

$$v'_1=\tfrac{1}{2}(v_1+v_2)-\tfrac{1}{2}(v_1-v_2)=v_2$$
$$v'_2=\tfrac{1}{2}(v_1+v_2)+\tfrac{1}{2}(v_1-v_2)=v_1.$$

COR. 4. The velocity which one body communicates to another at rest, when perfectly elastic, is equal to twice the velocity of the former divided by one plus the ratio of the masses.

Making in (49) $m_2=rm_1$, $v_2=0$, and $\varepsilon=1$, we obtain

$$v'_2=\frac{2v_1}{r+1}.$$

246. PROP. *When, in a series of* n *perfectly elastic bodies whose masses are in geometrical progression, the first impinges directly against the second at rest, the second against the third, and so on, to find the velocity of the* n*th body.*

Let $m, rm, r^2m \ldots\ldots r^{n-1}m$ be the bodies, and v the velocity of m. By *Cor.* 4, *Art.* 245,

the velocity of the second will be $\dfrac{2v}{r+1}$,

" " third " $2.\dfrac{2v}{1+r}.\dfrac{1}{1+r}=\dfrac{2^2v}{(1+r)^2}$,

" " fourth " $2.\dfrac{2^2v}{(1+r)^2}.\dfrac{1}{1+r}=\dfrac{2^3v}{(1+r)^3}$,

. . . .

" " nth " $=\dfrac{2^{n-1}v}{(1+r)^{n-1}}$.

COR. 1. The momentum of the nth body is

$$r^{n-1}m.\frac{2^{n-1}v}{(1+r)^{n-1}}=mv.\left(\frac{2r}{1+r}\right)^{n-1}.$$

COR. 2. If the bodies are equal, $r=1$, and the velocity of the last equals v, the velocity of the first. If all are in contact except the first before impact, all except the last will remain in contact after impact, and the last will move off with the velocity of the first.

247. PROP. *The motion of the common center of gravity of two bodies after direct impact is the same as before impact.*

Let $\bar{v}$ be the velocity of the common center of gravity of the bodies m_1 and m_2, moving in the same direction with velocities v_1 and v_2 respectively before impact, and $\bar{v}'$ the velocity of their common center of gravity after impact. Let x_1, x_2, $\bar{x}$ be the distances respectively of the centers of gravity of m_1 and m_2, and of their common center of gravity from any fixed point in their line of motion at any instant; x'_1, x'_2, $\bar{x}'$ the same quantities after an interval t, so that (36)

$$x'_1 = x_1 + v_1 t,$$
$$x'_2 = x_2 + v_2 t,$$
$$\bar{x}' = \bar{x} + \bar{v} t.$$

By (29), *Art.* 105,

$$(m_1 + m_2)\bar{x} = m_1 x_1 + m_2 x_2, \qquad (a)$$
$$(m_1 + m_2)\bar{x}' = m_1 x'_1 + m_2 x'_2. \qquad (b)$$

Substituting in (*b*) the values of $\bar{x}'$, x'_2, x'_1 above, and subtracting (*a*) from the result, we have

$$\bar{v} = \frac{m_1 v_1 + m_2 v_2}{m_1 + m_2}, \qquad (c)$$

which is the same as the velocity of two inelastic bodies after impact, and therefore equal to the velocity of their common center of gravity after impact, since the masses move on together.

If the bodies are elastic, and v'_1, v'_2 are their velocities after impact, then

$$x'_1 = x_1 + v'_1 t,$$
$$x'_2 = x_2 + v'_2 t,$$
$$\bar{x}' = \bar{x} + \bar{v}' t;$$

from which we deduce, as before,

$$\bar{v}' = \frac{m_1 v'_1 + m_2 v'_2}{m_1 + m_2} \qquad (d)$$

for the velocity of the common center of gravity of elastic bodies after impact. Now the denominators of (*c*) and (*d*) are the same, and by (51) the numerators are equal;

$$\bar{v} = \bar{v}',$$

or the velocity of the common center of gravity is unchanged by impact.

248. SCHOL. This proposition is only a particular case of a general principle in Mechanics, denominated the *conservation of the motion of the center of gravity.* The principle consists in this, that the mutual action of the several bodies or parts of a system upon each other can produce no change in the motion of the center of gravity of the entire system.

249. DEF. If a body impinge on a surface, the angle which its path, before impact, makes with the *perpendicular* to the surface at the point of impact is called the *angle of incidence,* and the angle which its path, after impact, makes with the same perpendicular is called the *angle of reflection.*

250. PROP. *To determine the motion of a smooth inelastic body after oblique impact upon a smooth, hard, and fixed plane.*

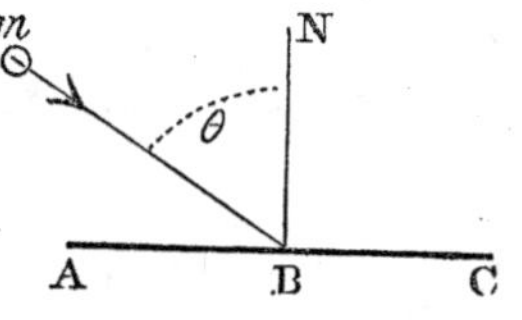

Let the body m impinge on the plane AC at B, with the velocity v, making the angle of incidence $m\text{BN}=\theta$.

The component of the velocity v parallel to the plane (39) is $v. \sin. \theta$; and this velocity will not be changed by impact, since the body and plane are smooth. The component of v perpendicular to the plane, viz., $v. \cos. \theta$, will be destroyed by the plane, and, since the body and plane are inelastic, there will be no vertical velocity after impact. Hence the body will slide along the plane with the velocity $v. \sin. \theta$.

251. PROP. *To determine the motion of an elastic body after oblique impact upon a smooth, hard, and fixed plane.*

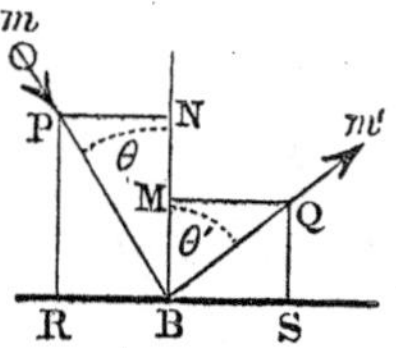

Let the body m impinge on the plane RS at B, with the velocity v, making the angle of incidence $\text{PBN}=\theta$. Let PB represent the velocity v before impact. Draw PR and PN perpendicular and parallel to the plane RS. $\text{PN}=\text{RB}=v. \sin. \theta$ is the component of the velocity parallel to the plane, and is not affected by the impact. $\text{PR}=\text{NB}=v. \cos. \theta$, the component of the velocity perpendicular to the plane, will be destroyed by the plane. But the body being elastic, the force of restitution will give it a velocity $\varepsilon.v.\cos.\theta$ in the direction BN. Take $BM=\varepsilon.v.\cos.\theta$, $BS=v.\sin.\theta$, and, completing the parallelogram, draw BQ. BQm' is the direction, and BQ the velocity of the body after impact. Now

$$BQ^2=v'^2=BM^2+MQ^2$$
$$=\varepsilon.^2v.^2\cos.^2\theta+v^2\sin.^2\theta$$
$$\therefore\ BQ=v'=v\sqrt{\varepsilon.^2\cos.^2\theta+\sin.^2\theta};$$

also, $$\tan.\theta'=\frac{MQ}{MB}=\frac{PN}{\varepsilon.NB}=\frac{\tan.\theta}{\varepsilon}.$$

Cor. If $\varepsilon=1$, $\theta=\theta'$, and $v=v'$, or if the body be perfectly elastic, the angle of incidence equals the angle of reflection, and the velocity is the same after as before impact. If $\varepsilon=0$, $\tan.\theta'=\infty$, and $\theta'=90°$.

252. Prop. *To determine the direction in which a body of given elasticity must be projected, in order that after reflection from a given plane it may pass through a given point.*

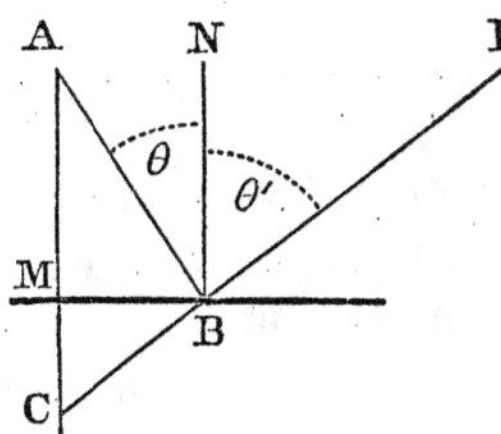

Let MB be the given plane, A the position of the body before projection, D the point through which it is required to pass after reflection, and ε the modulus of elasticity of the body, supposed known.

Draw AMC perpendicular to the plane, and take $AM : MC = 1 : \varepsilon$ Draw C D cutting the plane in B, and join A B. AB is the direction in which the body must be projected, and AB, BD will be its path.

Since $$\varepsilon.AM=MC,$$
$$\frac{MB}{\varepsilon.AM}=\frac{MB}{MC},$$

or $$\frac{\tan.MAB}{\varepsilon}=\tan.MCB;$$

or $$\frac{\tan.\theta}{\varepsilon}=\tan.\theta',$$

as in *Art.* 251.

253. PROP. *The modulus of elasticity is equal to the ratio of the relative velocities of the bodies after and before impact.*

For, eliminating v from (a) and (b) (*Art.* 245), we have

$$\varepsilon = \frac{v'_2 - v'_1}{v_1 - v_2}. \qquad (52)$$

254. SCHOL. Bodies suspended by fine cords, and made to oscillate like a pendulum, acquire velocities at the lowest point proportional to the chords of the arcs of descent, and will rise through arcs whose chords are proportional to the velocities impressed upon them at the lowest point of the arcs. Also, if the arcs be small, the times of descent will be equal, so that bodies descending through arcs of different lengths will impinge at the lowest points. If, therefore, two spherical bodies of the same material be made to descend, in the manner described, through arcs of given length, and the arcs through which they rise after impact be measured, the velocities v_1, v_2, v'_1, v'_2 will be known, and these, substituted in (52), will give ε the modulus of elasticity.

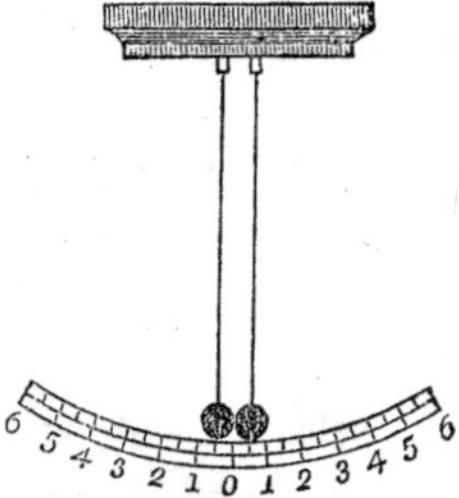

The following table exhibits the results of experiments, perfect elasticity being unity:

Substances.	Moduli.	Substances.	Moduli.
Glass	0.94	Bell-metal	0.67
Hard-baked clay	.89	Cork	.65
Ivory	.81	Brass	.41
Limestone	.79	Lead	.20
Steel, hardened	.79	Clay, just malleable by the hand	.17
Cast iron	.73		
Steel, soft	.67		

255. EXAMPLES.

Ex. 1. Two inelastic bodies, weighing 12 lbs. and 7 lbs. respectively, move in the same direction with velocities of 8 feet and 5 feet in a second. Find the common velocity after impact; also the velocity lost by one, and that gained by the other.

Ex. 2. A mass m_1, with a velocity 11, impinges on m_2 moving in the opposite direction with a velocity of 5; by impact m_1

loses one third of its momentum. What are the relative magnitudes of m_1 and m_2?

Ex. 3. m_1, weighing 8 lbs., impinges on m_2, weighing 5 lbs., and moving in m_1's direction with a velocity of 9; by impact m_2's velocity is trebled. What was m_1's velocity before impact?

Ex. 4. Two bodies m_1 and m_2 are moving in the same straight line with velocities v_1 and v_2. Find the velocity of each after impact when $6m_1=5m_2$, $v_1=7$, $4v_1+5v_2=0$, and $\varepsilon=\frac{2}{3}$.

Ex. 5. Two bodies m_1 and m_2 are perfectly elastic and move in opposite directions; m_1 is treble of m_2, but m_2's velocity is double that of m_1. Determine their motions after impact.

Ex. 6. There is a row of perfectly elastic bodies in geometrical progression whose common ratio is 3; the first impinges on the second, which transmits its velocity to the third, and so on; the last body moves off with $\frac{1}{64}$ the velocity of the first. What was the number of bodies?

Ex. 7. $m_1(=3m_2)$ impinges on m_2 at rest; m_1's velocity after impact is $\frac{3}{5}$ of its velocity before impact. Required the value of ε, the modulus of elasticity.

Ex. 8. Two bodies m_1 and m_2, whose elasticity is $\frac{2}{3}$, moving in opposite directions with velocities 25 and 16 respectively, impinge directly upon each other. Find the distance between them $4\frac{1}{2}$ seconds after impact.

Ex. 9. At what angle must a body whose elasticity is $\frac{1}{3}$ be incident on a perfectly hard plane, that the angle made by its path before and after impact may be a right angle?

Ex. 10. A ball whose elasticity is ε, projected from a given point in the circumference of a circle, after two reflections from the interior of the circle, returns to the same point. Required the angle θ made by the direction of projection with the radius at the given point.

Ans. Tan. $\theta=\frac{\varepsilon^{\frac{3}{2}}}{\sqrt{1+\varepsilon+\varepsilon^2}}$.

CHAPTER III.

MOTION FROM THE ACTION OF A CONSTANT FORCE.

256. By the definition of a *constant* force (*Art.* 222), the velocities generated in equal successive intervals of time by the action of the force are all equal, and the increment of velocity in a unit of time is a measure of the force. Hence the velocity is the same at no two successive instants, and, if the body or point move from rest, will *increase uniformly*, or be *uniformly accelerated.*

By the velocity acquired in t seconds is meant the space over which the body would pass in the second next succeeding the t seconds if the velocity should remain the same during this second as it was at the end of the t seconds; or the space described by the body in the interval between t seconds and $t+1$ seconds, if, at the end of t seconds, the force should cease to act. Putting f for the force, and v for the *velocity acquired* in the time t, we have, as before,

$$v=ft. \tag{53}$$

257. Prop. *To find the space in terms of the force and time when a body moves from rest by the action of a constant force.*

Let f be the force, and s the whole space described in the time t, and let t be divided into n equal parts, each $=\frac{t}{n}$. The intervals reckoning from the commencement of the time, will be

$$\frac{t}{n}, \frac{2t}{n}, \frac{3t}{n}, \frac{4t}{n}, \text{\&c.}, \ldots\ldots \frac{(n-1)t}{n}, \frac{nt}{n}.$$

By (53), the velocities at the end of these intervals will be

$$f\frac{t}{n}, f\frac{2t}{n}, f\frac{3t}{n}, f\frac{4t}{n} \text{ \&c.}, \ldots f.\frac{(n-1)t}{n}, f.\frac{nt}{n}.$$

If now the body moved *uniformly* during each interval $\frac{t}{n}$ with the velocity it had at the beginning or end of this interval, the spaces described during the intervals respectively, would be equal to the product of this uniform velocity by the interval (37), and the whole space described would be equal to the sum of these partial spaces.

If, therefore, the body moved *uniformly* during each interval $\frac{t}{n}$ with the velocity it had at the *beginning* of the interval, we should have

$$s=0+f.\frac{t^2}{n^2}+f\frac{2t^2}{n^2}+f\frac{3t^2}{n^2}+, \&c., \dots f.\frac{(n-1)t^2}{n^2},$$
$$=f.\frac{t^2}{n^2}\{1+2+3+, \&c., \dots (n-1)\},$$
$$=f.\frac{t^2}{n^2}.\frac{n(n-1)}{2},$$
$$=f.\frac{t^2}{2}-f.\frac{t^2}{2n}. \qquad (a)$$

But if the body moved *uniformly* during each interval $\frac{t}{n}$ with the velocity it had at the *end* of the interval, we should have

$$s=f.\frac{t^2}{n^2}+f.\frac{2t^2}{n^2}+f.\frac{3t^2}{n^2}+, \&c., \dots\dots f.\frac{(n-1)t^2}{n^2}+f.\frac{nt}{n^2}$$
$$=f.\frac{t^2}{n^2}(1+2+3+, \&c., \dots\dots n),$$
$$=f.\frac{t^2}{n^2}.\frac{n(n+1)}{2},$$
$$=f.\frac{t^2}{2}+f\frac{t^2}{2n}. \qquad (b)$$

Since the velocity is uniform during no sensible interval, the true value of s will lie between the two quantities (a) and (b), however small each interval may be, or however large n may be. But when n becomes indefinitely large the last terms in (a) and (b) vanish, and $(a)=(b)$.

$$\therefore\ s=\tfrac{1}{2}ft^2 \text{ and } s \propto t^2. \tag{54}$$

Hence the space described from rest by a body from the action of a constant force *is equal to half the product of the force by the square of the number of seconds, and the spaces vary as the squares of the times.*

258. Prop. *To determine the space in terms of the force and velocity; also in terms of the time and velocity.*

1°. Eliminating t from (53) and (54), we have

$$s=\frac{v^2}{2f}, \text{ or } s \propto v^2. \tag{55}$$

2°. Eliminating f from (53) and (54), we obtain

$$s=\tfrac{1}{2}vt, \text{ or } s \propto vt. \tag{56}$$

Cor. The space described in any time by a body moving from rest by the action of a constant force, is half that it would describe in the same time if it moved uniformly with the acquired velocity.

For the space s_1 described in the time t, with a uniform velocity v, is, by (37),

$$s_1=vt,$$

which, compared with (56), gives

$$s=\tfrac{1}{2}s_1.$$

259. Prop. *To find the space described by a body in the last* n *seconds of its motion.*

The space described in the time t (54) is

$$s_1=\tfrac{1}{2}ft^2. \tag{a}$$

The space described in the time $(t-n)$ is

$$s_2=\tfrac{1}{2}f(t-n)^2 \tag{b}$$

Subtracting (b) from (a), we have, for the space s described in the last n seconds,

$$s=s_1-s_2=\tfrac{1}{2}f(2nt-n^2). \tag{57}$$

Cor. If the space described in the m seconds next preceding the last n seconds is required, we have, for the space s_3, described in the time $t-(n+m)$ seconds, $s_3=\tfrac{1}{2}f(t-n-m)^2$, which, subtracted from (b), gives for the space s required,

$$s=s_2-s_3=\tfrac{1}{2}f(2mt-2mn-m^2). \tag{58}$$

260. Prop. *A body being projected with a given velocity in the direction in which a constant force acts, to find the velocity of the body at the end of a given time, and the space described in that time.*

By *Art.* 234, the velocity due to the joint action of the impulsive and constant forces will be equal to the sum or difference of the velocities due to each, according as they act in the same or opposite directions. If, therefore, v_1 be the velocity of projection, the whole velocity v, at the end of the time t, will be

$$v=v_1\pm ft. \qquad (59)$$

In the same manner, the space due to the joint action of the forces will be equal to the sum or difference of the spaces due to each, or

$$s=v_1t\pm\tfrac{1}{2}ft^2. \qquad (60)$$

261. Prop. *A body being projected with a given velocity in the direction in which a constant force acts, to find its velocity when it has passed through a given space.*

Let s be the given space, and h the space through which the body must pass to acquire the velocity v_1 by the action of the constant force. Then (55)

$$v_1^2=2fh,$$

and for the space $(h\pm s)$

$$\begin{aligned} v^2&=2f(h\pm s),\\ &=2fh\pm 2fs,\\ &=v_1^2\pm 2fs, \end{aligned} \qquad (61)$$

the signs to be taken as in the last proposition.

262. Prop. *When a body is projected in a direction opposite to that in which a constant force acts, the velocity acquired in returning to the point of departure is equal to the velocity of projection.*

If v_1 be the velocity of projection in the direction AB, from (61) we have

$$s=\frac{v_1^2-v^2}{2f}. \qquad (a)$$

Now the actual velocity v of the body is continually diminished by the action of the force f, and when the body has reached its greatest distance from A, as B, $v=0$. Hence (a) gives

$$s=\frac{v_1^2}{2f}, \text{ or } v_1=\sqrt{2fs}.$$

But the velocity v acquired in moving from rest through BA$=s$ is, by (55),

$$v=\sqrt{2fs}.$$

$$\therefore\ v=v_1.$$

B

C

A

COR. 1. The velocity of the body at any given distance from A is the same in going and returning.

For in the expression (a), since v_1 and $2f$ are constant, v is always the same for the same value of $s=$AC.

If $v>v_1$, s is negative, or on the opposite side of A from B.

COR. 2. The whole time of flight $T=\frac{2v_1}{f}$. Making $v=0$ in (59), we have $t=\frac{v_1}{f}$, when the body is at B.

But, in returning, it acquires the velocity $v_1=ft$.

Hence the time of return is $t=\frac{v_1}{f}$,

and the whole time $\quad 2t=T=\frac{2v_1}{f}$.

263. SCHOL. 1. There is no known instance in nature of a force which is constant. The law of Universal Gravitation is, that every particle of matter attracts every other particle with a force which varies directly as the mass of the attracting particle, and inversely as the square of the distance. A sphere of uniform density, or one whose density is the same at equal distances from the center, attracts a body *exterior* to it as if the matter of the sphere were collected at its center, and with a force varying *inversely as the square of the distance* of the body from the center, but a body or particle in the *interior* with a force varying *directly as its distance* from the center.

Regarding the earth as a sphere, this is the law of the *earth's* attraction for bodies exterior to it; but for all small distances above the surface the intensity of gravity may be considered constant, since at a distance of one mile above the surface the actual diminution of gravity is only $\frac{1}{1977.291}$, or about the 2000th part of that at the surface; a variation too small to affect sensibly the circumstances of the motion of a falling body computed on the hypothesis that the force suffers no variation at all.

In reality, the force by which a body is drawn toward the earth is equal to the sum of the attractions of each for the other; but when the mass of the body is inconsiderable with regard to the mass of the earth, the effect of the former is insensible, and the accelerations of all bodies of moderate size are the same. Within these limits, then, gravity may be taken as a constant or uniformly accelerating force, whatever be the mass.

264. Schol. 2. Representing the intensity of gravity at the surface of the earth, as before, by g, we have, from (54), $s=\frac{1}{2}gt^2$, and if $t=1$, $s=\frac{1}{2}g$, or $g=2s$ from which it appears that the acceleration is equal to twice the space described in the unit of time. It has been found by experiment that the space through which a body falls freely in one second in the latitude of New York is equal 16.0799 feet, or $16\frac{1}{12}$ feet nearly. Hence $g=32.1598$ feet, or $32\frac{1}{6}$ feet nearly. Hence, also, all the relations between the space, time, and velocity due to the action of a constant force are true in relation to the action of gravity near the earth's surface. Collecting these, and substituting g for f, we have the following relations, supposing no resistance from the air:

$$s=\frac{1}{2}gt^2=\frac{v^2}{2g}=\frac{1}{2}tv, \qquad (54')$$

$$v=gt=\frac{2s}{t}=\sqrt{2gs}, \qquad (53')$$

$$t=\frac{v}{g}=\frac{2s}{v}=\sqrt{\frac{2s}{g}}, \qquad (56')$$

$$g=\frac{v}{t}=\frac{2s}{t^2}=\frac{v^2}{2s}. \qquad (55')$$

Also, when a body is projected with a velocity v_1 vertically upward or downward, (260) and (261),

$$s=v_1t\pm\tfrac{1}{2}gt^2, \qquad (60')$$
$$v^2=v_1^2\pm2gs. \qquad (61')$$

265. EXAMPLES.

Ex. 1. A body has been falling 11 seconds. Find the space described and the veloctiy acquired.

By (54′), $s=\frac{1}{2}gt^2=16\frac{1}{12}\times121=1946\frac{1}{12}$ feet.
By (53′), $v=gt=32\frac{1}{6}\times11=353\frac{5}{6}$ feet.

Ex. 2. Find the time in which a falling body would acquire a velocity of 500 feet, and the height from which it must fall.

By (56′), $t=\dfrac{v}{g}=\dfrac{500}{32\frac{1}{6}}=15.544$ seconds.

By (54′), $s=\dfrac{v^2}{2g}=\dfrac{250000}{64\frac{1}{3}}=3886$ feet nearly,

or $s=\frac{1}{2}tv=\frac{1}{2}\times15.544\times500=3886$ feet, as before.

Ex. 3. What is the velocity acquired by a body in falling 450 feet? and if the body weigh 10 tons, what is the momentum acquired?

Ans. $v=170.14$ feet.
$M=3811136$ lbs.

Ex. 4. A body had fallen through a height equal to one quarter of a mile. What was the space described by it in the last second?

Ans. $s=275$ feet.

Ex. 5. A body had been falling 15 seconds. Compare the spaces described in the seventh and last seconds.

Ex. 6. A body had been falling 12.5 seconds. What was the space described in the last second but 5 of its fall?

Ex. 7. The space described by a body in the fifth second of its fall was to the space described in the last second but 4, as 1 to 6. What was the whole space described?

Ans. $s=15958.69$ feet.

Ex. 8. A body is projected vertically downward with a velocity of 100 feet. What is its velocity at the end of 5 seconds?

Ex. 9. A body is projected vertically upward with a velocity of 100 feet. Find its velocity at the end of 5 seconds, and its position at the end of 8 seconds.

Ans. $v=-60\frac{5}{6}$ feet.
$s=-229\frac{1}{3}$ feet.

Ex. 10. A body is dropped into a well and is heard to strike the bottom in 4 seconds. What is the depth of the well, the velocity of sound being 1130 feet?

Ans. 231 feet.

Ex. 11. A body is thrown vertically upward with a velocity v_1. Find the time at which it is at a given height h in its ascent.

If t be the time required, (60) gives

$$h=v_1t-\tfrac{1}{2}gt^2;$$

whence

$$t^2-t\frac{2v_1}{g}+\frac{2h}{g}=0,$$

or

$$t=\frac{v_1\pm\sqrt{v_1^2-2gh}}{g}.$$

The lower sign gives the time when the body is at the height h in its ascent, and the upper in its descent.

Ex. 12. A body is projected vertically upward, and the interval between the times of its passing a point whose height is h in its ascent and descent is $2t$. Find the velocity v of projection, and the whole time T of its motion.

Ex. 13. A body whose elasticity is ε is projected vertically upward to the height h above a hard plane, to which it returns, and from which it rebounds till its motion is destroyed. What is the whole space described by the body?

CHAPTER IV.

PROJECTILES.

In the preceding chapter we have discussed the motion of a body by the joint action of a projectile force and the force of gravity, when these forces were coincident or opposite in direction. We now proceed to determine the circumstances of the motion of a body, when the direction of the projectile force is other than vertically upward or downward, supposing, as heretofore, no resistance from the air.

266. Prop. *The path of a body, moving under the joint influence of a projectile force, and the force of gravity considered as a constant, accelerating force, is a parabola.*

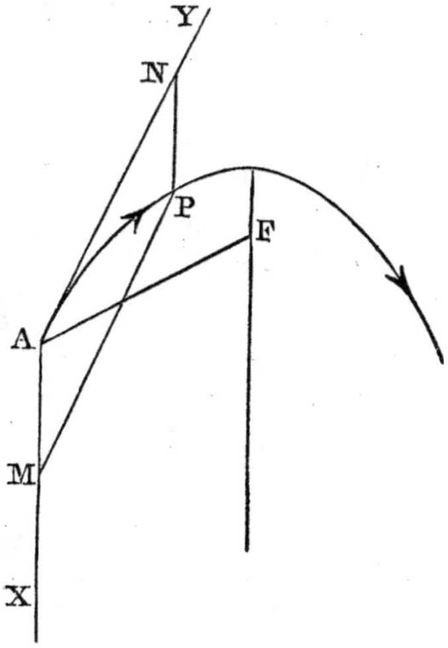

Let v be the velocity of projection from the point A, in the direction ANY, and t the time in which the body will describe AN$=y$, with the uniform velocity v, if gravity do not act. Let AM$=x$ be the space through which the body will fall in the time t by the action of gravity. Completing the parallelogram MN, the actual place of the body at the end of the time t is P.

By (37), AN$=y=vt$; and by (54), AM$=$NP$=x=\frac{1}{2}gt^2$. Eliminating t from these two equations, we get

$$y^2=\frac{2v^2}{g}x. \qquad (a)$$

If h be the space due to the velocity v, or the space through which the body must fall to acquire the velocity of projection (55), $v^2=2gh$, which, substituted in (a), gives

$$y^2=4hx, \qquad (62)$$

the equation of a parabola referred to the oblique axes AX, AY; and, since $4h$ is the parameter to the diameter through A, h is the distance from A, the point of projection to the focus F, or to the directrix.

COR. The velocity of the body *at any point of its path* is that which the body would acquire in falling vertically from the directrix to that point.

For if the body were projected from any point of its path in the direction, and with the velocity it has at that point, it would obviously describe the same path. Therefore the velocity at that point must be equal to that due to one fourth the parameter to the diameter at that point, which is the distance from that point to the directrix or to the focus.

267. PROP. *To find the equation to the path of a projectile when referred to horizontal and vertical co-ordinate axes.*

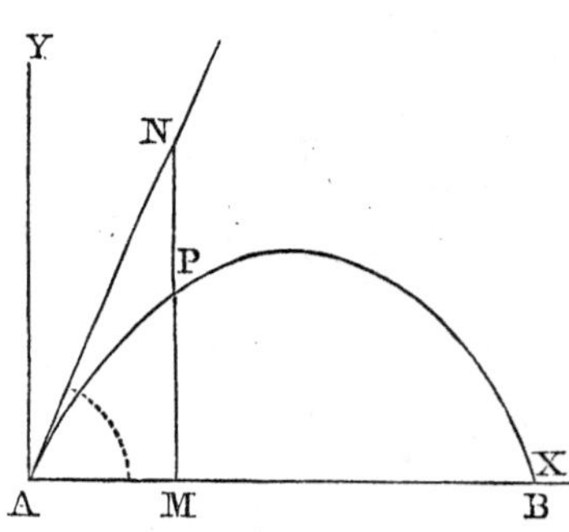

Let v be the velocity of projection in the direction AN, which makes with AX the *angle of elevation* NAX$=a$, APB the path of the body, AM$=x$, PM$=y$, the co-ordinates of the point P, t the time in which the body describes the arc AP, and produce MP to meet the direction of projection in N.

Then AN$=vt$, NP$=\frac{1}{2}gt^2$, and MN$=vt \sin. a$.

Also, AM$=x=vt. \cos. a$, (a)

and PM$=y=vt \sin. a-\frac{1}{2}gt^2$. (b)

Eliminating t from (a) and (b), we obtain

$$y=x \tan. a-x^2\frac{g}{2v^2 \cos.^2 a}, \qquad (63)$$

and, substituting the value of $v^2=2gh$,

$$y=x \tan. a-\frac{x^2}{4h \cos.^2 a}. \qquad (64)$$

268. Def. The *horizontal range* of a projectile is the distance AB from the point of projection to the point where it strikes the horizontal plane in its descent. The *time of flight* is the time occupied in describing APB. The height through which the body must fall to acquire the velocity of projection is called the *impetus*.

269. Prop. *To find the time of flight of a projectile on a horizontal plane.*

At the points A and B the ordinate $y=0$. This value of y, substituted in (b), *Art.* 267, gives

$$vt \sin. a-\tfrac{1}{2}gt^2=0.$$

$$\therefore\ t=0, \text{ and } t=\frac{2v \sin. a}{g}. \qquad (65)$$

The former value of t applies to the point A, and the latter to the point B, which is, therefore, the time of flight required. Or, since $v \sin. a$ is the vertical component of the velocity, or the velocity of projection estimated vertically, if this value of the velocity be substituted for v_1 in (60), and we make $s=0$, we get, as before,

$$0=vt \sin. a-\tfrac{1}{2}gt^2,$$

$$t=0, \text{ and } t=\frac{2v \sin. a}{g}.$$

270. Prop. *To find the range of a projectile on a horizontal plane.*

Put $y=0$ in (64), and we have

$$0=x \tan. a-\frac{x^2}{4h \cos.^2 a};$$

from which we obtain

for the point A, $x=0$,

for the point B, $x=4h \sin. a \cos. a$,

or $\quad AB=R=2h \sin. 2a, \qquad (66)$

the horizontal range required.

Cor. 1. The horizontal range is greatest when $a=45°$. For in this case $2a=90°$, and $\sin. 2a=1.\ \therefore\ x=2h.$

or the greatest horizontal range is equal to twice the height due to the velocity of projection, or twice the distance from the point of projection to the focus of the trajectory.

COR. 2. The range is the same for any two angles of elevation, the difference between which and 45° is the same, or for $(45 \pm \theta)$.

For $\quad \sin.(90° + 2\theta) = \sin.(90° - 2\theta)$,

or $\quad \sin.2(45° + \theta) = \sin.2(45° - \theta)$.

If, therefore, we put either $45° + \theta$ or $45° - \theta$ for a in (66), the value of R remains the same. Thus, if AB be the range of a projectile when the angle of elevation is 45°, and APB its path, the range AB′ will be that due to the elevation $45° - \theta$, for which the path is AP′B′, and to $45° + \theta$, for which the path is AP″B′.

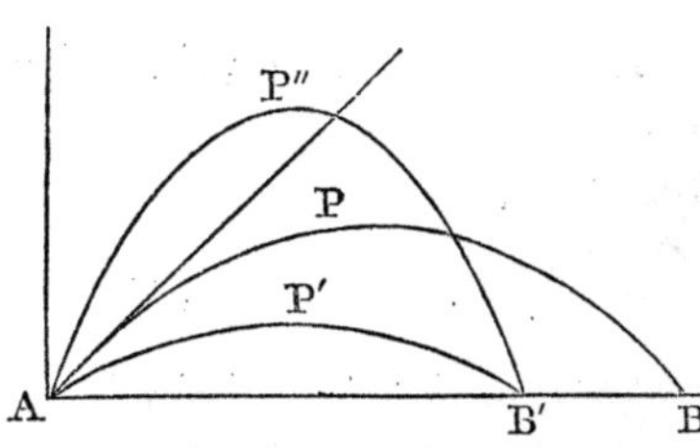

COR. 3. When the velocity of projection is given, and we know the range R due to the elevation a, we can readily find the range R′ due to any other elevation a'.

For $\quad R = 2h \sin. 2a$,

and $\quad R' = 2h \sin. 2a'$.

$$\therefore\ R' = \frac{\sin. 2a'}{\sin. 2a}.R. \qquad (67)$$

COR. 4. Since the horizontal velocity of a projectile is uniform, the range is equal to the horizontal component of the velocity into the time of flight, or

$$R = v.\cos.a \times \frac{2v \sin. a}{g} = \frac{2v^2 \sin. a. \cos. a.}{g} = 2h \sin. 2a, \text{ as before.}$$

271. PROP. *To find the greatest height which a projectile attains.*

The greatest height H is evidently the value of the ordinate at the middle of AB, or when the time is one half the time of flight. Putting $t = \frac{v \sin. a}{g}$ in (*b*), *Art.* 267, we have

$$H=\frac{v^2 \sin.^2 a}{g}-\frac{1}{2}\frac{v^2 \sin.^2 a}{g}=\frac{\frac{1}{2}v^2 \sin.^2 a}{g},$$

$$=h \sin.^2 a. \qquad (68)$$

272. PROP. *To find the co-ordinates of the point where a projectile will strike an inclined plane passing through the point of projection, the range on the inclined plane, and the time of flight.*

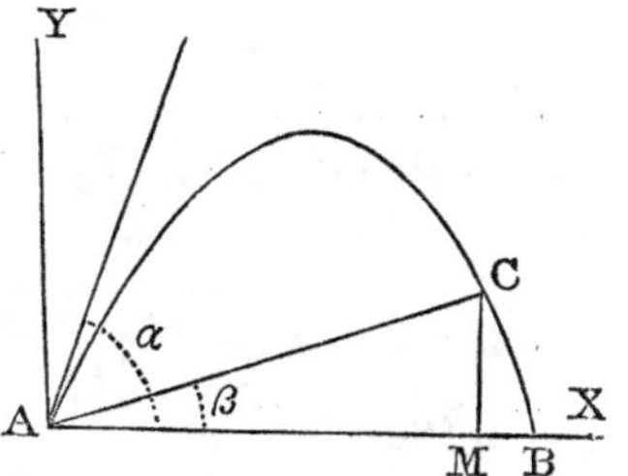

Let $y=x \tan.\beta$ be the equation of the line AC, which is the intersection of the inclined plane, with the vertical plane of the path of the body.

Substituting this value of y in (64), we obtain, after reduction, for the abscissa of the point C,

$$x=\frac{4h \cos. a. \sin.(a-\beta)}{\cos. \beta}. \qquad (a)$$

By substituting in (64) for x, its value $y \cot.\beta$, and reducing, we get for the ordinate of C,

$$y=\frac{4h. \cos. a. \sin. \beta. \sin. (a-\beta)}{\cos.^2\beta}. \qquad (b)$$

To find AC=R′, multiply (a) by $\sec. \beta=\frac{1}{\cos. \beta}$, and we have

$$AC=R'=x \sec. \beta=\frac{4h. \cos. a. \sin. (a-\beta)}{\cos.^2 \beta}. \qquad (69)$$

If the inclined plane cut the path of the projectile below the axis AX, β will be negative.

The time of flight is equal to the time of describing the abscissa AM with the horizontal component of the velocity. Hence (37), dividing (a) by $v \cos. a.$, we get

$$t=\frac{4h \sin. (a-\beta)}{v. \cos. \beta},$$

$$=\frac{2v \sin. (a-\beta)}{g. \cos. \beta}. \qquad (70)$$

273. SCHOL. It has been found by experiment, that if w be the weight of a ball or shell, p the weight of the gunpowder used in discharging the ball or shell from a mortar, and v the velocity generated by the powder,

$$v=1600\sqrt{\frac{3p}{w}} \text{ feet.} \qquad (71)$$

274. EXAMPLES.

Ex. 1. A body is projected at an angle of elevation of 15° with a velocity of 60 feet. Find the horizontal range, the greatest altitude, and time of flight.

From (54′), $h=\frac{v^2}{2g}=\frac{3600}{64\frac{1}{3}}=55.96.$

From (66), $R=2h \sin. 2a=2h. \sin. 30°=2.h.\frac{1}{2}=h=55.96$

From (68), $H= h \sin.^2 a.$ Log. sin. a= 9.4129962
" " = 9.4129962
" h = 1.7478777

∴ H=3.7486 . . . . . " H=.0.5738701

From (65), $T=\frac{2v \sin. a}{g}$. Log. $2v$= 2.0791812

Log. sin. a= 9.4129962
a.c. " g= 8.4925940

∴ T=0.96554 . . . . . " T= 9.9847714

Ex. 2. A body is projected at an angle of elevation of 45° and descends to the horizon at a distance of 500 feet from the point of projection. Required the velocity of projection, the greatest altitude, and time of flight.

Ans. v =126.82 feet
H=125. "
T =5.58 seconds.

Ex. 3. The horizontal range of a projectile is 1000 feet and the time of flight is 15 seconds. Required the angle of elevation, velocity of projection, and greatest altitude.

Ans. a =74°.33′.09″.
v =250.29 feet.
H=904.69 "

Ex. 4. If a body be projected at an angle of elevation of 60°, with a velocity of 850 feet, find the parameter to the axis of the parabola described, and the co-ordinates of the focus.

Ans. $p=11230.57$ feet.
$x=9725.67$ "
$y=5615.28$ "

Ex. 5. Find the velocity and angle of elevation of a ball tha it may be 100 feet above the ground at the distance of one quarter of a mile, and may strike the ground at the distance of one mile.

Ans. $a=5°.46'.04''.6$.
$v=921.566$ feet.

Ex. 6. What must be the angle of elevation of a body in order that the horizontal range may be equal to three times the greatest altitude? What, that the range may be equal to the altitude?

Ex. 7. A body is projected at an angle of elevation of 60° with a velocity of 150 feet. Find the co-ordinates of its position, its direction, and velocity at the end of 5 seconds.

Ex. 8. A body is projected from the top of a tower 200 feet high, at an angle of elevation of 60°, with a velocity of 50 feet. Find the range on the horizontal plane passing through the foot of the tower, and the time of flight.

Ex. 9. A body, projected in a direction making an angle of 30° with a plane whose inclination to the horizon is 45°, fell upon the plane at the distance of 250 feet from the point of projection, which is also in the inclined plane. Required the velocity of projection, and the time of flight.

Ex. 10. The heights of the ridge and eaves of a house are 40 feet and 32 feet respectively, and the roof is inclined at 30° to the horizon. Find where a sphere rolling down the roof from the ridge will strike the ground, and also the time of descent from the eaves.

Ex. 11. How much powder will throw an eight-inch shell, weighing 48 lbs., 1500 yards on an inclined plane, the angle of elevation of the plane being 28°.45′, and that of the mortar being 48°.30′?

Ex. 12. Find the velocity and angle of elevation that a projectile may pass through two points whose co-ordinates are $x=300$ feet, $y=60$ feet, $x'=400$ feet, and $y'=40$ feet. Also find the horizontal range, greatest altitude, and time of flight.

CHAPTER V.

CONSTRAINED MOTION.

§ I. MOTION ON INCLINED PLANES.

275. PROP. *To determine the relations of the space, time, and velocity when a body descends by the action of gravity down an inclined plane.*

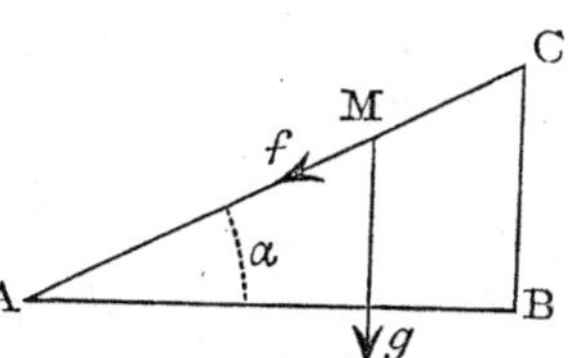

Let the body fall from C down the inclined plane CA, whose inclination is a, M be the position of the body at any time t from rest at C, CM$=s$, and $v=$ the velocity at M.

The force of gravity g, acting in the direction Mg, may be resolved into two forces, one $f=g$ sin. a acting in the direction CA, the other g cos. a acting perpendicularly to CA, and wholly ineffective in producing motion; the body is, therefore, urged down the inclined plane by the constant accelerating force

$$f=g.\ \sin.\ a.$$

If, therefore, this value of f be substituted

in (53), we have $v=gt$ sin. a, (72)

in (54), we have $s=\frac{1}{2}gt^2$ sin. a, (73)

in (55), we have $v^2=2gs$ sin. a, (74)

from which all the circumstances of the motion may be determined.

276. PROP. *The velocity acquired by a body in falling down an inclined plane is equal to that acquired in falling freely through the height of the plane.*

If $s=$AC, the length of the plane, and $h=$BC, the height, by (74),

$$v^2=2g.s.\ \sin.\ a$$
$$=2g.\text{CB}.$$

$$\therefore\ v=\sqrt{2gh},$$

which, by (53′), is the velocity due to h, the height of the plane.

277. Prop. *The times of descent down different inclined planes of the same height vary as the lengths of the planes.*

By (73), $$s=\tfrac{1}{2}gt^2 \sin. a,$$

or $$AC=\tfrac{1}{2}gt^2\frac{BC}{AC}.$$

$$\therefore\ t=AC\sqrt{\frac{2}{g.BC}}$$

$$\propto AC, \text{ when BC is constant.}$$

278. Prop. *To find the relations of the space, time, and velocity when a body is projected down or up an inclined plane.*

Substituting for f, its value, g sin. a, on an inclined plane,

in (59), we have $$v=v_1\pm gt \sin. a, \qquad (75)$$
in (60), we have $$s=v_1t\pm\tfrac{1}{2}gt^2 \sin. a, \qquad (76)$$
in (61), we have $$v^2=v_1^2\pm 2gs \sin. a, \qquad (77)$$

which give all the circumstances of the motion of the body.

279. Prop. *The times of descent down all the chords of a circle in a vertical plane, drawn from either extremity of a vertical diameter, are the same, and equal to that down the vertical diameter.*

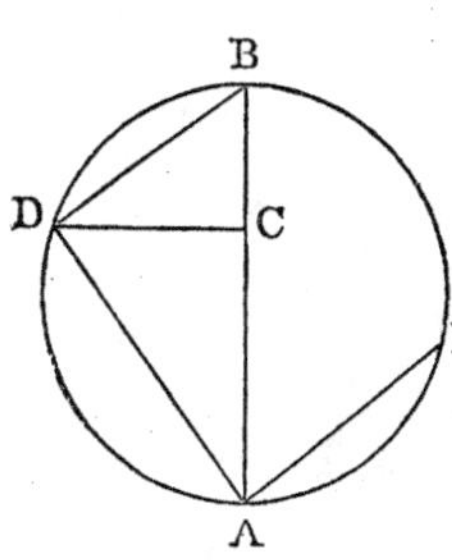

Let AB be the vertical diameter, BD and AD any chords drawn from its extremities, and DC perpendicular to AB. To find the time of descent down BD we have, from (73),

$$BD=\tfrac{1}{2}gt^2 \sin. BDC$$

$$=\tfrac{1}{2}gt^2\frac{BC}{BD}.$$

$$\therefore\ t^2=\frac{2BD^2}{g.BC}$$

$$=\frac{2.BA}{g}.$$

or $$t=\sqrt{\frac{2.BA}{g}},$$

which, by (56′), equals the time down the diameter
in the same manner,

$$DA=\tfrac{1}{2}gt^2\frac{CA}{DA},$$

from which we obtain, as before,

$$t=\sqrt{\frac{2.BA}{g}}.$$

280. Prop. *To find the straight line of quickest descent from a given point within a vertical circle to its circumference.*

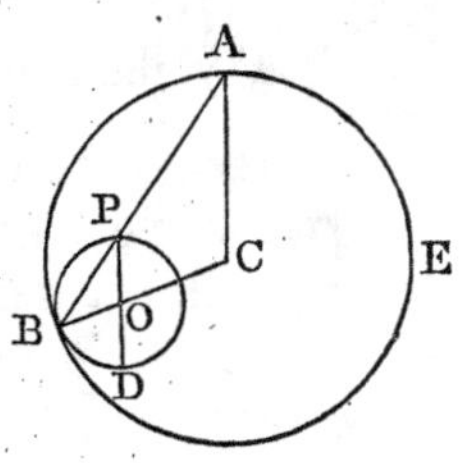

Let P be the given point. Draw the vertical radius CA, join AP and produce it to meet the circumference at B; PB will be the line required.

Join BC and draw POD parallel to AC. Since BC=AC, BO=PO. With O as a center, and radius PO, describe the circle PBD, which will be tangent to the circle C at B.

Now the time down PB will equal the time down the vertical diameter PD, and the time down every other chord drawn from P (*Art.* 279). But any other chord from P, produced to the circumference of C, will be partly without the circumference of O, and, therefore, the time down it will be greater than the time down PB, which is therefore the line of quickest descent to the circumference of C.

281. Prop. *To find the straight line of quickest descent from a given point to a given inclined plane.*

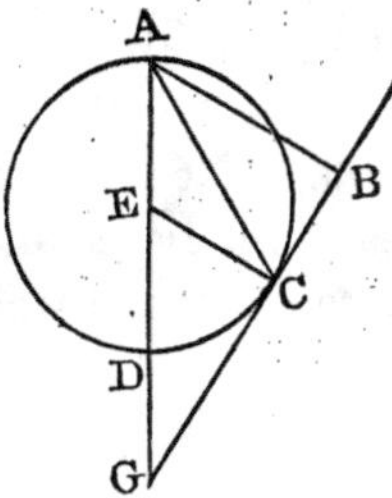

Let A be the given point, AG a vertical line passing through this point, and AB a perpendicular to the inclined plane BG. The line AC, bisecting the angle BAG, will be the line required.

Draw CE parallel to BA, and therefore perpendicular to BG. Since EAC=CAB

=ECA, the triangle EAC is isosceles. With E as a center, and EA or EC as a radius, describe the circle ACD, to which BCG will be a tangent.

Now the time down AC will be equal to the time down AD, and to the time down every other chord drawn from A. But any other chord of the circle must be produced to meet the plane BG. Therefore the time down any other line, drawn from A to the plane, will be greater than the time down AC, which must be the line of quickest descent required.

282. Prop. *Two bodies are suspended from the extremities of a cord passing over a fixed pulley: to determine the circumstances of their motion.*

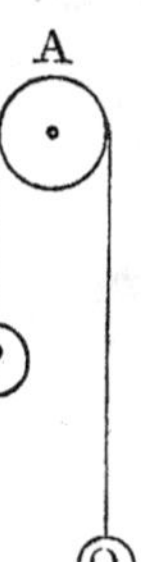

Let P and Q be the weights of the bodies, of which P is the greater. By (22) their masses are $\frac{P}{g}$ and $\frac{Q}{g}$ respectively. Neglecting the rigidity of the cord, the inertia and friction of the pulley, if P=Q, they will counterpoise, and no motion will ensue; but P>Q, P will descend, and Q rise through equal spaces by a force equal to the difference P−Q of their weights. But this being a moving force (*Art.* 225), is equal to the accelerating force into the mass moved, $=f\left(\frac{P+Q}{g}\right)$, where f represents the accelerating force.

Hence
$$P-Q=f.\frac{P+Q}{g},$$
or
$$f=g.\frac{P-Q}{P+Q}.$$

This being a constant accelerating force, by substituting its value in (53), (54), and (55), we have expressions for the space, time, and velocity.

283. Schol. If the inertia I of the pulley be taken into consideration, Ig is an additional force to be overcome by the difference of the weights, and in this case

$$f=g.\frac{P-Q}{P+Q+Ig}.$$

This is the formula for Atwood's machine, an instrument fo. illustrating the laws of falling bodies. The friction of the pulley is reduced by friction wheels, and the rigidity of the cord by employing a fine flexible thread. By making the difference between P and Q small, the motion is made so slow as to render the time, space, and velocity easily determinable.

§ II. MOTION IN CIRCULAR ARCS.

284. PROP. *When a body descends by the action of gravity down any smooth arc of a circle in a vertical plane, the velocity at the lowest point is proportional to the length of the chord of the arc.*

Let a body descend from the point D (*Fig., Art.* 279) down the arc DA. Since the reaction of the curve is always perpendicular to the path of the body, it can neither accelerate nor retard the motion of the body. Its velocity, therefore, at A will be that which it would acquire in falling through the vertical height CA of the arc.

Hence, by (55).

$$v^2 = 2g\text{CA}$$
$$= 2g\frac{\text{AD}^2}{\text{AB}}.$$
$$\therefore\ v = \text{AD}\sqrt{\frac{2g}{\text{AB}}}$$
$$\propto \text{AD}.$$

When the arcs are small, the velocities are nearly proportiona! to the arcs, the principle referred to in *Art.* 254.

285. PROP. *When a body is constrained to describe the sides of a polygon successively, to find the velocity lost in passing from one side to the succeeding one.*

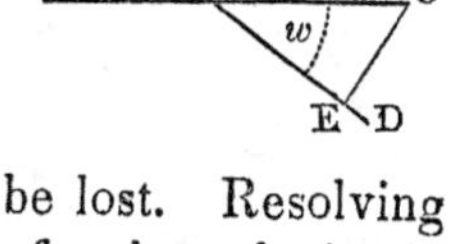

Let AB and BD be two adjacent sides of the polygon, w the angle made by their directions, and v the velocity of the body at the point B. In passing from AB to BD some of the velocity will necessarily be lost. Resolving the velocity in the direction BD, we have, for the velocity in BD,

$$v \cos. w.$$

This, subtracted from the primitive velocity v, will give for the velocity lost,

$$\begin{aligned} v_1 &= v - v \cos. w \\ &= v(1-\cos. w) \\ &= v \text{ versin. } w \\ &= \frac{v \sin.^2 w}{2-\text{versin. } w}. \end{aligned}$$

286. Cor. If the sides of the polygon be increased in num ber, the angle w diminishes; and when the polygon becomes a circle, w, and consequently its sine, becomes indefinitely small. In this case, versin. w, small in comparison with 2, may be rejected, and

$$v_1 = \frac{v}{2} \sin.^2 w.$$

But if sin. w is infinitely small, $\sin.^2 w$ is infinitely smaller, and the velocity lost

$$v_1 = 0.$$

When, therefore, a point or body is constrained to describe a curve, no velocity is lost by the reaction of the curve.

287. Prop. *If a material point move through one side of a regular polygon with a uniform velocity, to find the direction and intensity of the impulse which must be given to the material point at each angle of the polygon in order that it may describe the entire polygon with the same uniform velocity.*

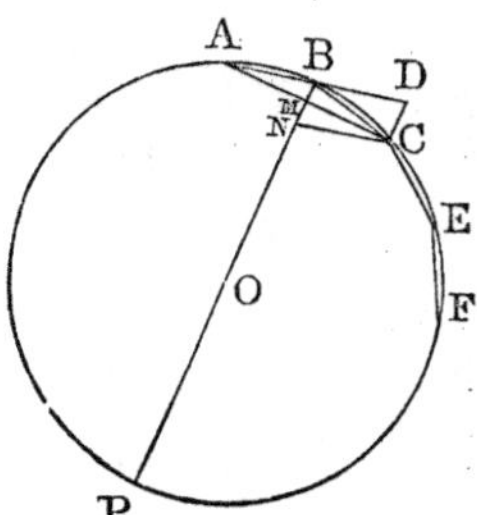

Let ABC ... be the polygon, and let the material point describe AB with the velocity v in the unit of time. If, when at B, no other force act upon the point, it will describe BD=AB in the same time. Join DC, and draw BN equal and parallel to DC. If, when at B, the point receive an impulse in the direction BN, such as to cause it to describe BN in the same time that it would describe BD, the point will describe the diagona BC, the succeeding side of the polygon.

in the same time. But since BC=BD, the velocity in BC is the same as that in AB, and the point will describe the two adjacent sides with the same uniform velocity. Now, since the triangle BCD is isosceles, BDC=BCD=CBN. But ABN=BDC; ∴ ABN=CBN, and BN bisects the angle ABC. Hence the direction of the impulse will pass through the center of the circumscribing circle, and, if similar impulses be applied at each of the angles of the polygon, the point will describe the entire polygon, with the original velocity unchanged.

To find the magnitude of the component force BN=f, let r be the radius of the circle; and, since BN is perpendicular to the chord AC,

$$\mathrm{BM}=\frac{\mathrm{BC}^2}{2r},$$

and

$$f=\mathrm{BN}=\frac{\mathrm{BC}^2}{r}.$$

But BC=AB is the space described in the unit of time, and is therefore represented by v. Hence

$$f=\frac{v^2}{r}; \qquad (78)$$

or, *the intensity of the impulse is equal to the square of the velocity in the polygon divided by the radius of the circumscribing circle.*

288. Cor. The force which must continually urge a material point toward the center of a *circle,* in order that it may describe the circumference with a uniform velocity, *is equal to the square of the velocity divided by the radius.*

Since the reasoning in the proposition is independent of the number of sides in the polygon, the sides of the polygon may be increased in number, and the frequency of the impulse increased in the same ratio, without affecting the relation of the intensity of the impulse to the velocity in the polygon. But when the number of sides becomes infinite, or the polygon becomes a circle, the impulses will no longer be successive, but an incessant action of the same intensity.

The force is, therefore, a constant accelerating force, and

$$f=\frac{v^2}{r}. \tag{79}$$

If the mass of the body be taken into consideration, the moving force (*Art.* 225) will be

$$F=mf=\frac{mv^2}{r}. \tag{80}$$

It is obviously immaterial whether the body be retained in its path by the reaction of a smooth curve, or by an inextensible cord without weight, by which it is connected with the center of the circle. F will be a measure of the resistance of the curve in the one case, and of the tension of the cord in the other.

289. Def. The force which constantly urges a body toward the center of its circular path is called a *centripetal force.* The tendency which the body has to recede from the center, in consequence of its inertia, or the resistance which it offers to a deflection from a rectilinear path, the resistance being estimated in the direction of the radius, is called a *centrifugal force.*

290. Prop. *To discuss the circumstances of the motion of a body constrained to move in a circle by the action of a central force.*

1°. When the masses are equal, by (79),

$$f=\frac{v^2}{r}.$$

If r be constant,

$$f \propto v^2;$$

or, when a body moves in the circumference of a circle by means of a cord fixed at the center, *the tension of the cord,* or *the centrifugal force, will vary as the square of the velocity.*

2°. If v be constant,

$$f \propto \frac{1}{r};$$

or, if equal bodies describe different circles with the same velocity, *the centrifugal forces will be inversely as the radii of the circles.*

3°. Let T be the *periodic time*, or time of one revolution.

Then (37) $$vT=2\pi r,$$

or $$v^2=\frac{4\pi^2r^2}{T^2}.$$

(79), $$=fr.$$

$$\therefore f=\frac{4\pi^2 r}{T^2}, \qquad (81)$$

$$\propto \frac{r}{T^2};$$

or, *the centrifugal force varies directly as the radius of the circle, and inversely as the square of the periodic time.*

4°. If $T^2 \propto r^3$, by substitution in (81),

$$f \propto \frac{r}{r^3} \propto \frac{1}{r^2};$$

or, when the squares of the periodic times are as the cubes of the distances from the center, *the centrifugal force will be inversely as the square of the distance.*

5°. If ω be the angle subtended by the arc described by the body in the unit of time, ω is called the *angular velocity*. Assuming for the *angular unit* the angle whose arc is equal in length to the radius, $v=r\omega$. Substituting this value of v in (79),

$$f=r\omega^2; \qquad (82)$$

or, *the centrifugal force varies as the product of the radius and the square of the angular velocity.*

6°. If in each of the above cases the masses be not the same, the centrifugal forces will vary directly as the masses in addition to the other causes of variation.

The foregoing principles admit of a very simple and satisfactory illustration by means of an instrument called the *whirling table.*

291. Prop. *To find the relation of the centrifugal force in a circle to the force of gravity.*

Let h be the height due to the velocity v which the body has in the circle. By (53′) $v^2=2gh$. This value of v^2, substituted in (79), gives

$$f=\frac{2gh}{r}$$

or

$$\frac{f}{g}=\frac{2h}{r}.$$

Hence *the centrifugal force is to the force of gravity as twice the height due to the velocity in the circle is to the radius of the circle.*

Cor. If $f=g$, $\qquad r=2h$,

or, in a circle whose radius is equal to twice the height due to the velocity in the circle, the centrifugal force is equal to the force of gravity.

292. Prop. *To find the centrifugal force at the equator.*

By (81), $$f=\frac{4\pi^2 R}{T^2}.$$

The equatorial radius R of the earth is 3962.6 miles=20,922,528 feet. $\pi=3.1415926$. And, since the earth revolves on its axis in 0.997269 of a day, $T=0.997269\times 86400^s=86164^s$. These values, substituted in the above, give

$$f=0.111255 \text{ feet.}$$

293. Schol. Since the force of gravity g at the equator has been found to be 32.08954 feet, if G be the force of gravity on the supposition that the earth does not revolve on its axis, then

$$g=G-f,$$

or

$$G=g+f,$$
$$=32.08954+0.111255=32.200795,$$

and

$$\frac{f}{G}=\frac{0.111255}{32.200795}=\frac{1}{289} \text{ nearly;} \qquad (83)$$

or the centrifugal force at the equator is $\frac{1}{289}$ the force of gravity.

294. Prop. *To find the time in which the earth must revolve on its axis, in order that the centrifugal force at the equator may equal the force of gravity.*

Let T′ be the required time, and f' the corresponding centrifugal force. From (79) we have

$$f : f' = \frac{R}{T^2} : \frac{R'}{T'^2}.$$

But in this case R=R′, and f' is to become equal to G.

$$\therefore f : G = \frac{1}{T^2} : \frac{1}{T'^2},$$

or $$T'^2 = \frac{f}{G}T^2 = \frac{1}{289}T^2,$$

and $$T' = \frac{1}{17}T \text{ nearly.}$$

Hence the earth must revolve in $\frac{1}{17}$th of its present period.

295. Prop. *The centrifugal force diminishes gravity at different places on the earth's surface in the ratio of the square of the cosine of the latitude.*

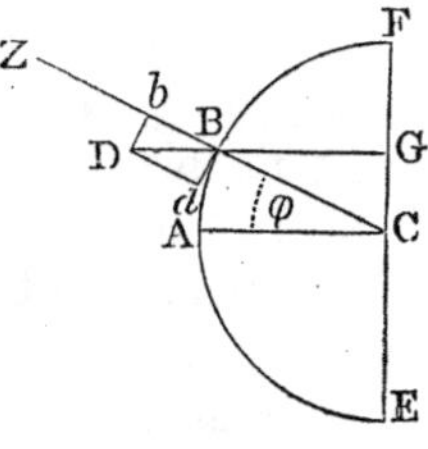

Regarding the earth as a sphere, from which it differs but very little, let EF be the axis, AC=R the equatorial radius, B any point whose latitude is the arc AB, measured by the angle ACB=ϕ, and BG=R′ the radius of the parallel of latitude passing through B. Let the centrifugal force at B, which acts in the direction GB, be represented by BD.

By (81), $$BD = \frac{4\pi^2 R'}{T^2}.$$

Resolving this in the direction of the vertical CZ, opposite to that in which gravity acts, and calling this component f', we have

$$f' = Bb = \frac{4\pi^2 R}{T^2} \cos. \phi.$$

But R′=R cos. ϕ,

$$\therefore f' = \frac{4\pi^2 R}{T^2} \cos.^2 \phi, \qquad (84)$$

in which the coefficient of cos.$^2 \phi$ is constant for all latitudes.

Hence $$f' \propto \cos.^2 \phi,$$

f' being the diminution of gravity by the action of the centrifugal force.

The latitude of Middletown being 41°.33'.10", the centrifugal force f', in a vertical direction at this place, will be found to be

$$f'=0.062305.$$

If this be added to the observed gravity $g'=32.16208$, we shall have for the whole gravity, undiminished by the centrifugal force,

$$G'=g'+f'=32.224385.$$

296. Cor. Resolving the centrifugal force at B in a direction perpendicular to the radius CB, we have

$$f'=\text{B}d=\frac{4\pi^2\text{R}}{\text{T}^2}\cos.\phi\sin.\phi,$$

$$=\frac{2\pi^2\text{R}}{\text{T}^2}\sin.2\phi.$$

If, therefore, the matter of the earth were susceptible of yielding to this component of the centrifugal force, it would necessarily cause the earth to deviate from a spherical form. The fluid portions of the surface are therefore urged toward the equatorial regions, thereby increasing the equatorial diameter and diminishing the polar. If the solid portion did not also partake of the same general form as the fluid, we should expect to find vast equatorial oceans and polar continents, with polar mountains far exceeding the equatorial in height. But the actual distribution of the waters and mountains on the surface of the earth is widely different. The amount of deviation from a spherical form, which the earth must take from the action of the forces developed by its motion, will depend in part on the law of variation of its density from the surface to the center. By measurement the equatorial diameter is found to exceed the polar about 26 miles. The ratio of the difference of the equatorial and polar radii to the equatorial radius, called the *compression* or *ellipticity* of the earth, is about $\frac{1}{300}$.

297. Prop. *To find the centrifugal force of the moon in its orbit.*

Let R be the radius of the earth, nR the mean radius of the moon's orbit, and P the periodic time of the moon in seconds. Then the velocity of the moon will be

$$v=\frac{2\pi n\mathrm{R}}{\mathrm{P}}.$$

This, substituted in (79), gives, for the centrifugal force,

$$f=\frac{4\pi^2 n\mathrm{R}}{\mathrm{P}^2}. \qquad (a)$$

If the earth were an exact homogeneous sphere, the intensity of gravity would be the same at all points on the surface. Being a spheroid, differing little from a sphere, the theory of attraction of spheroids shows that it is necessary to use the intensity of gravity at a latitude ϕ, of which the sine is the $\sqrt{\frac{1}{3}}$; whence $\phi=35°.15'.52''$. Supposing the compression of the earth to be 0.00324, in the latitude ϕ,

$$\mathrm{R}=20897947 \text{ feet.}$$

To find n, conceive two lines drawn from the center of the moon, one to the center of the earth, and the other tangent to the earth at the equator. The angle made by these lines at the center of the moon, and subtended by the equatorial radius of the earth, called the moon's *equatorial horizontal parallax*, is found by astronomical observations to be at a mean $57'.1''$. This angle, when subtended by the radius of the earth in the latitude of $35°.15'.52''$, is $\pi=56'.57''.33$. Now $\frac{\mathrm{R}}{n\mathrm{R}}=\sin.\pi=\frac{1}{n}$, from which we find $n=60.3612$. $\mathrm{P}=2360585^{\mathrm{s}}$. These values, substituted in (a), give

$$f=0.00894.$$

298. Cor. The moon is retained in its orbit by the gravity of the earth. For, since the intensity of gravity is inversely as the square of the distance from the earth's center, its intensity g' at the moon may be found by the proportion

$$g:g'=\frac{1}{\mathrm{R}^2}:\frac{1}{n^2\mathrm{R}^2};$$

whence

$$g'=\frac{g}{n^2}.$$

The intensity of gravity in the latitude of $35°.15'.52''$, un diminished by the centrifugal force of the earth in its diurnal motion, is

$$g=32.24538.$$

Substituting for n and g their values, we find

$$g'=0.00885.$$

The difference between f and g' is less than one ten thousandth of a foot, a difference which may be attributed to errors of observation.

§ III. PENDULUM.

299. Def. A *simple pendulum* is a material point suspended by a right line, void of weight, and oscillating about a fixed point by the force of gravity. The path of the point is the arc of a vertical circle, of which the fixed extremity of the line is the center.

300. Prop. *To find the force by which the pendulum is urged in the direction of its path.*

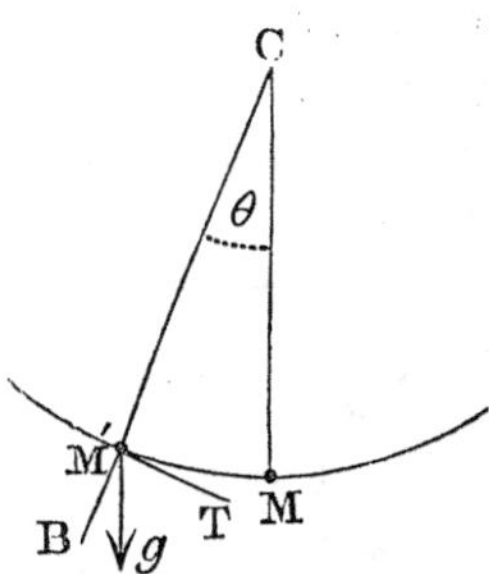

Let M be the material point suspended from C by the line CM. When removed from its vertical direction CM to the inclined position CM′, gravity will cause the point to descend and describe the arc M′M. Resolving the force of gravity g in the directions M′B and M′T, parallel and perpendicular to the radius CM′, the component g cos. θ, in the direction of CM′, will be counteracted by the fixed point C. The component g sin. θ, in the direction of the tangent M′T, will be the only effective force to produce motion. Hence the accelerative force is

$$f=g \sin. \theta.$$

Let CM$=l$, the length of the pendulum and the arc M′M$=s$ the semi-arc of the vibration; then $s=l\theta$.

$$\therefore f=g. \sin. \frac{l}{s}.$$

Now f x be any arc of a circle, it is shown in trigonometry that

$$\sin. x = x - \frac{x^3}{1.2.3} + \frac{x^5}{1.2.3.4.5} -, \text{\&c.}$$

Hence $$f = g\left(\frac{s}{l} - \frac{s^3}{1.2.3.l^3} + \frac{s^5}{1.2.3.4.5.l^5} - \right), \text{\&c.}$$

But when the arc s is small compared with l, the cube and higher powers of $\frac{s}{l}$ may be neglected, and

$$f = \frac{g}{l}s \qquad (85)$$

$$\propto s,$$

or the force varies as the distance from the lowest point measured on the arc.

301. Prop. *When a body is urged toward a fixed point by a force varying directly as its distance from that point, the times of descent to that point from all distances will be the same.*

For the space described will obviously depend on the intensity of the force and the time of its action, or will be measured by the product of the force by some function of the time.

Hence $$s = f.\phi(t).$$

Since s varies as f, let $s = nf$, n being the constant ratio of the space to the force; then, by substitution,

$$n = \phi(t);$$

or the function of the time is the same whatever be the distance, and, in the case of the pendulum, $n = \frac{l}{g}$. (85).

302. Schol. It will be shown hereafter that the time of descent to the center of force, or the point toward which the force is directed, is always equal to $\frac{\pi}{2}\sqrt{\frac{1}{\mu}}$, when μ is the accelerating force. In the case of the pendulum, $\mu = \frac{g}{l}$, and we shall have, for the time of descent to the lowest point, or the time of a semi-vibration.

$$T' = \frac{\pi}{2}\sqrt{\frac{l}{g}}.$$

When the material point has reached its lowest pos tion, its momentum will cause it to rise on the other side of the vertical line; and, since its velocity at the lowest point will be that due to the vertical height fallen through, and the force will require the same time to destroy the velocity that was required to generate it, the whole time of a vibration will be

$$T = \pi\sqrt{\frac{l}{g}}. \qquad (86)$$

As the circumstances of the material point at the end of the time T are the same as at the commencement, it will then perform another vibration in the same time, and so continuing, all its vibrations will be *isochronal.*

It should be recollected that these results are obtained on the supposition that the arcs of vibration are small.

303. Prop. *To discuss the circumstances of the vibration of different pendulums.*

1°. Since in (86) π is constant,

$$T \propto \frac{\sqrt{l}}{\sqrt{g}};$$

or, *the time of vibration of a pendulum varies directly as the square root of the length, and inversely as the square root of the accelerating force.*

2°. If g be constant, which is the case in the same latitude, and at the same elevation above the surface of the earth,

$$T \propto \sqrt{l};$$

or, *the time will vary as the square root of the length.*

3°. If l be constant, or the length of the pendulum remain the same, then

$$T \propto \frac{1}{\sqrt{g}};$$

or, *the time of a vibration will vary inversely as the square root of the intensity of gravity.*

4°. If T be constant, since

$$gT^2=\pi^2 l,$$
$$l \propto g;$$

or, *the lengths of pendulums vibrating in the same time vary as the accelerating force.*

304. Cor. Hence the force of gravity in different latitudes, and at different elevations above the surface of the earth, will vary as the length of the pendulum vibrating seconds. If, therefore, the length of the seconds pendulum be ascertained at various places on the earth, we shall have the relative intensities of gravity at those places. Since these intensities of gravity at different places are dependent upon the figure of the earth, they will serve to determine it. The pendulum, therefore, becomes an instrument for ascertaining the form of the earth.

305. Prop. *The lengths of pendulums at the same place are inversely as the square of the number of vibrations performed by each in the same time.*

Let N and N′ be the number of vibrations performed by each respectively in the time H. Then the duration of a vibration by each will be

$$T=\frac{H}{N} \text{ and } T'=\frac{H}{N'}.$$

But
$$T : T'=\sqrt{l} : \sqrt{l'}=\frac{H}{N} : \frac{H}{N'}.$$

Hence
$$l : l'=N'^2 : N^2.$$

Also,
$$l=\frac{l'N'^2}{N^2}. \qquad (87)$$

306. Schol. To determine experimentally the length of the seconds pendulum, or a pendulum which will oscillate 86400 times in a mean solar day, let a clock, the length l of whose pendulum is required, be regulated to vibrate seconds. Suspend in front of the clock a pendulum of known length l' $(l'<l)$, and observe the exact second when the two simultaneously commence a vibration. As the vibration of l' will be more rapid than that of l, count the number of returns to coincidence of vibration in a period of five or six hours; and, finally, note

the exact second of the simultaneous termination of a vibra tion. If now to the whole number of seconds N shown by the clock, there be added twice the number of coincidences of vibration, we shall have the number N′ of vibrations of l'. These values, substituted in (87), will give the length l of the seconds pendulum.

The length of the seconds pendulum in the latitude of New York has been determined to be equal to

$$39^{\text{in.}}.10168=3^{\text{ft.}}.25847.$$

307. PROP. *To find the value of* g, *the measure of the intensity of gravity.*

From (86), we deduce

$$g=\frac{\pi^2 l}{T^2}.$$

Making T=1, and using the length of the seconds pendulum above, we find, for the value of g in the latitude of New York,

$$g=385^{\text{in.}}.9183=32^{\text{ft.}}.1598.$$

308. PROP. *To find the correction in the length of a pendulum which gains or loses a known number of seconds in a day.*

Let l be the length of the seconds pendulum, $n=86400$, $y=$ the number of seconds gained or lost in a day, and $x=$ the corresponding correction in the length.

By *Art.* 305, and recollecting that a diminution of length corresponds to an increase in the number of vibrations,

$$n^2 : (n+y)^2=l-x : l,$$

from which we deduce

$$x=\frac{2ny+y^2}{n^2+2ny+y^2}.l;$$

or, rejecting y^2, as small in comparison with $2ny$,

$$x=\frac{2y}{2y+n}.l$$

$$=\frac{l}{1+\frac{n}{2y}}. \qquad (88)$$

Hence, *divide the length of the seconds pendulum by one plus the ratio of the number of seconds in a day to twice the gain or loss, and the quotient will be the correction in the length.*

x will have the same sign as y.

309. PROP. *To determine the rate of clock when carried to a given height above the surface of the earth.*

Let $N=86400$, $N'=$ the number of vibrations in a day when the clock is carried to the height h above the surface of the earth, and $r=$ the radius of the earth. The length of the pendulum remaining the same,

$$T : T'=\frac{1}{\sqrt{g}} : \frac{1}{\sqrt{g'}}=\frac{1}{N} : \frac{1}{N'}.$$

But
$$g : g'=\frac{1}{r^2} : \frac{1}{(r+h)^2}.$$

$$\therefore\ N : N'=r+h : r,$$

and
$$N'=\frac{rN}{r+h}$$

$$N-N'=\frac{hN}{r+h}$$

$$=\frac{N}{1+\frac{r}{h}}. \qquad (89)$$

$N-N'$, the loss in a day, is the rate of the clock. Hence, *divide the number of seconds in a day by one plus the ratio of the radius of the earth to the height, and the quotient will be the rate.* The quantities r and h must be in the same denomination.

310. COR. If the loss in a day by a seconds pendulum be determined by observation, we may find the height to which it is carried, for (89) gives

$$h=\frac{r(N-N')}{N'}. \qquad (90)$$

311. DEF. A body suspended by a cord and performing revolutions in a horizontal circle is called a *conical pendulum.*

312. PROP. *To determine the motion of a conical pendulum.*

Let $l=$ the length of the cord AP, fixed at A and attached to the body at P, $PC=r$ the radius of the circle which the body

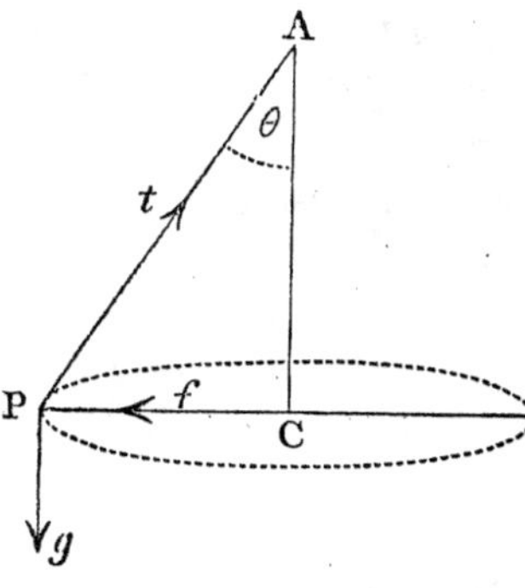

describes with the uniform velocity v, $m=$ the mass of the body, and the angle PAC$=\theta$. Since the body is in the same circumstances at each point of its path, the forces acting upon it must be in equilibrium. These are the tension t of the cord in the direction PA, the centrifugal force $f=\frac{mv^2}{r}$, in the direction CP, and the weight of the body mg downward.

Resolving horizontally and vertically, we have

$$t \sin. \theta - \frac{mv^2}{r} = 0. \qquad (a)$$

$$t \cos. \theta - mg = 0. \qquad (b)$$

From (b), we have the tension in the cord,

$$t = \frac{mg}{\cos. \theta}.$$

Eliminating t from (a) and (b),

$$v^2 = \frac{gr \sin. \theta}{\cos. \theta}$$

$$= gl \frac{\sin.^2 \theta}{\cos. \theta}.$$

The time t of performing one revolution is

$$t = \frac{2\pi r}{v}$$

$$= 2\pi \sqrt{\frac{l \cos. \theta}{g}}$$

$$= 2\pi \sqrt{\frac{\text{AC}}{g}}.$$

313. Dr. Bowditch, in the second volume of his translation of the *Mécanique Céleste*, gives the following formula for computing the length l of the seconds pendulum in any latitude ϕ

$$l = \lambda + \omega \sin.^2 \phi,$$

in which $\lambda = 39.01307$ in. and $\omega = 0.20644$ in.

314. EXAMPLES.

Ex. 1. The length of an inclined plane is 400 feet, its height 250 feet: a body falls from rest from the top of the plane. What space will it fall through in $3\frac{1}{2}$ seconds? in what time will it fall through 300 feet? and what velocity will it have when it arrives within 50 feet of the bottom of the plane?

Ex. 2. The angle of elevation of a plane is 25°.30′. A body, in falling from the top to the bottom, acquires a velocity of 450 feet. What is the length of the plane?

Ex. 3. The length of a plane is 240 feet, and its elevation is 36°. Determine *that* portion of it, equal to its height, which a body, in falling down the plane, describes in the same time it would fall freely through the height.

Ex. 4. A body descending vertically draws an equal body, 25 feet in $2\frac{1}{2}$ seconds, up a plane inclined 30° to the horizon by means of a string passing over a pulley at the top of the plane. Determine the force of gravity.

Ex. 5. The time of descent of a body down an inclined plane is thrice that down its vertical height. What is the inclination of the plane to the horizon?

Ex. 6. If a body be projected down a plane inclined at 30° to the horizon with a velocity equal to $\frac{3}{4}$ths of that due to the vertical height of the plane, compare the time of descent down the plane with that of falling through its height.

Ex. 7. A given weight P draws another given weight W up an inclined plane of known height and length, by means of a string parallel to the plane. When and where must P cease to act that W may just reach the top of the plane?

Ex. 8. Divide a given inclined plane into three parts such that the times of descent down them successively may be equal.

Ex. 9. At the instant that a body begins to descend down a given inclined plane from the top, another body is projected upward from the bottom of the plane with a velocity equal to that acquired in falling down a similar inclined plane n times its length. Where will they meet?

Ex. 10. A ball having descended to the lowest point of a circle through an arc whose chord is a, impels an equal ball up an arc whose chord is b. Find ε the modulus of elasticity of the balls.

Ex. 11. Determine that point in the hypothenuse of a right angled triangle whose base is parallel to the horizon, from which the time of a body's descent in a straight line to the right angle may be the least possible.

Ex. 12. Two bodies fall from two given points in the same vertical line down two straight lines to any point of a curve in the same time, all the lines being in the same vertical plane. Find the equation of the curve.

Ex. 13. 193 oz. is so distributed at the extremities of a cord passing over a pulley, that the more loaded end will descend through 3 inches in one second. What is the weight at each end of the cord?

Ex. 14. If an inelastic body be constrained to move on the interior of a regular hexagon, describing the first side with a uniform velocity in one second, find the time of describing the last side.

Ex. 15. A stone is whirled round horizontally by a string 2 yards long. What is the time of one revolution, when the tension of the string is 4 times the weight of the stone?

Ex. 16. What is the length of a pendulum which oscillates twice in one second?

Ex. 17. A seconds pendulum, carried to the top of a mountain, lost 48.6 seconds in a day. What was the height of the mountain?

Ex. 18. The length of a pendulum vibrating sidereal seconds, being 38.926 inches, what is the length of the sidereal day? How much must it be lengthened that it may measure mean solar time?

CHAPTER VI.

ROTATION OF RIGID BODIES.

315. PROP. *When a rigid body, containing a fixed axis, is acted upon by a given force in a plane perpendicular to that axis, to determine its motion.*

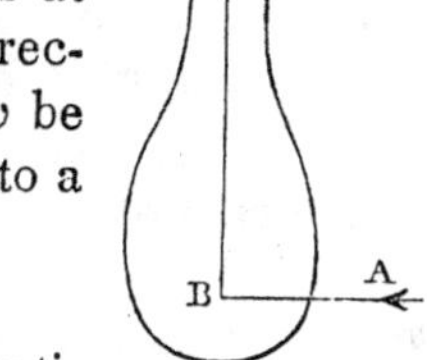

Let the annexed figure be a section of the body by a plane perpendicular to the axis at C, and the given force F act at B, in the direction AB, at a given distance from C. If v be the velocity which the force F can impart to a mass M when free,

$$F = Mv.$$

When the force acts on the body, each particle is constrained to move in a circle whose center is in the axis through C. If ω be the angle made by the body from the action of the force in a unit of time, the linear velocity of any particle m will be $r\omega$, and its momentum $f = mr\omega$. The moment of this force, in reference to the axis through C, is $rf = mr^2\omega$. The moment of any other particle will have the same form, and hence

$$rf + r'f' +, \text{\&c.}, = mr^2\omega + m'r'^2\omega +, \text{\&c.},$$

or

$$\Sigma.rf = \Sigma.mr^2\omega = \omega\Sigma mr^2,$$

since ω is the same for each particle.

But the moment of F is $rF = Mv.CB$, and, as the former is a measure of the effect of the latter,

$$\omega\Sigma.mr^2 = Mv.CB.$$

$$\therefore \omega = \frac{Mv.CB}{\Sigma.mr^2}. \qquad (91)$$

Thus, if the body consisted of two particles m and m' at the distances r and r' from the axis, and the force F would impart to m, when free, the velocity v, then the angular velocity

$$\omega = \frac{mvr}{mr^2 + m'r'^2},$$

and the linear velocity of m will be

$$r\omega = \frac{mvr^2}{mr^2 + m'r'^2}.$$

If the two particles were equal, and at equal distances from the axis, then $mr^2 = m'r'^2$, and

$$r\omega = \frac{v}{2}.$$

If $m = m'$, and $r' = 2r$,

$$r\omega = \frac{v}{5}.$$

Cor. If the body were in motion before the force acted on it, then the *change* of angular velocity will be given by (91).

316. Def. *The moment of inertia* of a rigid body about an axis is the sum of the products of the mass of each particle by the square of the distance of that particle from the axis.

Thus, if m be the mass of a particle of a rigid body, r its distance from a fixed straight line, $\Sigma . mr^2$ is the moment of inertia of the body about that line. The definition given above is to be regarded as a verbal enunciation of this analytical expression, which has required nomenclature by the frequency of its occurrence in dynamical investigations.

317. Prop. *The moment of inertia of a body about any fixed axis exceeds its moment of inertia about a parallel axis passing the center of gravity, by the product of the mass into the square of the distance between the axes.*

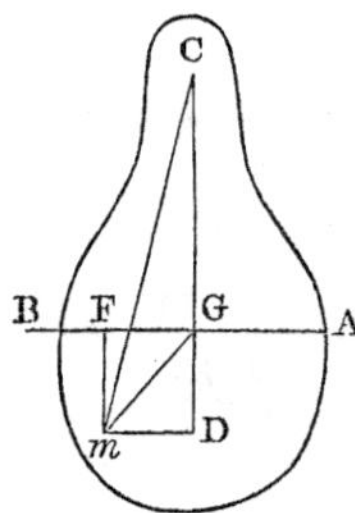

Let m be a particle of the body, and let the plane mCG, passing through m, cut the two axes at C and G. Through the axis at G pass a plane perpendicular to CG, intersecting the plane of the figure in AB. Let $m\text{C} = r$, $m\text{G} = r_1$, and $\text{CG} = h$. Draw mF perpendicular and mD parallel to AB. Then

$$r^2 = r_1^2 + h^2 \pm 2h.\text{GD},$$

according as the angle mGC shall be obtuse or acute. Hence

$$mr^2 = mr_1^2 + mh^2 \pm 2mh.\text{GD},$$

and
$$\Sigma.mr^2 = \Sigma.mr_1^2 + h^2\Sigma m \pm 2h\Sigma.m.\text{GD}.$$

Now, since the plane through the axis at G passes through the center of gravity, by (29),

$$\Sigma.m.\text{GD} = \Sigma.m.\text{F}m = 0.$$

$$\therefore\ \Sigma.mr^2 = \Sigma.mr_1^2 + h^2\Sigma m$$
$$= \Sigma mr_1^2 + \text{M}h^2,$$

M being the whole mass of the body.

318. Schol. Assuming k, such that $\text{M}k^2 = \Sigma.mr_1^2$, we have

$$\Sigma.mr^2 = \text{M}(k^2 + h^2).$$

$\text{M}k^2$ is the moment of inertia of a body about an axis through the center of gravity, and $\text{M}(k^2 + h^2)$ is the moment of inertia about a parallel axis at a distance h from the former. The moment of inertia about *any* axis will, therefore, be easily determined when the moment of inertia about an axis through the center of gravity is known. Since $k^2 = \frac{\Sigma.mr_1^2}{\text{M}}$, the determination of k will, in general, require the aid of the integral calculus.

The length $\sqrt{k^2 + h^2}$ is called the *radius of gyration* about the axis considered, and, similarly, k is called the radius of gyration about an axis through the center of gravity. Since k is the least value of $\sqrt{k^2 + h^2}$, it is sometimes called the *principal radius of gyration.*

319. Prop. *If a body oscillate about a fixed horizontal axis not passing through its center of gravity, there is a point in the right line, drawn from the center of gravity perpendicular to the axis, whose motion is the same as it would be if the whole mass were collected at that point and allowed to vibrate as a pendulum about the fixed axis.*

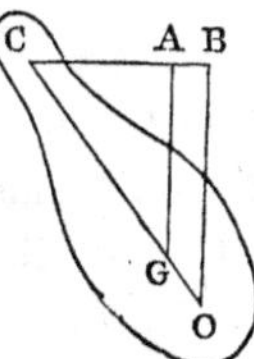

Let the horizontal axis be perpendicular to the plane of the figure at C, G be the center of gravity, CA be horizontal, and GA vertical. Then, if M be the mass of the body, and v the velocity which gravity can impart to the body if free in each instant of time, the moment of gravity will be $Mv.CA$. By *Art.* 315, *Cor.*, the change of angular velocity produced in each instant of time is

$$\omega=\frac{Mv.CA}{\Sigma.mr^2}.$$

Produce CG to some point O and draw OB vertically, meeting CA in B. Now, if the whole mass of the body were collected at O, the moment of inertia of the body would be $M.CO^2$, and the change of angular velocity in the same instant of time would be

$$\omega'=\frac{Mv.CB}{M.CO^2}.$$

But the position of the point O being arbitrary, CO, and therefore CB, will vary at pleasure. We may, therefore, put

$$\Sigma.mr^2 : M.CO^2=CA : CB=CG : CO, \qquad (a)$$

which gives

$$\frac{CA}{\Sigma.mr^2}=\frac{CB}{M.CO^2},$$

or

$$\omega=\omega';$$

that is, the change of angular velocity is the same in the two cases, and, therefore, the motion from rest will be the same.

320. Cor. The point O may be found from (*a*), which gives

$$CO=\frac{\Sigma.mr^2}{M.CG}=\frac{M(k^2+h^2)}{M.h}$$
$$=\frac{k^2+h^2}{h}$$
$$=h+\frac{k^2}{h}. \qquad (92)$$

321. Def. The point O is called *the center of oscillation* of the body with respect to the axis through C. It is thus defined: when a rigid body moves about a fixed horizontal axis under the action of gravity, in the straight line drawn through the center of gravity perpendicular to the axis, a point can be found such that, if the mass of the body were collected there and hung by a thread from the axis, the angular motion of the point would, under the same initial circumstances, be the same with that of the body, and this point is called the center of oscillation of the body with respect to the axis.

322. Cor. Hence, when a body makes small oscillations about a fixed horizontal axis, it is only necessary to calculate the value of the expression $h+\frac{k^2}{h}=l$, and the time of a small oscillation will be $\pi\sqrt{\frac{l}{g}}$. (86).

323. Def. A body of any form suspended from a fixed axis, about which it oscillates by the force of gravity, is called a *compound pendulum.*

324. Prop. *The centers of oscillation and suspension are convertible.*

Let h' be the distance of the center of oscillation from the center of gravity, when h is the distance of the axis from the center of gravity. Then, by (92),

$$CO-h=h'=\frac{k^2}{h},$$

and

$$h=\frac{k^2}{h'}; \qquad (93)$$

so that if h' be the distance from the center of gravity to the axis, h will be the distance from the center of gravity to the center of oscillation, and a body will oscillate in the same time about an axis, through the center of oscillation, as it oscillates about the original axis, the extent of vibration being the same.

325. Cor. 1. If in a straight line through the center of gravity, perpendicular to the axis of motion, points be taken at distances h and h' from the center of gravity, and on opposite sides of it, then the length of the equivalent *simple* pendulum is $h+h'$; so that the time of vibration about the axes through each of these two points is the same, and the length of the equivalent simple pendulum is in each case the distance between the two axes. Each of these points is the center of oscillation in reference to the other as a center of suspension.

326. Cor. 2. From (93), we have

$$hh'=k^2;$$

or, *the principal radius of gyration is a mean proportional be-*

tween the distances of the centers of oscillation and suspension from the center of gravity.

327. Cor. 3. Since $h : \sqrt{k^2+h^2} = \sqrt{k^2+h^2} : h + \frac{k^2}{h} = CO$, we infer that *the distances of the centers of gravity, of gyration, and of oscillation from the axis of motion, are continued proportionals.*

328. Schol. The convertibility of the centers of oscillation and suspension was employed by Captain Kater in finding the length of a simple pendulum vibrating seconds, and consequently the force of gravity at the place of observation.

A bar of brass one inch and a half wide and one eighth of an inch thick was pierced by two holes, through which triangular wedges of steel, called knife edges, were inserted, so that the pendulum could vibrate on the edge of either of these as an axis, resting on two fixed horizontal plates of agate, between which the bar was suspended. The axes were about 39 inches apart. Weights were attached to the bar and rendered capable of small motions by screws, by which means the position of the center of gravity of the bar could be changed. These weights were so adjusted by trial that the time of a small vibration through an angle of about 1° was the same when either knife edge was the axis, so that each gave the center of oscillation belonging to the other. The distance of the knife edges was obtained by placing the pendulum so that the edges were viewed by two fixed microscopes, each furnished with a micrometer of ascertained value, and afterward placing a scale of known accuracy in a similar position under the same microscopes. This method, combined with that referred to in *Art.* 306, served to determine the length of the seconds pendulum, and thence the force of gravity.

329. The relation of the simple to the compound pendulum will be illustrated by one or two examples in which the principal radius of gyration is supposed to have been previously determined.

Ex. 1. A material straight line vibrates about an axis perpendicular to its length: required the length of the isochronal simple pendulum.

Let $2a$ be the length of the line, and h the distance of the point of suspension from its center of gravity, which is its middle point. The radius of gyration of a straight line about an axis through its center of gravity, perpendicular to its length, is $\frac{a}{\sqrt{3}}$. Since the radius of gyration is a mean proportional between the distances of the centers of oscillation and of suspension from the center of gravity (93),

$$h'=\frac{k^2}{h}=\frac{a^2}{3h},$$

and (*Art.* 325), $$l=h+h'=h+\frac{a^2}{3h}.$$

1°. Let the point of suspension be at the extremity of the line, in which case $h=a$.

Then $$h'=\tfrac{1}{3}a,$$
and $$l=a+\tfrac{1}{3}a=\tfrac{2}{3}.2a,$$
that is, the center of oscillation is two thirds of its length below the axis of motion.

2°. Let $h=\frac{1}{3}a$; then

$$h'=\tfrac{1}{2}a$$
and $$l=\tfrac{2}{3}.2a,$$
the same as before. Hence the time of a small vibration is the same, whether the line be suspended from one extremity, or from a point one third of its length from the extremity. This also illustrates the convertibility of the centers of oscillation and of suspension.

3°. If $h=\frac{1}{4}a$,

$$h'=\tfrac{4}{3}a,$$
and $$l=\tfrac{19}{12}a=\tfrac{5}{8}.2a+\tfrac{1}{6}.2a;$$
or, when the center of suspension is three eighths of its length from one end, the center of oscillation is one sixth of its length below the other end.

4°. If $h=0$, $$h'=\infty,$$
and $$l=\infty;$$
or, when the center of suspension is at the center of gravity, the length of the equivalent simple pendulum is infinite, and

therefore the time of one vibration is infinite. This is obvious, also, from the fact that, when the line is suspended from the center of gravity, no motion can result from the action of gravity.

Ex. 2. A sphere being made to oscillate about a given axis; required the length of the equivalent simple pendulum.

Let r be the radius of the sphere, and h the distance of the axis of motion from its center. The principal radius of gyration of a sphere is $r\sqrt{\frac{2}{5}}$.

Hence $$h'=\frac{2r^2}{5h}.$$

1°. Let $h=r$, or the sphere vibrate about an axis tangent to its surface. Then

$$h'=\tfrac{2}{5}r,$$

and $$l=r+\tfrac{2}{5}r;$$

or, the center of oscillation is two fifths of the radius distant from the center of the sphere.

2°. If $h=10r$, $$h'=\frac{r}{25},$$

and $$l=10r+\frac{r}{25}.$$

3°. If $h=\frac{1}{5}r$, $$h'=2r,$$

and $$l=2r+\frac{r}{5};$$

or, the center of oscillation is without the sphere and at a distance r from the surface.

MOMENT OF INERTIA.

330. Since (*Art.* 316) the expression for the moment of inertia of a system consisting of a finite number of points is $\Sigma.mr^2$, m being the portion of the mass which is at the distance r from the axis of rotation; when the number of points becomes indefinite, the expression will evidently become $\int r^2 d\text{M}$, $d\text{M}$ being an element of the mass at the distance r.

Hence (*Art.* 318), $$\text{M}k^2=\int r^2 d\text{M}. \qquad (a)$$

We shall now illustrate the method of determining Mk^2 in a few simple cases.

Ex. 1. To find the moment of inertia of a straight line revolving about an axis perpendicular to it at any point of it.

Let the axis be at a distance a from one end and b from the other, and let r be any distance from the axis; since the thickness and density of the line are supposed to be uniform, each may be taken $=1$.

$$\therefore \ M=a+b \text{ and } dM=dr.$$

Hence (a) $$(a+b)k^2=\int r^2dr=\frac{r^3}{3}+c,$$

and the integral being taken between the limits $-a$ and $+b$.

$$k^2=\frac{a^3+b^3}{3(a+b)}.$$

If $a=b$, or the axis be at the center of gravity, we have, for the principal radius of gyration,

$$k=\frac{a}{\sqrt{3}},$$

and $$Mk^2=\frac{Ma^2}{3}=\frac{2}{3}a^3,$$

whatever be the thickness and density of the line.

Ex. 2. To find the moment of inertia of the circumference of a circle about an axis perpendicular to its plane through its center.

In this case $M=2\pi r$, and since all the points are at the same distance from the axis, r is constant.

$$\therefore \ Mk^2=\int r^2dM=r^2\int dM=2\pi r^3,$$

and $$k=r.$$

Ex. 3. To find the moment of inertia of a circle about an axis through the center perpendicular to its plane.

Putting $a=$ the radius of the circle, its area will be πa^2, and at a distance r from the axis the area will be πr^2. Hence, in this case, $dM=2\pi rdr$ and $Mk^2=\pi a^2k^2=2\pi\int r^3dr=\frac{1}{2}\pi r^4=\frac{1}{2}\pi a^4$, when $r=a$, and

$$k=\frac{a}{\sqrt{2}}=\frac{1}{2}a\sqrt{2}.$$

Ex. 4. To find the moment of inertia of the circumference of a circle about a diameter.

Taking the proposed diameter for the axis of x, the distance of any point (x, y) from the axis of rotation is y.

From the equation of the circle, $y^2=a^2-x^2$, we find

$$\sqrt{dx^2+dy^2}=d\mathrm{M}=\frac{a}{y}dx.$$

$$\therefore\ r^2d\mathrm{M}=y^2d\mathrm{M}=aydx,$$

and $$\mathrm{M}k^2=2\pi ak^2=a\int ydx=a\times \text{area of circle}=\pi a^3.$$

$$\therefore\ k=\frac{a}{\sqrt{2}}.$$

Ex. 5. To find the moment of inertia of a sphere about a diameter.

Let the axis of rotation be the axis of x, and conceive the sphere to be generated by a circle of variable radius, y, whose center moves along the axis. By *Ex.* 3 the moment of this generating circle is $\frac{1}{2}\pi y^4$. Hence the whole moment will be the sum of the moments of the generating circle in all its positions.

$$\therefore\ \mathrm{M}k^2=\tfrac{1}{2}\pi\int y^4dx.$$

But $y^2=a^2-x^2$. $\quad\therefore\ \mathrm{M}k^2=\frac{1}{2}\pi\int(a^2-x^2)^2dx,$

and this integral, taken between the limits $+a$ and $-a$, is

$$\mathrm{M}k^2=\tfrac{8}{15}\pi a^5.$$

But $\mathrm{M}=\frac{4}{3}\pi a^3$. $\quad\therefore\ k=a\sqrt{\tfrac{2}{5}}.$

CHAPTER VII.

331. The methods employed in the preceding chapters to determine the motion of a body are partial in their application and limited to the simplest cases. When the force is variable, or the motion of a body is due to the action of several forces, varying, in direction as well as in intensity, with the varying position of the body, some more general method is required. We now proceed to show, to a limited extent, how the circumstances of the motion of a body or point, in such cases, may be determined. In order to this the fundamental formulæ will need some modification and extension.

332. By *Art.* 214, variable velocity at any instant is measured by the limit of the ratio of the space to the time, that is, from the nature of the differential coefficient, it is the differential coefficient of the space regarded as a function of the time Hence

$$v=\frac{ds}{dt}. \qquad [\text{VII.}]$$

In like manner, it appears from *Art.* 223, that the force, when variable, is measured at any instant by the differential coefficient of the velocity regarded as a function of the time. Hence

$$\phi=\frac{dv}{dt}. \qquad [\text{VIII.}]$$

From [VII.], by differentiation, we obtain

$$\frac{dv}{dt}=\frac{d^2s}{dt^2}.$$

$$\therefore\ \phi=\frac{d^2s}{dt^2}. \qquad [\text{IX.}]$$

From [VII.] and [VIII.] we obtain, by eliminating dt,

$$vdv=\phi ds. \qquad [\text{X.}]$$

§ I. RECTILINEAR MOTION OF A FREE POINT.

333. If a free point at rest is acted on by a single force, or forces whose resultant is equivalent to a single force, the motion must be entirely in the direction of that force, and may at once be determined from [VII.] and [VIII.]. But in order that we may integrate these expressions, the law of variation of the force must be given as a function, either of the space, time, or velocity; and as the intensity of a variable force would depend on the position of the body, the force is naturally and generally given as a function of the distance of the body from some fixed point.

In order that we may obtain an exact expression for the value of the force at any variable distance, its value at some given distance must be known. This given distance is most conveniently assumed as the unit of distance, and if μ be the intensity of the force at this unit of distance, it is usually called the *absolute force.*

If, then, the force be given to vary as the nth power of the distance x from the fixed point, we have

$$\phi=\mu x^n. \qquad (94)$$

We shall first consider the case of a constant force, for the purpose of showing with what facility the calculus enables us to determine the relations of the time, space, and velocity, although these relations have already been deduced in chapter III.

334. PROP. *To determine the space described by a point acted on by a constant force in terms of the time, the velocity of the point in the direction of the force at the commencement of the time being given.*

The velocity due to the action of the force in the time t is, by [III.],

$$v'=\phi t,$$

where ϕ represents the velocity generated by the force in the unit of time.

If v_1 be the given velocity, or the velocity when the time commences, we have, for the whole velocity, as in (59),

$$v=v_1\pm\phi t.$$

Substituting this value of v in [VII.], we have

$$ds=v_1dt\pm\phi tdt,$$

and, integrating,

$$s=v_1t\pm\tfrac{1}{2}\phi t^2+c;$$

in which c is an arbitrary constant depending on the position of the point when $t=0$.

If the space be reckoned from the position of the point when $t=0$, then $c=0$, and

$$s=v_1t\pm\tfrac{1}{2}\phi t^2. \quad (60).$$

If the point move from rest, $v_1=0$, and

$$s=\tfrac{1}{2}\phi t^2. \quad (54).$$

All the other relations are readily deduced from these.

335. Prop. *To determine generally the velocity of a particle moving in a right line to or from a fixed point by the action of a variable force.*

By [X.], $$vdv=\phi ds;$$

or, measuring the line on the axis of x,

$$vdv=\phi dx.$$

Hence, by integration,

$$v^2=2\int\phi dx+c; \qquad (95)$$

and if the force varies as some function of the distance x, the integration may in general be effected.

If the motion is toward the fixed point, dx will be negative; if from it, positive.

To determine the constant, we must know the velocity at some given point.

336. Prop. *To determine generally the time of the motion.*

From [VII.] we have at once

$$t=\int\frac{dx}{v}+c, \qquad (96)$$

and since v is determined by (95) in terms of x, the integration may in general be effected.

To determine c, we must know the position of the particle at some given time.

337. Prop. *To determine the velocity of a particle attracted to a fixed point by a force varying directly as the distance of the particle from that point.*

Assuming the axis of x in the direction of the motion, and the origin at the center of force, by (94) we have

$$\phi=-\mu x;$$

the sign being negative, because the force which is directed to the origin tends to diminish the distance x.

Substituting this value of ϕ in (95), and integrating, we get

$$v^2=c-\mu x^2.$$

If a be the distance of the particle from the origin when the motion commences, when $x=a$, $v=0$. Hence

$$c=\mu a^2,$$

and $$v^2=\mu(a^2-x^2), \qquad (97)$$

from which it appears that x can not pass the limits $\pm a$.

338. Cor. Since the force by which a material point, at liberty to move along a perforation from the surface to the center of the earth, varies directly as the distance from the center, (97) will serve to determine the velocity of the particle at any distance x from the center.

In this case, if $a=r$, the radius of the earth, and g the force of gravity at the surface, μ will be found from the proportion

$$r:1=g:\mu.$$

Hence $$v^2=\frac{g}{r}(r^2-x^2).$$

As this velocity must be spent before the particle stops, if we make $v=0$, we get $x=r$, or the particle will go to the opposite point of the earth's surface.

339. Prop. *To find the time when the force varies directly as the distance.*

From [VII.], we have $$dt=\frac{dx}{v}$$

$$=\frac{1}{\sqrt{\mu}}\cdot\frac{dx}{\sqrt{a^2-x^2}}.$$

To integrate this, let $x=a\sin.\theta$, whence $a^2-x^2=a^2\cos.^2\theta$.

$$\therefore\ dt = \frac{1}{\sqrt{\mu}} \cdot \frac{a.d \sin.\theta}{a \cos.\theta}$$

$$= \frac{1}{\sqrt{\mu}}\, d\theta,$$

and $$t = \frac{1}{\sqrt{\mu}} . \theta + c.$$

Hence $$\theta = \sqrt{\mu}(t - c),$$

and $$x = a.\sin.\sqrt{\mu}(t - c).$$

Suppose the time to commence when the particle is at the origin, or $x=0$; then

$$0 = -a.\sin.\sqrt{\mu}.c,$$

or $$c = 0.$$

$$\therefore\ t = \frac{1}{\sqrt{\mu}}.\theta. \qquad (a)$$

If $x=a$, $\sin.\theta=1$, and $\theta=\frac{1}{2}\pi$ or $\frac{5}{2}\pi$, &c.,

and $$t = \tfrac{1}{2}\pi\frac{1}{\sqrt{\mu}} \text{ or } \tfrac{5}{2}\pi\frac{1}{\sqrt{\mu}}, \text{ \&c.,}$$

from the commencement of the motion.

If $x=-a$, $\sin.\theta=-1$, and $\theta=\frac{3}{2}\pi$ or $\frac{7}{2}\pi$, &c.,

and $$t = \tfrac{3}{2}\pi\frac{1}{\sqrt{\mu}} \text{ or } \tfrac{7}{2}\pi\frac{1}{\sqrt{\mu}}, \text{ \&c.}$$

Hence the particle moves from $+a$ to $-a$, while t changes from $\frac{1}{2}\pi\frac{1}{\sqrt{\mu}}$ to $\frac{3}{2}\pi\frac{1}{\sqrt{\mu}}$, or the time of one vibration is

$$T = \pi\frac{1}{\sqrt{\mu}}, \qquad (98)$$

and this is true whether a be large or small. Compare *Art.* 302.

COR. 1. When a particle is attracted to a fixed point by a force varying directly as the distance, the time of descent to that point will be the same for all distances.

COR. 2. The time of descent to the center of the earth, in which case $\mu=\frac{g}{r}$, is

$$T' = \frac{\pi}{2}\sqrt{\frac{r}{g}},$$
$$= 21'.5''.8.$$

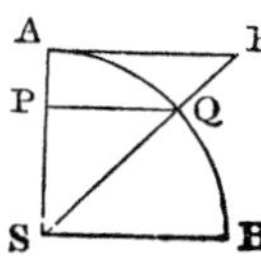

COR. 3. Let S be the center of force, AS$=a$ the distance of the particle at the commencement of motion, P the position of the particle at any instant, and t the time of describing PS$=x$. On AS describe a quadrant, and draw PQ perpendicular to AS. Then

$$x = a.\sin. QSB.$$
$$\therefore\ QSB = \theta \text{ and } QB = a.\theta.$$

By (97), $\quad v = \sqrt{\mu}.\sqrt{a^2 - x^2} = \sqrt{\mu}.PQ \propto PQ,$

and by (a), $\quad t = \frac{1}{\sqrt{\mu}}.\theta = \frac{1}{\sqrt{\mu}}.\frac{QB}{a} = \frac{QB}{\text{vel. at S}} \propto QB.$

When the particle is at A, arc BQ becomes BA$=\frac{1}{2}\pi a$,

and $\quad \therefore T' = \frac{\pi}{2\sqrt{\mu}}.$

340. PROP. *To determine the velocity when a particle is attracted by a force varying inversely as the square of the distance.*

In this case (94), $\quad \phi = \mu x^{-2},$
and [X], $\quad vdv = -\mu x^{-2}dx.$
Integrating, $\quad v^2 = 2\mu x^{-1} + c.$
If $v=0$ when $x=a$, $\quad c = -2\mu a^{-1}.$

$$\therefore\ v^2 = 2\mu(x^{-1} - a^{-1}) = 2\mu\frac{a-x}{ax}. \qquad (99)$$

341. COR. Since the intensity of gravity above the earth's surface varies directly as the mass of the body, and inversely as the square of its distance from the center, (99) will give the velocity of a body falling from any height to the surface. When the body arrives at the surface $x=r$, and since $\mu : g = \frac{1}{1^2} : \frac{1}{r^2}$, $\mu = gr^2$. Hence

$$v^2 = 2gr.\frac{a-r}{a}.$$

If $a=\infty$, $\quad v = \sqrt{2gr} = 6.9428$ miles,

from which it appears that the velocity can never equal seven miles, and if a body be projected upward with the above velocity, supposing no resistance from the air, it would never return.

If we make $x=0$ in (99),

$$v=\infty;$$

or the velocity at the center would be infinite if the same law of force continued.

342. PROP. *To find the time when the force varies inversely as the square of the distance.*

We have, from [VII.],

$$dt=\frac{-dx}{v}=\frac{-dx}{\sqrt{2\mu}\sqrt{\frac{a-x}{ax}}}.$$

$$\therefore\ t=\left(\frac{a}{2\mu}\right)^{\frac{1}{2}}\int\frac{-xdx}{\sqrt{ax-x^2}}.$$

By adding $\frac{1}{2}adx$ to the numerator, and then subtracting the same quantity from it, we have

$$t=\left(\frac{a}{2\mu}\right)^{\frac{1}{2}}\int\left\{\frac{\frac{1}{2}adx-xdx}{\sqrt{ax-x^2}}-\frac{\frac{1}{2}adx}{\sqrt{ax-x^2}}\right\}$$

$$=\left(\frac{a}{2\mu}\right)^{\frac{1}{2}}\left\{(ax-x^2)^{\frac{1}{2}}-\frac{a}{2}\text{versin.}^{-1}\frac{2x}{a}+c\right\}.$$

When $t=0$, $x=a$, and $\text{versin.}^{-1}\frac{2x}{a}=\pi$.

$$\therefore\ c=\pi,$$

and
$$t=\left(\frac{a}{2\mu}\right)^{\frac{1}{2}}\left\{(ax-x^2)^{\frac{1}{2}}+\frac{a}{2}\left(\pi-\text{versin.}^{-1}\frac{2x}{a}\right)\right\}. \quad (100)$$

COR. 1. If $x=0$, we have

$$t^2=\frac{\pi a^3}{8\mu}\propto a^3,$$

or the square of the time of falling to the center varies as the cube of the distance.

COR. 2. On AS$=a$ describe a semicircle, and draw PQ perpendicular to AS. Then PQ$=\sqrt{ax-x^2}$, arc AQS$=\frac{1}{2}\pi a$, arc SQ$=$SC.$\angle$ SCQ$=\frac{a}{2}\text{versin.}^{-1}\frac{2x}{a}$.

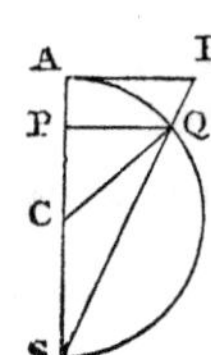

Hence the time through AP is, by (100),

$$t=\left(\frac{a}{2\mu}\right)^{\frac{1}{2}}(\text{PQ}+\text{arc AQS}-\text{arc QS})$$

$$=\left(\frac{a}{2\mu}\right)^{\frac{1}{2}}(\text{PQ}+\text{arc AQ}).$$

If SQ be produced to meet in R a tangent at A,

$$\text{AR}=\frac{\text{SA.PQ}}{\text{PS}}=\frac{a\sqrt{ax-x^2}}{x}=a^{\frac{3}{2}}\left(\frac{a-x}{ax}\right)^{\frac{1}{2}}$$

$\therefore$ by (99), $\quad v=\left(\frac{2\mu}{a^3}\right)^{\frac{1}{2}}.\text{AR}\propto\text{AR}.$

When P coincides with S, AR is infinite.

343. Prop. *To find the velocity and time when the force varies inversely as the cube of the distance.*

In this case, $\qquad \phi=\mu x^{-3},$

and [X], $\qquad vdv=-\mu x^{-3}dx.$

Integrating, $\qquad v^2=\mu x^{-2}+c.$

But when $x=a$, $v=0$; $\therefore\ c=-\mu a^{-2},$

and $\qquad v^2=\mu(x^{-2}-a^{-2}),$

or $$v=\sqrt{\mu}.\frac{\sqrt{a^2-x^2}}{ax}. \qquad (101)$$

By [VII.], $$dt=\frac{-dx}{v}=\frac{a}{\sqrt{\mu}}.\frac{-xdx}{\sqrt{a^2-x^2}}.$$

Integrating, $$t=\frac{a}{\sqrt{\mu}}.\sqrt{a^2-x^2}+c.$$

When $x=a$, $t=0$; $\qquad \therefore\ c=0,$

and $$t=\frac{a}{\sqrt{\mu}}.\sqrt{a^2-x^2}. \qquad (102)$$

If $x=0$, $$t=\frac{a^2}{\sqrt{\mu}},$$

the time to the center of force.

Cor. (See *Fig.*, *Art.* 339) $\text{AR}=\frac{\text{AS.PQ}}{\text{PS}}=\frac{a\sqrt{a^2-x^2}}{x}.$

$$\therefore\ v=\frac{\sqrt{\mu}}{a^2}.\text{AR}\propto\text{AR}.$$

and $$t=\frac{a}{\sqrt{\mu}}\ \text{PQ}\propto\text{PQ}.$$

§ II. CURVILINEAR MOTION OF A FREE POINT.

344. If several forces in different directions in the same plane act *continually* on a material point, it will have a resulting motion at any instant, whose direction and rate must be determined by the relations between the directions and intensities of the forces which act upon it. The motion may or may not be rectilinear, and, in order to investigate the circumstances of the resulting motion generally, we shall employ the method of resolution to two rectangular axes, as in Statics, using X and Y, instead of Σ.X and Σ.Y, to denote the sums of the resolved forces.

345. PROP. *To determine, generally, the motion of a point acted upon by any number of forces in the same plane.*

Resolve the forces in the direction of two rectangular axes, and let X and Y represent the sums of the resolved forces in each axis, x and y being the co-ordinates of the position of the point. The point may be regarded as acted upon by the two forces X and Y, independently of each other. Hence, by [IX.],

$$X=\frac{d^2x}{dt^2},\ Y=\frac{d^2y}{dt^2}. \qquad (103)$$

If ds represent any small element of the path of the point, by [VII.], $\frac{ds}{dt}$ will represent the velocity of the point in its path. Hence, if a and β represent the angles which ds makes with the axes, the velocities in the direction of the axes (39) are

$$\frac{ds}{dt}.\cos. a=\frac{dx}{dt},\ \frac{ds}{dt}.\cos. \beta=\frac{dy}{dt}. \qquad (104)$$

Such are the general equations by which the motion of a point in a plane may be determined.

346. If the velocity is given, we must, by resolution, obtain the velocity in each axis, and thence, by differentiation, the force in each axis, and, by composition, the resultant force may be determined.

If the forces are given, we may, by integration, find the ve locity in each axis, and thence, by composition, the velocity of the point in its path.

If the path is required, each of equations (103) must undergo two integrations, at each of which operations a constant must be introduced. The constants introduced at the first integration will depend on the velocity of the point when the time commences, or at some given time; those introduced at the second integration will depend on the position of the point. We thus have two equations involving the co-ordinates x and y and the time t, and, by elimination of t, an equation between the co-ordinates will be obtained, which is the equation of the path of the point.

347. Prop. *To determine the velocity of a point in its path.*

The velocity may be found as above indicated, but the following is the preferable method.

Multiply the first of equations (103) by $2dx$, the second by $2dy$, and, adding the results, we have

$$\frac{2dxd^2x+2dyd^2y}{dt^2}=2(\mathrm{X}dx+\mathrm{Y}dy).$$

The first member of this equation is the differential of

$$\frac{dx^2+dy^2}{dt^2}=\frac{ds^2}{dt^2}=v^2.$$

Therefore, integrating, we find

$$v^2=2\int(\mathrm{X}dx+\mathrm{Y}dy)+c. \qquad (105)$$

348. Cor. If this expression $\mathrm{X}dx+\mathrm{Y}dy$ is integrable, the velocity may be found, provided we can correct the integral, or know the velocity at some given point.

Thus, if the expression is a differential of the co-ordinates of the position of the point, so that its integral is a function of these co-ordinates, or

$$v^2=2f(x, y)+c,$$

and if v_1 is the velocity at the point whose co-ordinates are a and b, so that

$$v_1^2=2f(a, b)+c;$$

then
$$v^2-v_1^2=2f(x, y)-2f(a, b).$$

Hence it appears that the velocity acquired by the particle

in passing from one point (a, b) to any other (x, y) is the same, whatever be the curve described between these points, since the change of velocity is independent of the co-ordinates of any intermediate point.

349. PROP. *The expression* Xdx+Ydy *is always integrable whenever the forces are directed to fixed centers, and their intensity is a function of the distance from those centers.*

Let F, F_1 . . . be two forces directed to fixed centers,
" a, b, a_1, b_1 be the co-ordinates of the centers,
" x, y . . . . be the co-ordinates of any position of the particle,
" r, r_1 . . be the distances of the particle from the centers,
" $\alpha, \beta, \alpha_1, \beta_1$ be the angles which r and r_1 make with the axes.

Resolving F parallel to each axis, we have for the components

$$F.\cos.\alpha,\ F\cos.\beta.$$

But
$$\cos.\alpha=\frac{x-a}{r} \text{ and } \cos.\beta=\frac{y-b}{r}.$$

Hence the components are

$$F.\frac{x-a}{r},\ F\frac{y-b}{r}.$$

The other forces may be resolved similarly.
Therefore the expression

$$Xdx+Ydy=F\left(\frac{x-a}{r}dx+\frac{y-b}{r}\right)$$
$$+F_1\left(\frac{x-a_1}{r_1}dx+\frac{y-b_1}{r_1}\right).$$

But
$$r^2=(x-a)^2+(y-b)^2,$$

and, by differentiation, $dr=\frac{x-a}{r}dx+\frac{y-b}{r}dy.$

Similarly,
$$dr_1=\frac{x-a_1}{r_1}dx+\frac{y-b_1}{r_1}dy.$$

$$\therefore\ Xdx+Ydy=Fdr+F_1dr_1.$$

And since, by hypothesis, each force F is a function of the distance r, each of the terms of the second member of the equation is an exact differential, and therefore the first is also.

In the same manner, the reasoning may be extended to any number of centers of force.

350. PROP. *To determine the motion of a point acted upon by a force directed to a fixed center.*

Assuming the fixed point as the origin of co-ordinates, let F be the force, whether a single one or the resultant of several, and r the distance of any position of the particle from the center, called the *radius vector.*

Multiply the first of equations (103) by y, and the second by x, and subtract the former from the latter. This gives

$$\frac{xd^2y-yd^2x}{dt^2}=x\mathrm{Y}-y\mathrm{X}. \qquad (106)$$

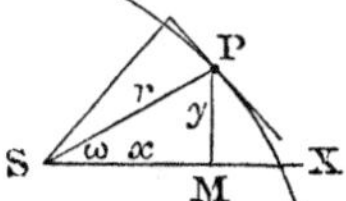

Resolving F in the direction of the axis, we have

$$\mathrm{X}=\mathrm{F}\frac{x}{r},\ \mathrm{Y}=\mathrm{F}\frac{y}{r}.$$

Multiplying the first by y, and the second by x, and taking the difference, we get

$$x\mathrm{Y}-y\mathrm{X}=0.$$

Therefore the first member of (106) is zero, and since it is the differential of $\frac{xdy-ydx}{dt}$, by integration we obtain

$$\int\frac{xd^2y-yd^2x}{dt^2}=\frac{xdy-ydx}{dt}=c,$$

or

$$xdy-ydx=cdt. \qquad (107)$$

Let ω be the angle made by r with the axis of x; then

$$x=r\cos.\ \omega, \qquad (a)$$

$$y=r\sin.\ \omega; \qquad (b)$$

also,

$$dx=\cos.\ \omega dr-r.\sin.\ \omega.d\omega, \qquad (c)$$

$$dy=\sin.\ \omega dr+r\cos.\ \omega.d\omega. \qquad (d)$$

Multiplying (c) by (b), and (d) by (a), and subtracting, we get

$$xdy-ydx=r^2d\omega=cdt.$$

$$\therefore\ \frac{r^2d\omega}{dt}=c. \qquad (108)$$

But $\frac{1}{2}r^2d\omega$ is the differential of the area of a plane curve referred to polar co-ordinates. Hence the ratio of the element of the area to the element of the time is constant, and *the area described by the radius vector in any time is proportional to the time.*

By integrating $\frac{1}{2}r^2d\omega=\frac{1}{2}cdt$, we have for the area described in any time

$$A=\tfrac{1}{2}ct, \qquad (109)$$

and if $t=1$, we have

$$c=2A,$$

or the constant c is twice the area described in the unit of time.

351. Prop. *Conversely, if a material point describe by its radius vector around a fixed center areas proportional to the times, the force is directed to that point.*

When the radius vector describes areas proportional to the times, we have (107)

$$\frac{xdy-ydx}{dt}=c.$$

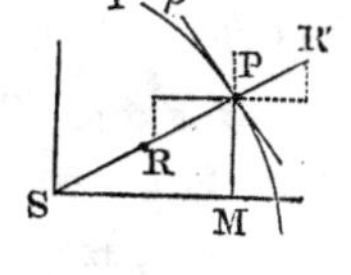

Differentiating, we get

$$\frac{xd^2y-yd^2x}{dt^2}=0.$$

∴ (106)

$$xY-yX=0,$$

and

$$X:Y=x:y.$$

Hence the forces X and Y, which are parallel respectively to the co-ordinates x and y of the point P, are proportional to these co-ordinates, and the resultant PR of X and Y must take the direction of PS, which is the hypothenuse of the right-angled triangle PMS, in which $PM=y$ and $SM=x$.

As an application of the general equations of motion of a free point (103), we may take the following cases.

352. Prop. *To determine the motion of a point moving from the action of an impulsive force.*

Since, after the action of the impulse, the point is abandoned to itself, there are no accelerating forces acting on the point, and we have (103)

$$\frac{d^2x}{dt^2}=X=0, \quad \frac{d^2y}{dt^2}=Y=0.$$

Integrating, $$\frac{dx}{dt}=v_1,\ \frac{dy}{dt}=v_2; \qquad (a)$$
v_1 and v_2 being constants, added to complete the integral.

Compounding these velocities (*Art.* 236), we have for the resultant velocity

$$v=\sqrt{\frac{dx^2+dy^2}{dt^2}}=\frac{ds}{dt}=\sqrt{v_1^2+v_2^2}.$$

The velocity is therefore uniform.

Integrating equations (a),

$$x=v_1t+s_1,\ y=v_2t+s_2, \qquad (b)$$

the constants s_1 and s_2 being the co-ordinates of the point when the time commences.

Eliminating t from equations (b), we have

$$y=\frac{v_2}{v_1}x+v_2(s_2-s_1),$$

the equation of a straight line.

Hence the path of the point is a straight line and the motion uniform, which accords with the first law of motion.

353. Prop. *To determine the motion of a projectile acted upon by gravity regarded as a constant force.*

In this case, taking the axis of x horizontal and that of y vertical, (103) give

$$\frac{d^2x}{dt}=X=0,\ \frac{d^2y}{dt^2}=Y=-g.$$

Multiplying by dt, and integrating,

$$\frac{dx}{dt}=v_1,\ \frac{dy}{dt}=v_2-gt.$$

Multiplying by dt, and integrating again, we have

$$x=v_1t+s_1,\ y=v_2t-\tfrac{1}{2}gt^2+s_2. \qquad (a)$$

But if the point of projection be at the origin of co-ordinates, and t be reckoned from the commencement of motion, s_1 and s_2 will each be zero. Putting $v_2=av_1$, $s_1=0$, $s_2=0$, and eliminating t from the equations (a), we get

$$y=ax-x^2\frac{g}{2v_1^2}.$$

Compare this with (63).

§ III. CONSTRAINED MOTION OF A POINT.

354. Prop. *To determine the velocity of a point moving on a given curve.*

If a point is constrained to move on a curve, the reaction of the curve will be a new force normal to the curve, and may be resolved in the direction of the axes and combined with the other components, as in Statics, *Art.* 77.

Let N be this normal force, a and β the angles which the normal makes with the axes, and X and Y the sums of the components of all the other forces in the direction of the axes.

Resolving N in the same directions, we shall have from (103)

$$\frac{d^2x}{dt^2}=\mathrm{X}+\mathrm{N}\cos. a$$

$$\frac{d^2y}{dt^2}=\mathrm{Y}+\mathrm{N}\cos. \beta$$

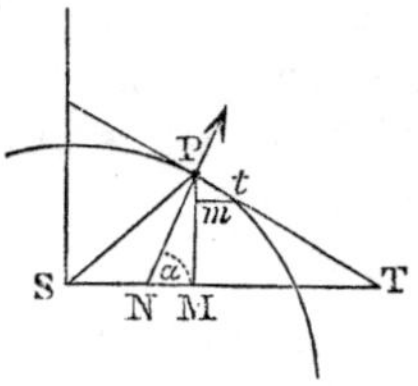

But $\cos. a=\sin. \mathrm{P}tm=+\frac{dy}{ds}$, supposing the motion in the direction tP, and $\cos. \beta=\sin. a=\cos. \mathrm{P}tm=-\frac{dx}{ds}$.

$$\therefore\ \frac{d^2x}{dt^2}=\mathrm{X}+\mathrm{N}.\frac{dy}{ds} \qquad (a)$$

$$\frac{d^2y}{dt^2}=\mathrm{Y}-\mathrm{N}\frac{dx}{ds}. \qquad (b)$$

Multiplying (a) by $2dx$, and (b) by $2dy$, and adding

$$\frac{2dxd^2x+2dyd^2y}{dt^2}=2(\mathrm{X}dx+\mathrm{Y}dy).$$

The first member is the differential of

$$\frac{dx^2+dy^2}{dt^2}=\frac{ds^2}{dt^2}=v^2.$$

Whence, integrating,

$$v^2=2\int(\mathrm{X}dx+\mathrm{Y}dy)+c. \qquad (110)$$

Cor. 1. Since this result is exactly the same as that obtained for unrestrained motion (105), we may conclude.

1°. That, if no accelerating force act on the point, or $X=0$ and $Y=0$, its velocity will remain constant, and be not at all retarded by the action of the curve. See *Art.* 284.

2°. If any accelerating forces do act on the point, the velocity is independent of the curve on which it is constrained to move. See *Art.* 286.

Cor. 2. If the accelerating forces are all parallel, they may be assumed parallel to the axis of x, and in this case (110) becomes

$$v^2=2\int Xdx+c. \qquad (111)$$

Cor. 3. If in this last case the force, or resultant of the forces, is constant, and equal to f, we have

$$v^2=\pm 2fx+c,$$

where the upper or lower sign is to be used according as the force tends to increase or diminish x.

If the distance of the point from the origin, $x=a$, when $v=0$, $c=2fa$, and

$$v^2=2f(a-x)\,; \qquad (112)$$

and since only the ordinates on the axis of x are involved, the velocity of the point on the curve depends not on the curve described, but on the difference of the ordinates on the axis of x.

355. Prop. *To find the time of motion of a point on a given curve.*

In all cases [VII.], $\quad dt=\frac{ds}{v}.$

Hence, when the nature of the curve and the velocity at any point of it is known, the value of v being found from (110) and substituted in that of dt, the time may be found by integration

Cor. If the forces act in parallel lines, and their resultant is constant, by (112)

$$dt=\frac{ds}{\sqrt{2f(a-x)}}. \qquad (113)$$

356. Prop. *To find the reaction of the curve.*

Multiplying (a), *Art.* 354, by $\frac{dy}{ds}$, and (b) by $\frac{dx}{ds}$, and subtracting the latter from the former, we get

$$N=Y\frac{dx}{ds}-X\frac{dy}{ds}+\frac{dyd^2x-dxd^2y}{dsdt^2}.$$

Eliminating dt by the equation $v=\frac{ds}{dt}$,

$$N=Y\frac{dx}{ds}-X\frac{dy}{ds}+v^2.\frac{dyd^2x-dxd^2y}{ds^3}.$$

But if ρ be the radius of the osculating circle,

$$\rho=\frac{ds^3}{dyd^2x-dxd^2y}.$$

$$\therefore\ N=Y\frac{dx}{ds}-X\frac{dy}{ds}+\frac{v^2}{\rho}$$

$$=Y\cos.\beta-X\cos.a+\frac{v^2}{\rho}.$$

The first two terms of this value of N are equal and opposite to the forces X and Y resolved in the direction of the normal. They give, therefore, the pressure on the curve due to the action of these forces, while the other term $\frac{v^2}{\rho}$ is the reaction of the curve or pressure due to the motion.

Otherwise. If we suppose no accelerating forces to act on the point, then X and Y are each equal to zero in (*a*) and (*b*), *Art.* 354. Hence

$$N^2\left(\frac{dx^2+dy^2}{ds^2}\right)=N^2=\frac{(d^2x)^2+(d^2y)^2}{dt^4};$$

or, eliminating dt by the equation $dt=\frac{ds}{v}$,

$$N=v^2.\frac{\sqrt{(d^2x)^2+(d^2y)^2}}{ds^2}.$$

But the coefficient of v^2 is the reciprocal of the radius of curvature when s is the independent variable.

$$\therefore\ N=\frac{v^2}{\rho} \qquad (114)$$

Compare this with (79).

As an application of the formulæ of constrained motion, we may take the following cases:

357. Prop. *To determine the motion of a body descending by the force of gravity down the arc of a vertical circle.*

1°. To find the velocity. Formula (112) applies to this case, and if h=SA be the height from which the body at P descends, and x any distance from the lowest point on SA, the axis of x, we have

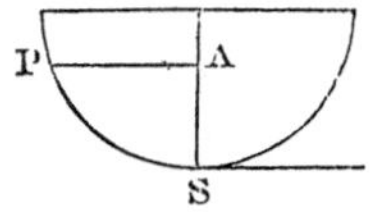

P A S

$$v=\sqrt{2g(h-x)}.$$

When $x=0$, we have for the velocity at the lowest point,

$$v=\sqrt{2gh},$$

which is the same as that due to the vertical height h.

2°. To find the time, we have from (113)

$$dt=\frac{ds}{\sqrt{2g(h-x)}}.$$

The equation of the circle is

$$y^2=2ax-x^2.$$

Hence
$$dy^2=\frac{(a-x)^2dx^2}{2ax-x^2}.$$

But
$$ds^2=dx^2+dy^2=dx^2\left(1+\frac{(a-x)^2}{2ax-x^2}\right)$$
$$=\frac{a^2dx^2}{2ax-x^2},$$
$$\therefore\ ds=\frac{adx}{\sqrt{2ax-x^2}}.$$

This, taken negative, because the arc is a decreasing function of the time, and substituted in the above, gives

$$dt=-\frac{a}{\sqrt{2g}}\cdot\frac{dx}{\sqrt{(h-x)(2ax-x^2)}}$$
$$=-\frac{a}{\sqrt{2g}}\cdot\frac{dx}{\sqrt{(hx-x^2)(2a-x)}}. \qquad (a)$$
$$=-\frac{a}{\sqrt{2g}}\cdot\frac{(2a-x)^{-\frac{1}{2}}dx}{\sqrt{hx-x^2}}$$
$$=-\tfrac{1}{2}\sqrt{\frac{a}{g}}\cdot\frac{\left(1-\frac{x}{2a}\right)^{-\frac{1}{2}}dx}{\sqrt{hx-x^2}}.$$

Expanding $\left(1-\frac{x}{2a}\right)^{-\frac{1}{2}}$ by the binomial theorem, we have

$$dt=-\frac{1}{2}\sqrt{\frac{a}{g}}\cdot\frac{dx}{\sqrt{hx-x^2}}\left\{1+\frac{1}{2}\left(\frac{x}{2a}\right)+\frac{1}{2}\cdot\frac{3}{4}\left(\frac{x}{2a}\right)^2+\frac{1}{2}\cdot\frac{3}{4}\cdot\frac{5}{6}\left(\frac{x}{2a}\right)^3+,\right.$$

$\left.\&c.\right\}$.

Thus the terms to be integrated are of the form $\frac{-x^n dx}{\sqrt{hx-x^2}}$ the exponents n being natural numbers beginning with 0. Performing the operations and taking the integral between the limits $x=h$ and $x=0$, we find

$$t=\frac{\pi}{2}\sqrt{\frac{a}{g}}\left\{1+\left(\frac{1}{2}\right)^2\cdot\left(\frac{h}{2a}\right)+\left(\frac{1}{2}\cdot\frac{3}{4}\right)^2\left(\frac{h}{2a}\right)^2+\left(\frac{1}{2}\cdot\frac{3}{4}\cdot\frac{5}{6}\right)^2\left(\frac{h}{2a}\right)^3+,\right.$$

$\left.\&c.\right\}$.

If h be small, the first term will give an approximate value for the time, viz.,

$$t=\tfrac{1}{2}\pi\sqrt{\frac{a}{g}}.$$

Otherwise. If in (a) x be rejected as small in comparison with $2a$, the equation reduces to

$$dt=-\tfrac{1}{2}\sqrt{\frac{a}{g}}\cdot\frac{dx}{\sqrt{hx-x^2}}.$$

$$\therefore\ t=-\tfrac{1}{2}\sqrt{\frac{a}{g}}\cdot\text{versin.}^{-1}\frac{2x}{h}+c.$$

But $t=0$ when $x=h$;

$$\therefore\ c=\tfrac{1}{2}\pi\sqrt{\frac{a}{g}},$$

and $$t=\tfrac{1}{2}\sqrt{\frac{a}{g}}\left(\pi-\text{versin.}^{-1}\frac{2x}{h}\right),$$

and when $x=0$, $$t=\tfrac{1}{2}\pi\sqrt{\frac{a}{g}}.\quad\text{(See (86)).}$$

358. Prop. *To determine the motion of a body descending by gravity down the inverted arc of a cycloid whose base is horizontal and axis vertical.*

The cycloid being inverted, let the origin be at the lowest

point, the axis of x being vertical, and that of y horizontal Let the axis $AB=2a$, $AM=x$, $PM=y$, and $AP=s$.

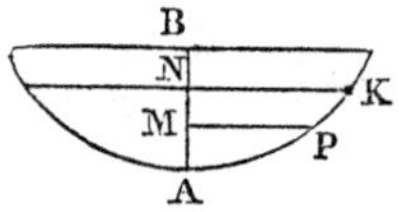

The velocity, as in the preceding proposition, is

$$v=\sqrt{2g(h-x)},$$

h being the abscissa of the point K from which the body begins to descend.

The differential equation of the curve, referred to the vertex as the origin, is

$$dy=\frac{(2ax-x)dx}{\sqrt{2ax-x^2}}.$$

But $$ds^2=dy^2+dx^2=\left(\frac{(2a-x)^2}{2ax-x^2}+1\right)dx^2.$$

$$\therefore\ ds=dx\sqrt{\frac{2a}{x}}.$$

Substituting this value of ds with its proper sign in [VII.], we find

$$dt=-\sqrt{\frac{a}{g}}.\frac{dx}{\sqrt{hx-x^2}}.$$

Integrating, $$t=\sqrt{\frac{a}{g}}.\left(\pi-\text{versin.}^{-1}\frac{2x}{h}\right),$$

and therefore for the time of descent to the lowest point, where $x=0$,

$$t=\pi\sqrt{\frac{a}{g}}=\frac{\pi}{2}\sqrt{\frac{4a}{g}}. \qquad (115)$$

Cor. The time t being independent of the abscissa h, will be the same, from whatever point in the curve the body begins to descend; in this respect differing from the motion in circular arcs. On account of this remarkable property the cycloid is called *tautochronal.*

359. Schol. The property of *tautochronism* which attaches to the cycloid was supposed to give peculiar advantages to a pendulum made to oscillate in this curve. The method of accomplishing this is naturally suggested by the property of the *evolutes;* and as an evolute of a cycloid is another equal cy-

cloid, we have only to take two equal semicycloids and place the extremities of their bases contiguous in the same horizontal line as in the figure. If we then suspend a body at the point S, common to the two semicycloids, by a flexible string equal in length to either semicycloid, or, which is the same thing, to twice the axis, and make it oscillate between the two, it will generate the involute, or another equal cycloid.

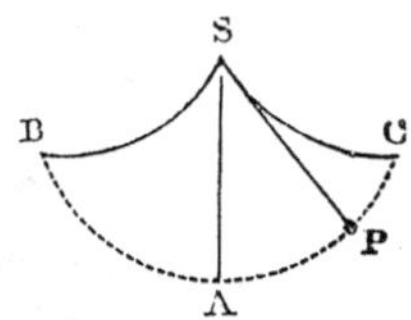

The constant change, however, in the center of motion arising from the contact of the string with the two curves or checks, which changes also the relative velocity of the different parts of the vibrating body, renders this contrivance, although beautiful in theory, yet useless in practice; independently of the difficulty of obtaining a string sufficiently flexible, and of ensuring accuracy in the plates.

Vibrations, therefore, in *small* circular arcs, which are at the same time also most natural, have been adhered to in practice, and it has been shown (*Art.* 302 and 300) that, as long as the arc is small, these vibrations have all the advantages of the vibrations in cycloidal arcs.

§ IV. MOTION OF A POINT ACTED UPON BY A CENTRAL FORCE

360. The case of a point revolving in an orbit by the action of a force tending to a fixed center is of sufficient importance to justify a distinct discussion, especially as the formulæ are susceptible of considerable simplification, when the force is given as a function of the distance from the center. The fundamental equations

$$X=\frac{d^2x}{dt^2},\ Y=\frac{d^2y}{dt^2},$$

are components of the force in two rectangular directions But when the force is directed to or from a fixed center, and is given in functions of the distance of the particle from that center, it is most natural and convenient to introduce polar coordinates.

361. Prop. *To transform the expressions for the force from rectangular to polar co-ordinates.*

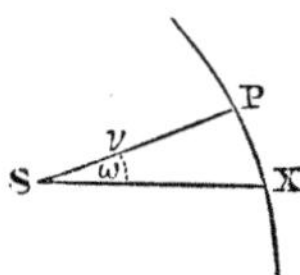

Suppose the origin to be at S, the center of force, and let r=SP, the radius vector or distance of the particle, and ω=PSX, the angle made by r with axis of x.

Then $\mathrm{X}=-\mathrm{F}\cos.\omega$, $\mathrm{Y}=-\mathrm{F}\sin.\omega$,

the signs being negative because the force is supposed to be attractive, or act toward the center in opposition to X and Y.

Multiplying the first by $\sin.\omega$, and the second by $\cos.\omega$, and subtracting, we have

$$\mathrm{X}\sin.\omega-\mathrm{Y}\cos.\omega=0. \qquad (a)$$

Similarly, multiplying the first by $\cos.\omega$, and the second by $\sin.\omega$, and adding, we get

$$\mathrm{X}\cos.\omega+\mathrm{Y}\sin.\omega=-\mathrm{F}. \qquad (b)$$

But x and y being the co-ordinates of the position of the particle

$$x=r\cos.\omega,\quad y=r\sin.\omega.$$

Differentiating each twice, we obtain

$$d^2x=d^2r.\cos.\omega-2dr.d\omega.\sin.\omega-rd\omega^2.\cos.\omega-rd^2\omega.\sin.\omega,$$
$$d^2y=d^2r.\sin.\omega+2dr.d\omega.\cos.\omega-rd\omega^2.\sin.\omega+rd^2\omega.\cos.\omega.$$

By dividing these expressions by dt^2, the values of X and Y will be obtained (103), and if we substitute these values in (a) and (b), we get

$$2.\frac{dr}{dt}.\frac{d\omega}{dt}+r.\frac{d^2\omega}{dt^2}=0, \qquad (116)$$

$$\frac{d^2r}{dt^2}-r.\frac{d\omega^2}{dt^2}+\mathrm{F}=0, \qquad (117)$$

which are two polar equations to determine the motion.

362. Cor. 1. *The radius vector of a point revolving about a center of force describes equal areas in equal times.*

For, multiplying (116) by rdt, we obtain

$$2rdr.\frac{d\omega}{dt}+r^2.\frac{d^2\omega}{dt}=0=d.\frac{r^2d\omega}{dt}.$$

Integrating, $$\frac{r^2 d\omega}{dt}=c, \tag{118}$$

and $$\int r^2 d\omega = ct + c_1.$$

But $\frac{1}{2}\int r^2 d\omega$ is the area described by the radius vector in the time t, and it varies as t.

363. Cor. 2. *The angular velocity of the radius vector varies inversely as the square of the distance of the particle.*

From (118) we obtain $$\frac{d\omega}{dt}=\frac{c}{r^2}. \tag{119}$$

But $\frac{d\omega}{dt}$ is the angular velocity of r, and it varies inversely as r^2.

364. Prop. *To determine the velocity of the particle in its orbit.*

Multiply (116) by $rd\omega$, and (117) by dr, and add the results. This gives

$$dr.\frac{d^2r}{dt^2}+rdr.\frac{d\omega^2}{dt^2}+r^2d\omega.\frac{d^2\omega}{dt^2}+\text{F}dr=0,$$

or $$d.\left(\frac{dr^2}{dt^2}+r^2\frac{d\omega^2}{dt^2}\right)+2\text{F}dr=0.$$

Therefore, integrating, we have

$$\frac{dr^2+r^2d\omega^2}{dt^2}+2\int \text{F}dr=c'. \tag{120}$$

But $$v^2=\frac{ds^2}{dt^2}=\frac{dr^2+r^2d\omega^2}{dt^2}. \quad \text{(See } Art. \text{ 367, 4°)}$$

$$\therefore v^2=c'-2\int \text{F}dr. \tag{121}$$

Cor. Since the value of v depends only on the distance of the particle from the center, the velocity of the particle will be the same at any two points equally distant from the center, and the velocity acquired in passing from one point to another will be the same, whatever be the path or curve between them, if the law of force remain the same.

365. Prop. *To find the time of describing any portion of the orbit.*

By (119),
$$\frac{d\omega^2}{dt^2}=\frac{c^2}{r^4},$$
which, substituted in (120), gives
$$\frac{dr^2}{dt^2}+\frac{c^2}{r^2}+2\int \mathrm{F}dr=c'. \qquad (122)$$
In this expression we have the relation of r to t, independent of the orbit.

366. Prop. *To find the equation of the orbit.*

Substituting in (122) for dt^2 its value $\frac{r^4 d\omega^2}{c^2}$, obtained from (118), we get
$$c^2\left(\frac{dr^2}{r^4 d\omega^2}+\frac{1}{r^2}\right)+2\int \mathrm{F}dr=c'; \qquad (123)$$
or, to give the equation a more convenient form, put $r=\frac{1}{u}$, from which we find $dr=-\frac{du}{u^2}$. These values of r and dr, substituted in (123), give
$$c^2\left(\frac{du^2}{d\omega^2}+u^2\right)-2\int \mathrm{F}\frac{du}{u^2}=c'. \qquad (124)$$
When the law of force is given, we can obtain from this expression, by integration, the relation between u and ω, or the polar equation of the orbit; or, if the relation between u and ω is given, the law of force may be determined.

367. Schol. A familiarity with the geometrical representatives of the terms and factors in the foregoing formulæ will conduce to facility in their application.

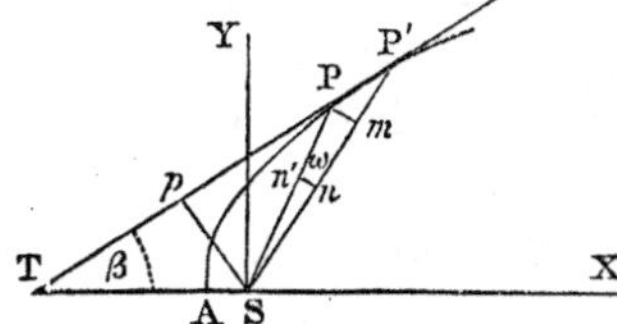

In the annexed figure let the center of force be at S, the origin of co-ordinates, SP$=r$ the radius vector of the particle at P, P$'$P$=ds$ an element of its path coinciding with the tangent PT, $\omega=$PST the angle made by the radius vector with the axis of x, P$'$SP$=d\omega$ the angle described by the radius vector in the indefinitely small time dt, and mP $=dr$ the increment or decrement of the radius vector in the same time. Let the arc Pm be described with S as a center, and radius SP, and the arc nn' with the radius Sn=1. Then,

1°. Since Sn : SP$=nn'$: Pm or $1 : r = d\omega$: Pm,

$$\text{P}m = r d\omega.$$

2°. The area of the sector SPm is

$$\text{A} = \frac{\text{SP.P}m}{2} = \tfrac{1}{2} r^2 d\omega.$$

3°. If the point P′ be indefinitely near to P, PP′ and Pm may be considered straight lines, and the triangle PmP′ a plane rectilinear triangle, right-angled at m. Hence

$$\text{P}'\text{P}^2 = \text{P}m^2 + \text{P}'m^2,$$
$$ds^2 = r^2 d\omega^2 + dr^2.$$

4°. If S$p=p$ be a perpendicular on the tangent, when the triangle PmP′ is in its nascent or evanescent state, the angle PP′m=pPS and the triangle PP′m is similar to pPS. Hence PP′ : Pm=PS : Sp, or $ds : r d\omega = r : p$. $\therefore p = \frac{r^2 d\omega}{ds}$, and [VII.], $p = \frac{r^2 d\omega}{v dt}$. $\therefore vp = \frac{r^2 d\omega}{dt} =$, by (118), twice the area described in the unit of time.

Also,
$$p^2 = \frac{r^4 d\omega^2}{ds^2} = \frac{r^4 d\omega^2}{dr^2 + r^2 d\omega^2}.$$

5°. Since Sp : pP : PS=Pm : mP′ : P′P=$r d\omega : dr : ds$,

$$\therefore \frac{\text{S}p}{\text{SP}} = \frac{\text{P}m}{\text{PP}'} = \frac{r d\omega}{ds} = \sin.\ \text{P},$$

$$\frac{\text{P}p}{\text{PS}} = \frac{\text{P}'m}{\text{P}'\text{P}} = \frac{dr}{ds} = \cos.\ \text{P},$$

and
$$\frac{p\text{S}}{p\text{P}} = \frac{m\text{P}}{m\text{P}'} = \frac{r d\omega}{dr} = \tan.\ \text{P}.$$

6°. If the force F be resolved in the direction of the curve, we shall have

$$\frac{d^2 s}{dt^2} = -\text{F} \cos.\ \text{P} = -\text{F}\frac{dr}{ds}.$$

Multiplying both sides by $2ds$, and integrating, we have

$$\frac{ds^2}{dt^2}=v^2=c'-2\int F dr, \text{ as before (121).}$$

By reference to (117), we see that the first term has the form of the differential expression [IX.] of a force in the direction of r; the second term $\frac{rd\omega^2}{dt^2}=\frac{r^2d\omega^2}{rdt^2}$. But $\frac{rd\omega}{dt}$ is the arc of a circle of radius r, divided by the time. It is, therefore, the linear velocity v_1 of a body in a circle. Hence $\frac{r^2d\omega^2}{dt^2}=\frac{v_1^2}{r}=f$, the centrifugal force in a circle. $\therefore$ (117) becomes $\frac{d^2r}{dt^2}=f-F$, F being the force by which the point describes any orbit, and f the centrifugal force in a circle whose radius is r, the distance of the point at the instant. $\frac{d^2r}{dt^2}$, the difference of these forces at any time, is that by which the radius vector is increased or diminished, and is called the *paracentric force.* In like manner, the integral of $\frac{d^2r}{dt^2}$ or $\frac{dr}{dt}$ is the velocity of approach or recession from the center, and is called the *paracentric velocity.* The paracentric force is the difference between the centrifugal and centripetal forces.

COR. 1. Since $\frac{rd\omega^2}{dt^2}=\frac{1}{r^3}\cdot\frac{r^4d\omega^2}{dt^2}=\frac{c^2}{r^3}$.

$$\therefore f=\frac{c^2}{r^3}. \tag{125}$$

COR. 2. The quantity by which c^2 is multiplied in (123) is, by (4°), equal to $\frac{1}{p^2}$. Hence

$$\frac{c^2}{p^2}+2\int F dr=c'.$$

Differentiating, p being variable,

$$-\frac{2c^2dp}{p^3}+2Fdr=0.$$

Hence $$F=\frac{c^2dp}{p^3dr},\qquad(126)$$

an expression most useful in finding the law of force by which a given orbit may be described.

COR. 3. Since $v^2=\frac{ds^2}{dt^2}=\frac{dr^2+r^2d\omega^2}{dt^2}=\frac{r^4d\omega^2}{p^2dt^2}=\frac{c^2}{p^2}$,

and (126), $$\frac{c^2}{p^2}=\frac{Fpdr}{dp}=2F.\frac{pdr}{2dp},\ \therefore\ v^2=2F.\frac{pdr}{dp}.\qquad(127)$$

But $\frac{pdr}{2dp}=\frac{1}{4}$ chord of curvature through the pole, and (if F be constant) is (55) the space through which a body must fall to acquire the velocity v. Hence the velocity of a body at any point of its orbit is equal to that acquired by falling through one fourth the chord of curvature through the pole.

Application of the preceding formulæ to the motions of the planets.

368. Observation has established three facts respecting the motions of the planets, which, from their discoverer, are called Kepler's Laws.

1°. *The areas described by the radius vector of a planet are proportional to the times.*

2°. *The orbit of a planet is an ellipse of which the center of the sun is one of the foci.*

3°. *The squares of the times of revolution of the different planets are proportional to the cubes of their mean distances from the sun, or the semi-major axes of their orbits.*

These laws relate only to the center of inertia of each planet, and conclusions from them must be limited to the motion of that point; in other words, to the motion of *translation* of the planets.

369. PROP. *The accelerating force by which a planet describes its orbit is directed to the center of the sun.*

For, by the first of Kepler's laws, the areas described by the radius vector are proportional to the times, and when this is the case, by *Art.* 351, the force must be directed to the point about which this equable description of areas takes place.

370. PROP. *The force by which a planet describes its orbit varies inversely as the square of its distance from the center.*

By the second law the orbit is an ellipse. The equation of an ellipse referred to the focus as a pole is,

$$r=\frac{a(1-e^2)}{1+e\cos.(\omega-\theta)}, \qquad (128)$$

in which a is the semi-major axis, called the *mean* distance, and e the eccentricity. The angle θ is an angle made by a fixed line, through the focus with the major axis, reckoned from the lower apsis, or least distance, and called the *longitude of the perihelion*, the lower apsis being called the *perihelion*, and the higher the *aphelion*. The angle $\omega-\theta$, which expresses the angular distance from the perihelion, is called the *true anomaly*.

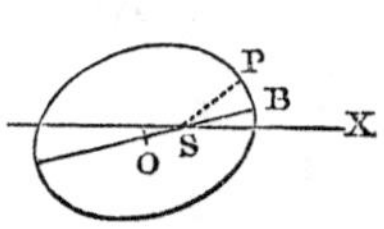

Putting $r=\frac{1}{u}$ in the above equation, we get

$$u=\frac{1+e\cos.(\omega-\theta)}{a(1-e^2)}. \qquad (a)$$

Differentiating in reference to ω,

$$\frac{du}{d\omega}=-\frac{e.\sin.(\omega-\theta)}{a(1-e^2)}.$$

Differentiating, again,

$$\frac{d^2u}{d\omega^2}=-\frac{e\cos.(\omega-\theta)}{a(1-e^2)}. \qquad (b)$$

Adding (b) to (a),

$$\frac{d^2u}{d\omega^2}+u=\frac{1}{a(1-e^2)}. \qquad (c)$$

Differentiating (124),

$$\frac{d^2u}{d\omega^2}+u=\frac{F}{c^2u^2}.$$

Hence $$F=\frac{c^2u^2}{a(1-e^2)}=\frac{c^2}{a(1-e^2)}\cdot\frac{1}{r^2}. \qquad (129)$$

The coefficient of $\frac{1}{r^2}$ being constant, the force varies inversely as the square of the distance.

If $r=1$, $$F=\frac{c^2}{a(1-e^2)}.$$

Hence the coefficient of $\frac{1}{r^2}$ is the intensity of the force at the unit of distance, or the absolute force.

371. Prop. *The intensity of the force is the same for all the planets at the same distance.*

The quantity c (109) is twice the area described in the unit of time. If, therefore, T be the time of one revolution of a planet, cT will be twice the area of the ellipse described by the radius vector.

But the area of an ellipse is $\pi ab=\pi a^2\sqrt{1-e^2}$.

$$\therefore\ cT=2\pi a^2\sqrt{1-e^2},$$

and $$\frac{c^2}{a(1-e^2)}=\frac{4\pi^2a^3}{T^2}.$$

Similarly, if a', e', c', T′ express the same quantities in any other orbit, we shall have

$$\frac{c'^2}{a'(1-e'^2)}=\frac{4\pi^2a'^3}{T'^2}.$$

By Kepler's third law,

$$T^2 : T'^2=a^3 : a'^3,$$

$$\therefore\ \frac{c^2}{a(1-e^2)}=\frac{c'^2}{a'(1-e'^2)}.$$

Hence the force does not vary from one planet to another except in consequence of change of distance.

372. Prop. *To find the velocity of a planet at any point in its orbit.*

By (121), $$v^2=c'-2\int F dr.$$

Substituting in this the value of F (129), we have

$$v^2=c'-\frac{2c^2}{a(1-e^2)}\int\frac{dr}{r^2}.$$

Integrating, $$v^2=c'+\frac{2c^2}{a(1-e^2)}.\frac{1}{r}.$$

To determine c', we must know the velocity at some given

distance, or at what distance and with what velocity the planet was originally projected into space.

Let v_1 be this primitive velocity, and r_1 the corresponding distance.

Then $$v_1^2=c'+\frac{2c^2}{a(1-e^2)}.\frac{1}{r_1},$$

or $$c'=v_1^2-\frac{2c^2}{a(1-e^2)}.\frac{1}{r_1}.$$

$$\therefore\ v^2=v_1^2+\frac{2c^2}{a(1-e^2)}.\left(\frac{1}{r}-\frac{1}{r_1}\right). \qquad (130)$$

Otherwise. From *Art.* 367, *Cor.* 3, we have

$$v^2=\frac{c^2}{p^2},$$

and (Analytic Geometry) $p^2=\frac{b^2r}{2a-r}=\frac{a^2(1-e^2)}{2a-r}$

$$\therefore\ v^2=\frac{c^2}{a^2(1-e^2)}.\frac{2a-r}{r}. \qquad (131)$$

Cor. 1. Hence the velocity is greatest when r is least or at perihelion, and least at aphelion, where r is the greatest.

Cor. 2. If a body move in a circle whose radius is r, v_1 being its velocity, the central force is

$$F=\frac{v_1^2}{r}=\frac{c^2}{a(1-e^2)}.\frac{1}{r^2}\ (129),$$

if the force is the same as that which retains the planet in its orbit

$$\therefore\ v^2:v_1^2=\frac{c^2}{a^2(1-e^2)}.\frac{2a-r}{r}:\frac{c^2}{a(1-e^2)}.\frac{1}{r}$$
$$=2a-r:a.$$

That is, the square of the velocity in the ellipse is to the square of the velocity in the circle as the distance of the planet from the unoccupied focus is to the semi-major axis.

Cor. 3. If r_1=perihelion distance, and r_2=aphelion distance then, if in (131) $r=r_1$, $v^2=\frac{c^2}{a^2(1-e^2)}.\frac{2a-r_1}{r_1}=\frac{c^2}{a^2(1-e^2)}.\frac{r_2}{r_1}$, (a)

" " $r=r_2$, $v^2=\frac{c^2}{a^2(1-e^2)}.\frac{2a-r_2}{r^2}=\frac{c^2}{a^2(1-e^2)}\frac{r_1}{r_2}$, (b)

" " $r=a$, $v^2=\frac{c^2}{a^2(1-e^2)}=\sqrt{(a).(b)}$;

or, the velocity at the extremity of the minor axis is a mean proportional between the velocities at the perihelion and aphelion.

373. Prop. *To determine the orbit which a body will describe when the force varies inversely as the square of the distance.*

If μ be the force at the unit of distance, then

$$F=\frac{\mu}{r^2}=\mu u^2.$$

$$\therefore \int F\frac{du}{u^2}=\int \mu du=\mu u.$$

This, substituted in (124), gives

$$c^2\left(\frac{du^2}{d\omega^2}+u^2\right)-2\mu u=c',$$

which is the differential equation of the orbit.

To integrate this, we find

$$d\omega^2=\frac{c^2du^2}{c'+2\mu u-c^2u^2}.$$

Put $u=z+\frac{\mu}{c^2}$, and $b^2=\frac{1}{c^4}(c'c^2+\mu^2)$, and we have

$$d\omega=\frac{dz}{\sqrt{b^2-z^2}};$$

or, integrating, $$\omega=\theta+\cos.^{-1}\frac{z}{b}.$$

$$\therefore\ z=b\cos.(\omega-\theta);$$

or, replacing the value of z,

$$u=\frac{\mu}{c^2}+b\cos.(\omega-\theta).$$

Hence $$r=\frac{\frac{c^2}{\mu}}{1+\frac{bc^2}{\mu}\cos.(\omega-\theta)}, \qquad (132)$$

which is the equation of a conic section referred to the focus. Hence the orbit which a body will describe when the force varies inversely as the square of the distance is a conic section.

To determine the constants, and therefore the dimensions and form of the orbit, some other circumstances must be given

Suppose we know v_1 the initial velocity or velocity of projection into space, ρ the distance of the planet from the sun, and ψ the angle made by the direction of projection with the distance.

By *Art.* 367, 5°, $\sin. \psi=\frac{\rho d\omega}{ds}$.

Hence (118), $c=\frac{\rho^2 d\omega}{dt}=\rho.\frac{ds}{dt}.\frac{\rho d\omega}{ds}=\rho.v. \sin. \psi.$ (*a*)

By (121), $v^2=c'-2\int F dr=c'+\frac{2\mu}{r}.$

$$\therefore\ c'=v_1^2-\frac{2\mu}{\rho}. \qquad (b)$$

Comparing (132) with (128), we see that

$$a(1-e^2)=\frac{c^2}{\mu} \text{ and } e=\frac{bc^2}{\mu}=\frac{1}{\mu}(c'c^2+\mu^2)^{\frac{1}{2}}.$$

$$\therefore\ a=-\frac{\mu}{c'}=\frac{\mu\rho}{2\mu-v_1^2\rho}, \qquad (c)$$

and

$$e^2=\frac{c'c^2}{\mu^2}+1=1+\frac{\rho^2 v_1^2 \sin.^2\psi\left(v_1^2-\frac{2\mu}{\rho}\right)}{\mu^2}. \qquad (d)$$

Cor. From (*c*) it appears that the semi-major axis a depends only on the distance ρ and the velocity of projection v_1, and is independent of the angle of projection ψ. Hence, in whatever direction the body is projected, the major axis of the orbit will be the same.

The orbit will be an ellipse, hyperbola, or parabola, according as the value of e is less than, greater than, or equal to unity, or (*d*) according as

$$v_1^2-\frac{2\mu}{\rho},$$

is negative, positive, or equal to zero.

Or, since (99), $v^2=\frac{2\mu}{x}\left(1-\frac{x}{a}\right)$, if $a=\infty$, $v^2=\frac{2\mu}{x}$. Therefore when a body is attracted from an infinite distance to a center of force which varies inversely as the square of the distance the square of the velocity at any distance ρ is

$$v_1^2=\frac{2\mu}{\rho} \text{ and } v_1^2-\frac{2\mu}{\rho}=0.$$

Hence the orbit will be an ellipse, parabola, or hyperbola, according as the velocity of projection is less than, equal to, or greater than, that acquired from an infinite distance.

374. PROP. *To find the time of describing any portion of the orbit.*

Substituting the value of $F=\frac{\mu}{r^2}$ in (122), we get

$$\frac{dr^2}{dt^2}+\frac{c^2}{r^2}-\frac{2\mu}{r}=c.$$

The values of the constants found in terms of the elements of the orbit in the preceding proposition are

$$c'=-\frac{\mu}{a} \text{ and } c^2=\mu a(1-e^2).$$

Substituting these values, and multiplying by $\frac{r^2a}{\mu}$, we obtain

$$\frac{a}{\mu}\cdot\frac{r^2dr^2}{dt^2}+(a-r)^2-a^2e^2=0.$$

Hence $$dt=\left(\frac{a}{\mu}\right)^{\frac{1}{2}}\cdot\frac{rdr}{\sqrt{a^2e^2-(a-r)^2}}.$$

In order to integrate this expression, let $(a-r)=aez$, then

$$dt=-a\left(\frac{a}{\mu}\right)^{\frac{1}{2}}\cdot\frac{(1-ez)dz}{\sqrt{1-z^2}}$$

$$=\left(\frac{a^3}{\mu}\right)^{\frac{1}{2}}\cdot\left\{\frac{-dz}{\sqrt{1-z^2}}+e\frac{zdz}{\sqrt{1-z^2}}\right\}.$$

Integrating, $$t=\left(\frac{a^3}{\mu}\right)^{\frac{1}{2}}\{\cos.^{-1}z-e(1-z^2)^{\frac{1}{2}}\}+k.$$

When $z=1$, or $r=a-ae$, and the body is at the nearer apsis, then $t_1=k=$ the time at the perihelion.

Hence the time through any portion of the arc from the perihelion is

$$t-t_1=\left(\frac{a^3}{\mu}\right)^{\frac{1}{2}}\{\cos.^{-1}z-e(1-z^2)^{\frac{1}{2}}\}. \quad (133)$$

If $z=-1, r=a+ae$, and the body is at the further apsis, in which case $t-t_1$ is the time of half a revolution, and

$$t-t_1=\pi\left(\frac{a^3}{\mu}\right)^{\frac{1}{2}}.$$

Also, if T be the periodic time,

$$T^2=\frac{4\pi^2a^3}{\mu}\propto a^3,$$

or the square of the periodic time varies as the cube of the mean distance.

375. Schol. The quantity μ which enters into the results of the preceding investigations is the value of the accelerating force at the unit of distance from the center of force. Now the attractive forces of the sun and planets vary directly as their masses, and if M be the number of units of mass of the sun, and m the same of any planet, and if we assume for the unit of force the attraction of a unit of mass at a unit of distance, M will express the attractive force of the sun at the unit of distance, and m that of any planet; and the whole force by which they will tend to approach each other, or the whole force which the sun, regarded as fixed, exerts on the planet at the unit of distance is $M+m=\mu$; and for any other distance r

$$\frac{M+m}{r^2}=\frac{\mu}{r^2}.$$

The intensity of the solar and planetary attractions may be expressed in terms of terrestrial gravity. For this, let r_1 be the radius of the earth, and m_1 its mass, and since the forces are directly as the masses, and inversely as the squares of the distances, we have

$$\frac{m_1}{r_1^2}:\frac{M}{r^2}=g:\frac{Mr_1^2}{m_1r^2}g,$$

the fourth term being the attractive force of the sun at any distance r. Similarly, the attractive force of the planet is $\frac{mr_1^2}{m_1r^2}g$; hence their joint influence is $\frac{M+m}{r^2}\cdot\frac{r_1^2}{m_1}g=\frac{\mu}{r^2}\cdot\frac{r_1^2}{m_1}g$, in which

$\frac{r_1^2}{m_1}.g$ is that which was assumed above for the unit of force, or the attraction of the unit of mass at the unit of distance.

376. EXAMPLES.

Ex. 1. Required the velocity and time when a material particle is attracted to a fixed point by a force varying inversely as the square root of the distance.

$$Ans.\ v=2\mu^{\frac{1}{2}}(a^{\frac{1}{2}}-x^{\frac{1}{2}})^{\frac{1}{2}}$$

$$t=\frac{2}{3\mu^{\frac{1}{2}}}(x^{\frac{1}{2}}+2a^{\frac{1}{2}})\ (a^{\frac{1}{2}}-x^{\frac{1}{2}})^{\frac{1}{2}}.$$

Ex. 2. Find the velocity and periodic time of a body revolving in a circle at a distance of n radii from the earth's center.

$$Ans.\ v=\left(\frac{gr}{n}\right)^{\frac{1}{2}}$$

$$\mathrm{P}=2\pi\left(\frac{n^3r}{g}\right)^{\frac{1}{2}}.$$

Ex. 3. Find the least velocity with which a body must be projected from the moon in the direction of a line joining the center of the earth and moon, so that it may reach the earth.

Ex. 4. Given the velocity, distance, and direction of projection, when the force is attractive and varies as the distance, to find the orbit.

Ex. 5. A body is projected in a direction which makes an angle of 60° with the distance, with a velocity which is to the velocity from infinity as $1:\sqrt{3}$: the force varying inversely as the square of the distance. Find the major-axis, the position of the apsis, the eccentricity, and the periodic time.

Ex. 6. A body is projected at a given distance from the center of force with a given velocity, and in a direction perpendicular to the distance, when the force is repulsive and varies inversely as the cube of the distance. Find the path of the body.

HYDROSTATICS.

CHAPTER I.

377. MATTER exists in the various states of solid, fluid, and aeriform. In solid bodies the homogeneous integrant particles cohere firmly, and do not admit of interchange of position. The homogeneous integrant constituents of fluids possess less cohesion, and change their position among each other on the application of a moderate force.

Fluids differ from each other in the degree of cohesion of their constituent parts and the facility with which they will yield to the action of an impressed force.

378. *A perfect fluid* is one in which the cohesion of the constituent particles is so feeble that the least force is capable of effecting a separation and causing motion among each other in all directions. Perfect fluids are the only ones which will be made the subject of investigation.

Fluids are divided into two classes, *incompressible* or *liquid*, and *compressible* or *aeriform*.

379. *Incompressible fluids* are those which retain the same volume under a variable pressure. They may be made to take an infinite variety in form, which form they will retain when acted upon by no external pressure nor accelerating force. Of this class water may be regarded as the type.

380. *Compressible fluids*, of which atmospheric air is the type, change at once in form or volume with every variation of pressure which they sustain, and return again to the same form and volume when the same circumstances of pressure and temperature are restored.

PRINCIPLE OF EQUALITY OF PRESSURE.

381. PROP. *A force or pressure applied to a given portion of the surface of a fluid is transmitted, undiminished in intensity, to every other equal portion of the surface.*

A force impressed on a solid is effective only in the direction of its action, and is sustained by an equal force impressed in an opposite direction. Applied to a fluid the force is effective in every direction, and can be sustained only by forces applied to every point in the surface of the fluid.

Let the annexed figure represent a section of a vessel filled with a fluid without weight. In this case, supposing the pressure of the atmosphere removed also, the surface of the fluid would be subject to no pressure, and would retain its form if the sides of the vessel were removed.

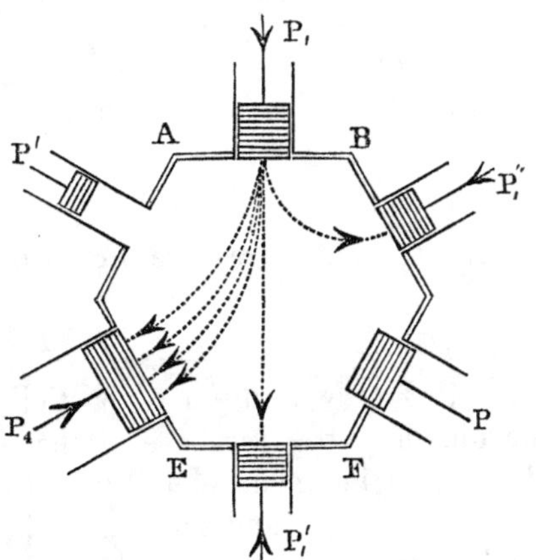

Let an aperture of a given size, as one square inch, be made in either side, AB, and a piston be accurately fitted to it, and supposed to move without friction. Now, since the particles are without friction, a force P_1 of one pound, applied to this piston, will act upon the stratum of particles in contact with its base, and this stratum upon the next, and so on. The force will therefore be transmitted to the opposite side, EF, and will cause motion in the equal piston P'_1, unless counterpoised by an opposite force of one pound. If this counterpoise be applied, then, since the particles move freely among each other without resistance from friction or cohesion, the particles in the direction of the applied forces will act on those lying without, and will communicate motion to them, unless resisted. A motion will therefore be communicated to any equal piston P''_1, unless sustained by a pressure of one pound. Hence both pistons P'_1 and P''_1 will sustain the same pressure by the action of P_1.

Further, if the pistons P'_1 and P''_1 were contiguous, or one of them twice the size of P_1, then it is obvious that a force of two pounds would be necessary to preserve the equilibrium; or, if the base of the piston P_4 be four times that of P_1, then a force of four pounds would be necessary to keep P_4 at rest; and if the base of one piston be n times that of the other, the pressure on the former will be n times that on the latter, or a force of one pound will produce a pressure of n pounds, where n may be the ratio less one, of the whole surface of the vessel to the base of the piston.

COR. 1. The pressures P_1 and P_n on any two portions A_1 and A_n of the surface will be proportional to their areas, or

$$\frac{P_n}{P_1}=\frac{A_n}{A_1}=n. \qquad (134)$$

Also, the normal pressure p on a unit of surface will be

$$\frac{P_1}{A_1}=\frac{P_n}{A_n}=p. \qquad (135)$$

COR. 2. Every stratum of particles in the interior, of the same dimensions as the base of the piston, wherever situated and however inclined, is subject to a pressure equal to that applied to the piston; and since the fluid is without weight, every particle presses and is pressed equally in every direction.

COR. 3. If all the pistons except P and P′ be firmly fixed, and P be forced in, since the fluid is incompressible, whatever fluid is displaced by P will be forced into the tube P′. If h and h' be the spaces through which the pistons move, a and a' the areas of their bases, then

$$ha=h'a'.$$

But P and P′ being the equilibrating pressures on these areas,

$$\frac{a}{a'}=\frac{P}{P'};$$

hence $$hP=h'P', \text{ or } hP-h'P'=0,$$

or the principle of virtual velocities applies to fluids in equilibrium.

SURFACES OF EQUILIBRIUM.

382. Prop. *The free surface of a fluid subjected to the action of an accelerating force is, when in equilibrium, perpendicular to the direction of the force.*

A liquid void of gravity will take and retain any form impressed upon it; but when subject to the action of an accelerating force, a containing vessel will be necessary in order to preserve a coherent mass.

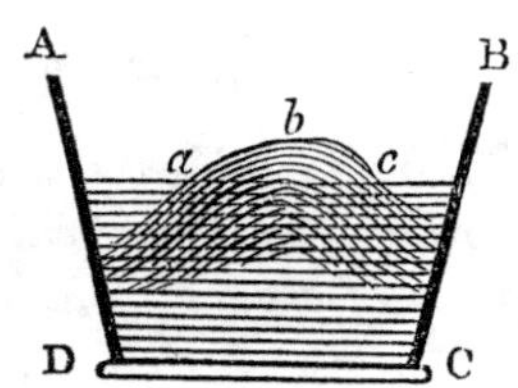

Let ABCD be the section of a vessel containing a fluid subject to the action of gravity, the base DC being horizontal. If any portion of the free surface, as *abc*, have a direction not perpendicular to the direction of gravity, this force may be resolved into two components, one of which is parallel to this surface; and since the particles are free to move in that direction as down an inclined plane, they will yield to this component. But whenever every portion of the surface is perpendicular to the direction of gravity this force will have no component in the direction of the surface, and every portion of it will be urged vertically downward with equal intensity.

Cor. Hence the surfaces of fluids at rest, and acted on only by gravity, are horizontal. But since the directions of gravity, acting on particles remote from each other, are convergent to the center of the earth nearly, the surfaces of large masses of fluid are not plane, but curved, and conform to the general figure of the earth.

383. Prop. *If a vessel containing a fluid be moved horizontally with a constant accelerating force, the surface will take the position of an inclined plane.*

Let the vessel ABCD containing a fluid be moved horizontally with a uniformly accelerating force P. Then any element, k, of the surface, whose weight is w, and mass $\frac{w}{g}$ (22), will be urged by its weight w in the direction kg, and by its

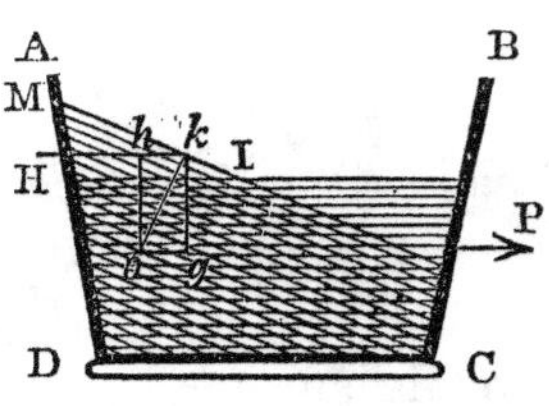

inertia $P\frac{w}{g}$ in the direction kh, and therefore by their resultant in the direction ko. As the same is true of every element of the surface, the whole surface (*Art.* 382) will be perpendicular to ko.

Let the angle of inclination to the horizon HIM=I. Then since HIM$=okg$,

$$\tan. okg=\tan. I=\frac{P\frac{w}{g}}{w}=\frac{P}{g}.$$

384. PROP. *If a vessel containing a fluid be made to revolve uniformly about a vertical axis, the surface of the fluid will take the form of a paraboloid of revolution.*

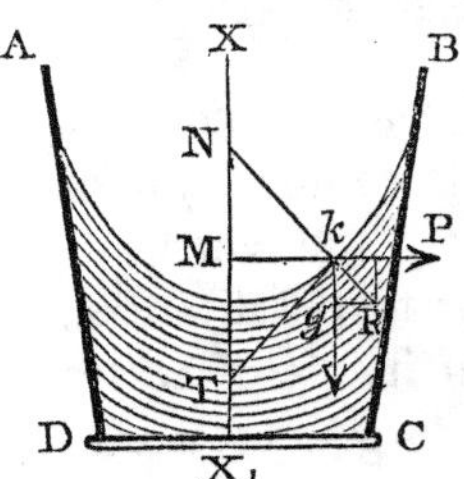

Let the vessel ABCD revolve uniformly about the vertical axis XX_1. Any element k of the surface will be urged horizontally by a centrifugal force directed from the axis, and vertically downward by gravity. Let ω be the angular velocity of the fluid, w the weight of the element k, and y its distance kM from the axis. Then (82) we have, for the centrifugal force P of this element,

$$P=\omega^2.y\frac{w}{g}.$$

Being urged vertically downward by its weight w, the surface of the element will be perpendicular to the resultant kR=R of these two forces. If this resultant be produced to meet the axis in N, we have

$$\angle Rkg=\angle kNM,$$

and

$$\frac{Rg}{kg}=\frac{kM}{MN},$$

or

$$\frac{\omega^2 y\frac{w}{g}}{w}=\frac{y}{MN}.$$

$\therefore$ the subnormal $MN=\frac{g}{\omega^2}=$ a constant, a property of the parabola. Hence the surface is a paraboloid of revolution.

NORMAL PRESSURES ON IMMERSED SURFACES.

385. Prop. *The pressure on the horizontal base of a vessel containing an incompressible fluid is proportional to its depth below the surface of the fluid, and independent of the form of the vessel.*

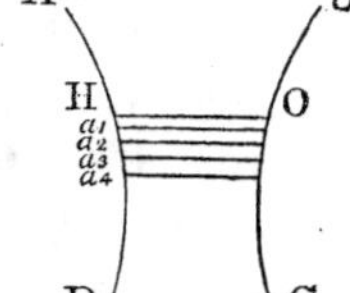

Let HO be the free surface of the fluid in the vessel ABCD, whose base DC is horizontal, and suppose the fluid divided into horizontal strata of small but equal thickness.

Let a_1, a_2, a_3, &c., denote the successive strata,

A_0, A_1, A_2, the units of surface in each,

P_0, P_1, P_2, the whole pressure on each,

p_0, p_1, p_2, the pressure on a unit of each,

a the common thickness of the strata, and ρ the density of the fluid. Since the thickness of each stratum is supposed indefinitely small, the upper and lower surfaces of each may be regarded as equal.

Now the weight of a_1 is $w_1=g\rho aA_0$ (24),

" " " " a_2 is $w_2=g\rho aA_1$,

. . . .

" " " " a_n is $w_n=g\rho aA_{n-1}$.

The pressure on a_1 is P_0,

on a unit of a_1 is $\frac{P_0}{A_1}=p_0$ (135),

on a_2 is $P_1=P_0+w_1=P_0+g\rho aA_0$,

on a unit of a_2 is $\frac{P_1}{A_1}=\frac{P_0}{A_1}+\frac{g\rho aA_0}{A_1}$ or $p_1=p_0+g\rho a$,

on a_3 is $P_2=P_1+w_2=P_1+g\rho aA_1$,

on a unit of a_3 is $\frac{P_2}{A_2}=\frac{P_1}{A_2}+\frac{g\rho aA_1}{A_2}$ or $p_2=p_1+g\rho.a=p_0+g\rho.2a$.

. . . .

on a_{n+1} is $P_n=P_{n-1}+w_n=P_{n-1}+g\rho aA_{n-1}$,

on a unit of a_{n+1} is $\frac{P_n}{A_n}=\frac{P_{n-1}}{A_n}+\frac{g\rho a A_{n-1}}{A_n}$ or $p_n=p_0+g\rho.na$.

But $na=h$ is the depth of a_{n+1} below the surface of the fluid, and if the upper surface of a_{n+1} represent the surface of the base, A the number of units in the base, then the pressure p on a unit of the base will be

$$p=p_0+g\rho h,$$

and the whole pressure on the base

$$P=Ap=Ap_0+g\rho hA=P_0+g\rho hA \propto h. \quad (136)$$

Cor. 1. Since hA is equal to a prism whose base is A and height h, and $g\rho hA$ is its weight, if we disregard P_0, which may represent the pressure of the atmosphere on the surface of the fluid, we have, for the pressure of the fluid on the base of the vessel, *the weight of a column of the fluid whose base is that of the vessel, and height the height of the surface of the fluid above the base.* It is obviously immaterial whether the surface pressed is that of the base of the vessel or a horizontal surface of an immersed solid.

Cor. 2. Since a cubic foot of water weighs 1000 oz.=62.5 lbs., we have, for the pressure on the base of any vessel containing water,

$$P=62.5\ hA \text{ lbs.}, \quad (137)$$

where h is the height in feet of the surface of the water above the base, and A the number of square feet in the base.

Cor. 3. The pressure on every portion of a horizontal stratum of the fluid will be the same, and since this pressure is transmitted equally in every direction, the pressure on every element of the sides of the vessel having the same depth will be equal to that on the surface of the stratum. If the sides of the vessel are inclined, this pressure will be the normal pressure on the sides at that depth. Let the annexed figures, A, B, C, D, represent vessels of different forms and capacities with equal bases. Then the pressures on the base of each will be equal when filled to a common height, and the points a, b, c, e, g, h, k, l, m, having a common depth, will be equally pressed normally to the surfaces. In B the horizontal surface df of the vessel will experience the same pressure vertically

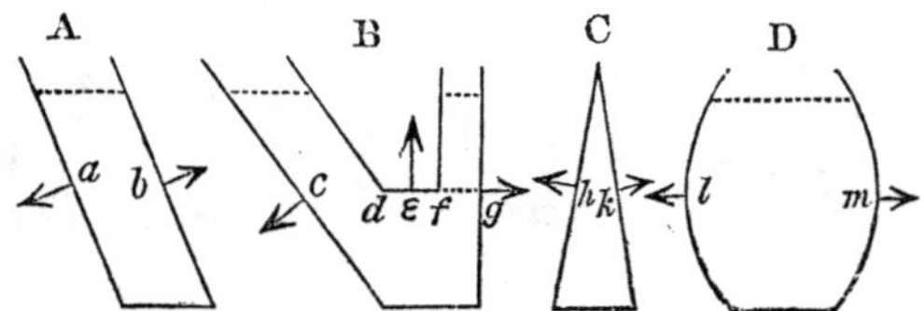

upward as that to which the stratum fg of the fluid is subject. Moreover, the stratum fg being in equilibrium, must be pressed equally upward and downward. The downward pressure is equal to the area of the section fg into the depth h of the section below the surface of the fluid in the vertical branch. But the upward pressure is due to the pressure of the fluid in the inclined branch; and since the area of the section fg is common to the measures of both pressures, the vertical heights of the fluid in both branches are necessarily equal.

386. Prop. *The normal pressure on any plane surface inclined to the horizon is proportional to the depth of its center of gravity below the surface of the fluid.*

Let k be any indefinitely small element of the immersed surface, x its depth, and p the pressure it sustains. By the preceding proposition,

$$p = g\rho xk;$$

and since the pressure is the same in every direction, p will be the pressure normal to this element, whatever be its position or inclination. The expression for the normal pressure on every other element will be of the same form. Hence the whole normal pressure P will be

$$P = \Sigma g\rho xk = g\rho\Sigma xk.$$

Let $\bar{x}$ be the depth of the center of gravity of the immersed surface below the surface of the fluid, and A its area. Then (29),

$$A\bar{x} = \Sigma xk;$$

$$\therefore P = g\rho A\bar{x} \propto \bar{x}. \qquad (138)$$

Cor. Since $g\rho A\bar{x}$ is the weight of a prism of the fluid whose base is A and altitude $\bar{x}$, the normal pressure on any immersed surface inclined to the horizon, or on any side of the containing vessel, is equal to the weight of a fluid prism whose base is the surface pressed, and height the depth of its center of gravity below the fluid surface, or, if the fluid be water,

$$P=62.5\ A\bar{x}\ \text{lbs.} \qquad (139)$$

387. Prop. *To find the pressure on any immersed surface or side of a vessel in a given direction.*

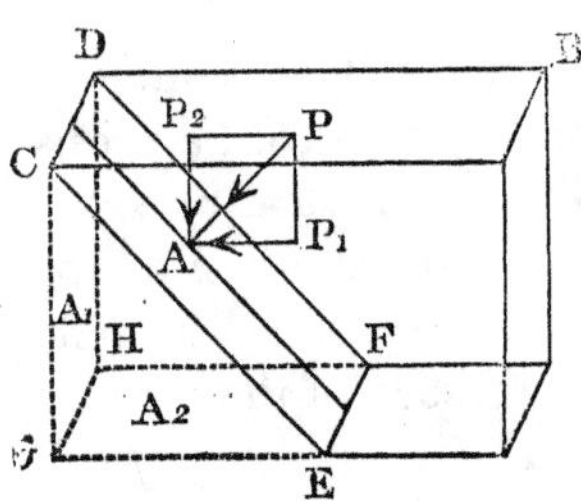

Let P be the normal pressure on the inclined side, CDFE=A of the vessel CBE, and let it be required to determine the horizontal and vertical pressures on A. Let a be the angle which the side A makes with a vertical plane, and β its inclination to the horizon. Then, since $PAP_1=a$ and $PAP_2=\beta$, resolving P horizontally and vertically, we have

$$P_1=P\cos. a=g\rho\bar{x}A\cos. a,$$

and $$P_2=P\cos. \beta=g\rho\bar{x}A\cos. \beta.$$

But $CDHG=A_1$ is the projection of the side A on a vertical plane, and $EFHG=A_2$ is the projection of the same side on a horizontal plane.

Therefore, $A_1=A\cos. a$ and $A_2=A\cos. \beta$,

and $P_1=g\rho A_1\bar{x}$ and $P_2=g\rho A_2\bar{x}$.

Hence the horizontal pressure on A is equal to the weight of a prismatic volume of the fluid whose base is the vertical projection of A and height the depth of its center of gravity; and vertical pressure, a volume whose base is the horizontal projection, and height the depth of the center of gravity of A. And generally the pressure on any immersed surface in a given direction is equal to the weight of a prismatic column of the fluid whose base is the projection of the surface on a plane perpendicular to the given direction, and height the depth of the center of gravity of the surface below the surface of the fluid.

388. Prop. *To find the resultant pressure of a fluid on the interior surface of the containing vessel.*

Let ABCD be a section of a vessel of any form filled with a fluid. When the sides of the vessel are curved, the normal pressures on the elements of the surface will not be parallel. But we may resolve the normal pressure on each element into vertical and horizontal components, and each horizontal component into two others parallel to two rectangular axes. We shall then have to find the resultants of three sets of parallel forces, and, finally, the resultant of these three resultants. We may, however, determine the resultant more simply, as follows:

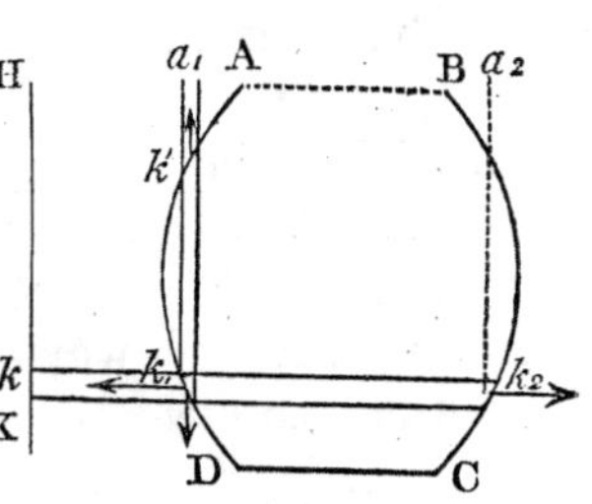

1°. The resultant of the horizontal pressures.

Let HK be the intersection of a vertical plane by the plane ABCD produced, and let k be the vertical projection of any element k_1 on this plane. Now the normal pressure on the element k_1 of depth $a_1k_1=h$ is hk_1, and the horizontal pressure (*Art.* 387) is hk. The horizontal pressure on each element being of the same form, the whole horizontal pressure perpendicular to HK of the surface convex toward HK will be $\Sigma.hk$. But to every element k_1 of the surface convex toward HK there is a corresponding element k_2 of the surface concave toward HK, the projection k of which is the same. Hence the horizontal pressure on k_2 is $-hk$, and the pressure on the whole surface concave toward HK is $-\Sigma hk$. The resultant, therefore, of all the horizontal pressures perpendicular to KH is $\Sigma.hk-\Sigma.hk=0$. Since the same is true on whatever side of the vessel the plane HK be drawn, the resultant of all the horizontal pressures will be zero.

2°. The resultant of the vertical pressures.

The pressure on the element k_1 vertically downward is the weight of a filament of the fluid whose base is k_1 and height a_1k_1. The pressure on k' vertically upward is equal to the weight of a filament of the fluid whose base is k' and height a_1k'. The resultant of the pressures on the corresponding elements k' and k_1 is therefore equal to the weight of the filament of fluid $k'k_1$. In the same manner, the resultant of the vertical pressures on any two corresponding elements will be the weight of the filament of fluid whose bases are these elements. The resultant of all the vertical pressures will then obviously be the sum of the weights of these filaments, or the weight of the fluid.

Hence the resultant of all the partial pressures on the interior surface of the vessel is equal to the weight of the fluid, directed vertically downward, and will pass through the center of all the parallel vertical pressures (*Art.* 44), or the center of gravity of the fluid (*Art.* 94).

Cor. Since the horizontal pressures balance, there will be no resultant pressure to cause motion in the vessel horizontally. If, however, an aperture be made on one side of the vessel, the pressure on this portion of the vessel will be removed, and the pressure on the corresponding opposite portion will tend to produce motion in that direction, and if the vessel be free to move, motion will actually ensue.

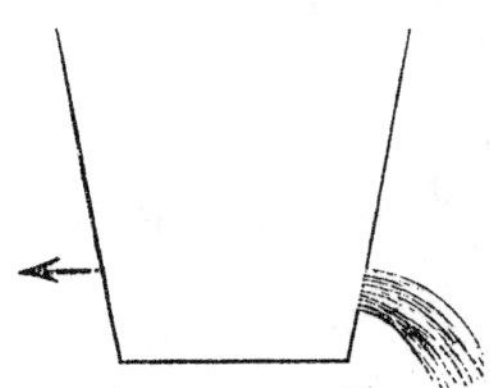

389. Prop. *To find the resultant pressure on any solid immersed in a fluid, and its point of application.*

1°. The horizontal pressure on any element of the immersed solid is the same as that on the vertical projection of the element (*Art.* 387), and the entire pressure on each of the opposite surfaces of the solid will be equal to that on their vertical projections. But the projections of any two opposite surfaces on the same vertical plane are equal. Hence the horizontal pressures on any two opposite surfaces are equal, and the resultant of the horizontal components of the pressures in every direction is zero.

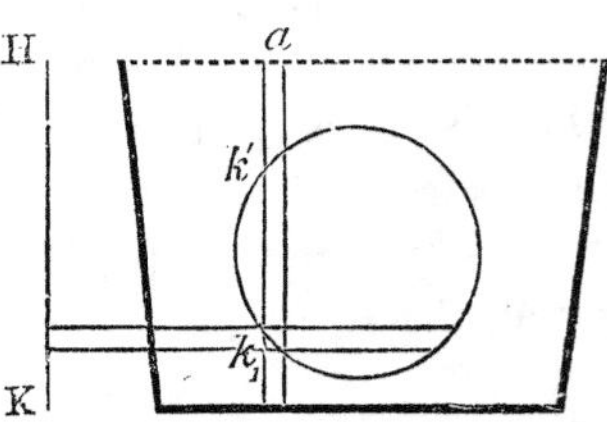

2°. The vertical downward pressure p' on any element k of the upper surface is equal to the weight of a filament of the fluid ak' whose base is k' and altitude $ak'=h'$, or if k be the horizontal transverse section of the filament through the cen ter of gravity of k', $p'=g\rho h'k$. But the sum of all the filaments resting on the upper surface of the solid is equal to the volume V′ of the fluid vertically above the solid, and the whole downward pressure P′ is the weight of this volume, or

$$P'=\Sigma.p'=g\rho\Sigma.h'k=g\rho V'.$$

In the same manner, the vertical upward pressure p_1 on the corresponding element k_1 of the lower surface is equal to the weight of a filament ak_1 of fluid whose base is k_1 and height $ak_1=h_1$, or $p_1=g\rho h_1k$. But the sum of all these filaments is the volume V_1 of the solid and fluid above the upper surface, and the whole upward pressure P_1 is the weight of this volume of fluid, or

$$P_1=\Sigma p_1=g\rho\Sigma.h_1k=g\rho V_1.$$

Now $h_1-h'=h$ is the length of the filament $k'k_1$ of the solid $p_1-p'=p$ its weight, and $V_1-V'=V$ is the volume of the solid.

Therefore, the difference P of the upward and downward pressure is

$$P=P_1-P'=\Sigma.p_1-\Sigma.p'=\Sigma.p,$$

and $$P=g\rho\Sigma.h_1k-g\rho\Sigma.h'k=g\rho\Sigma hk=g\rho V;$$

$$\therefore\ P=g\rho V=\Sigma p, \qquad (140)$$

or the resultant pressure is vertically upward, and equal to the weight of a mass of the fluid of the same volume as the solid.

3°. To find the point of application of this resultant. Let the distance of the point of application of $p=hk$ from the vertical plane HK be x, and the distance of the point of application of the resultant P be $\bar{x}$. Then, since the moment of the resultant equals the sum of the moments of the components,

$$P\bar{x}=\Sigma.px,$$

and $$\bar{x}=\frac{\Sigma.px}{P}=\frac{\Sigma.px}{\Sigma.p}. \qquad (141)$$

But (29) the point thus determined is the center of gravity of the displaced fluid.

Hence the resultant of all the pressures on the immersed solid is equal to the weight of the displaced fluid, acts vertically upward, and its point of application is the center of gravity of the displaced fluid.

390. Prop. *To find the conditions of equilibrium of an immersed solid.*

The conditions of equilibrium involve a consideration of the weight of the body. Let V be its volume, and σ its density. Its weight will be $g\sigma V$. The body will be urged downward by a force equal to $g\sigma V$ applied at its center of gravity, and by *Art.* 389 it is urged upward by a force equal to $g\rho V$, since the volumes of the solid and displaced fluid are equal.

In order to equilibrium, then, 1°. These forces must be equal; and, 2°. Their lines of direction coincident.

The first condition gives

$$g\sigma V = g\rho V,$$

or

$$\sigma = \rho;$$

that is, their densities must be the same.

Cor. 1. If σ be not equal to ρ, the body will ascend or descend by a force equal to $g(\rho - \sigma)V$, according as $\rho - \sigma$ is positive or negative. If $\rho > \sigma$, the body will rise to the surface, and be but partially submerged. Let V_1 be the displaced fluid, or the part of the solid immersed when the equilibrium is restored. Then

$$g\rho V_1 = g\sigma V, \qquad (142)$$

and

$$V : V_1 = \rho : \sigma;$$

or the whole volume of the solid is to the part immersed as the density of the fluid is to the density of the solid.

Cor. 2. If the centers of gravity of the solid and displaced fluid be not in the same vertical line, the body will be acted upon by two parallel forces in opposite directions, and will cause the body to turn round. The point of application of the resultant of these forces may be found by *Art.* 29.

391. Def. The section of a floating body made by a plane coincident with the surface of the fluid is called the *plane of flotation.* The line joining the centers of gravity of the solid and of the displaced fluid is called the *axis of flotation.*

DEPTH OF FLOTATION.

392. Prop. *To find the depth of flotation when the volume and density of the body are known.*

Let V be the volume of the body, σ its density, V_1 the volume of the displaced fluid, and ρ the density of the fluid. Then (142)

$$g\sigma V = g\rho V_1,$$

and

$$V_1 = \frac{\sigma}{\rho} V.$$

Now whenever V_1 can be determined in terms of the depth of flotation x, this expression will suffice to determine x.

If the solid be a right cylinder whose axis a is vertical, and the radius of whose base is r, we have

$$V_1 = \pi r^2 x,$$

and

$$V = \pi r^2 a;$$

$$\therefore\ x = \frac{\sigma}{\rho} a.$$

If the body be a right cone with the axis vertical and vertex downward, let r be the radius of the base, and a the altitude. Then the radius of the plane of flotation is $\frac{r}{a}x$. Hence

$$V_1 = \tfrac{1}{3}\pi \frac{r^2}{a^2} x^3,$$

and

$$V = \tfrac{1}{3}\pi r^2 a;$$

$$\therefore\ x = a\sqrt[3]{\frac{\sigma}{\rho}}.$$

If the vertex of the cone be upward,

$$x = a\sqrt[3]{1 - \frac{\sigma}{\rho}}.$$

393. Prop. *To find the positions of equilibrium of a right triangular prism when the axis is horizontal.*

Whenever a body can be conceived to be generated by the motion of a plane surface perpendicular to itself, and the body floats with its axis horizontal, it is evident that, in whatever position it be turned, its center of gravity and that of the part immersed will lie in the same vertical section; and, further, that the center of gravity of the body will be at the center of gravity of the section, and the center of gravity of the immersed part at the center of gravity of its section. Also, the ratio of the part immersed to the whole mass is the same as the ratio of their sections. We can therefore limit the investigation to the positions of equilibrium of one of the generating sections.

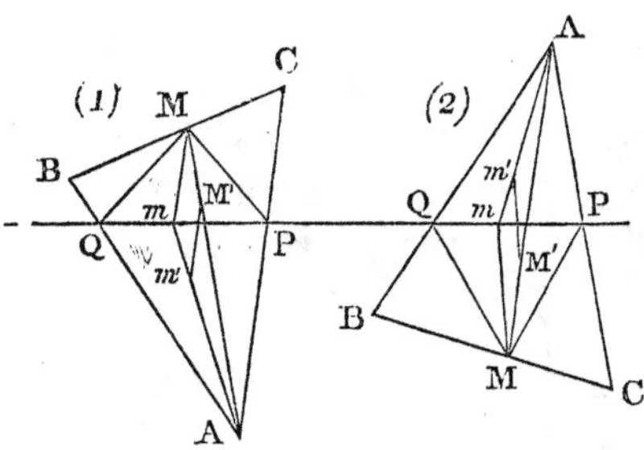

Let ABC, *Fig.* 1, be a generating section when only one angle is immersed, and *Fig.* 2 when two angles are immersed. Let $AB=c$, $AC=b$, $BC=a$, $AQ=x$, $AP=y$, and let σ be the ratio of the density of the solid to that of the fluid.

Now from (142) we have, in *Fig.* 1,

$$\sigma=\frac{AQP}{ABC}=\frac{\frac{1}{2}xy\sin. A}{\frac{1}{2}bc\sin. A}=\frac{xy}{bc}.$$

$$\therefore\ xy=\sigma.bc. \qquad (a)$$

In *Fig.* 2,

$$\sigma=\frac{BQPC}{ABC}=\frac{ABC-AQP}{ABC}=\frac{bc-xy}{bc}.$$

$$\therefore\ xy=bc(1-\sigma). \qquad (b)$$

Since (*b*) is derived directly from (*a*), by changing σ into $1-\sigma$, we may examine the case of *Fig.* 1, and deduce that of *Fig.* 2 from the result, by changing σ into $1-\sigma$.

Bisect BC in M, and QP in m. Join AM and Am. Take $MM'=\frac{1}{3}MA$ and $mm'=\frac{1}{3}mA$. M$'$ and m' are the centers of gravity of ABC and AQP. Join Mm and M$'m'$. It is obvious that the center of gravity of BQPC is in the line M$'m'$ produced. Now the second condition of equilibrium requires that M$'m'$ should be vertical, and since Mm is parallel to M$'m'$, Mm is perpendicular to QP and MQ=MP. Let $AM=h$, $MAQ=\theta$, and $MAP=\phi$. Then

$$MQ^2=x^2+h^2-2hx\cos.\theta,$$

and

$$MP^2=y^2+h^2-2hy\cos.\phi.$$

Hence

$$x^2-y^2-2hx\cos.\theta+2hy\cos.\phi=0. \qquad (c)$$

Substituting in (*c*) the value of y from (*a*), and reducing,

$$x^4-2h\cos.\theta.x^3+2\sigma bch\cos.\phi.x-\sigma^2b^2c^2=0, \qquad (d)$$

and changing σ into $1-\sigma$, we have for *Fig.* 2,

$$x^4-2h\cos.\theta.x^2+2(1-\sigma)bch\cos.\phi.x-(1-\sigma)^2b^2c^2=0. \quad (e)$$

The values of x deduced from (d) and (e), and the correspond ing values of y from (a) and (b), will give the positions of equi librium required.

Now, since the degree of the equation is even and the abso- ute term is negative, there are at least two possible roots, one positive and the other negative. The other two roots may be real or imaginary. If real, Descartes' rule of signs indicates that three will be positive and one negative. Hence, since the values of x and y, which are applicable to the question, are necessarily positive, equations (d) and (e) indicate no more than three positions of equilibrium. The values of x must also be less than b, and when subtituted in (a) and (b) give for y values less than c.

Let us take the case of an equilateral triangle.

Then $\theta=\phi=30$, $a=b=c$, and $h=b\cos.\theta=c\cos.\phi$.

These values, substituted in (d), give

$$x^4-\tfrac{3}{2}ax^3+\tfrac{3}{2}\sigma a^3x-\sigma^2a^4=0,$$

or

$$(x^2-\sigma a^2)\,(x^2-\tfrac{3}{2}ax+\sigma a^2)=0. \quad (f)$$

This equation is satisfied by putting

$$x^2-\sigma a^2=0;$$

whence

$$x=a\sqrt{\sigma},$$

the negative value of x being inapplicable.

Since $\sigma<1$, we have $x<a$, and the value of x indicates an actual position of equilibrium. But from (a) we have

$$y=\frac{\sigma.a^2}{x}=\frac{\sigma.a^2}{a\sqrt{\sigma}}=a\sqrt{\sigma}=x,$$

or one side of the triangle is parallel to the surface of the fluid.

Equation (f) is also satisfied by making

$$x^2-\tfrac{3}{2}ax+\sigma a^2=0;$$

whence

$$x=\frac{a}{4}\{3\pm\sqrt{9-16\sigma}\}. \quad (g)$$

Now, in order that these values of x may be real, we must have

$$16\sigma<9 \text{ or } 16\sigma=9.$$

Hence $$\sigma<\frac{9}{16} \text{ or } \sigma=\frac{9}{16},$$

or σ can not be greater than $\frac{9}{16}$.

But if $\sqrt{9-16\sigma}>1$, then $x>a$, which is inconsistent with the supposition that only one angle is immersed. The greatest value, therefore, which $\sqrt{9-16\sigma}$ can have is unity, and in this case the second angle will lie in the surface of the fluid.

Putting $$\sqrt{9-16\sigma}=1,$$

we get $$\sigma=\frac{1}{2},$$

for the least value σ can have when one angle only is immersed. The limits of σ for this case are, therefore,

$$\frac{1}{2} \text{ and } \frac{9}{16}.$$

If in (g) we change σ into $1-\sigma$, we get

$$x=\frac{a}{4}\{3\pm\sqrt{9-16(1-\sigma)}\}$$

$$=\frac{a}{4}\{3\pm\sqrt{16\sigma-7}\}. \qquad (h)$$

From which it may be shown that the limits of the value of σ for this case are

$$\frac{7}{16} \text{ and } \frac{1}{2}.$$

Since equations (d) and (e) may have three real positive roots each, there may be three positions of equilibrium for every single angle immersed, and three for every two angles immersed, and therefore eighteen in the whole. But in the particular form considered above these are not all possible, except $\sigma=\frac{1}{2}$, in which case $x=a$ or $y=a$; that is, either the angle B or C lies in the surface of the fluid. This would render six of the positions pertaining to *Fig.* 1 the same as six of those pertaining to *Fig.* 2, making but twelve really different ones. If $\sigma>\frac{1}{2}$ and $\sigma<\frac{9}{16}$, there will be nine positions with one angle immersed and three with two immersed; if $\sigma<\frac{1}{2}$ and $\sigma>\frac{7}{16}$, nine of the latter and three of the former; if $\sigma<\frac{7}{16}$ or $\sigma>\frac{9}{16}$, there will be only three of each.

CENTER OF PRESSURE.

394. Def. The *center of pressure* in any immersed surface is the point of application of the resultant of all the pressures upon it. It is therefore that point in an immersed surface or side of a vessel containing a fluid, to which, if a force equal and opposite to the resultant of all the pressures upon it be applied, this force would keep the surface at rest.

395. Prop. *To find the center of pressure on any immersed plane surface.*

Let the immersed surface be the inclined side AE of the vessel ABE supposed to be filled with a fluid. Let x and y be the co-ordinates of any element k referred to the rectangular axes AY and AX, and $hk=h$ the vertical distance of the element below the surface of the fluid.

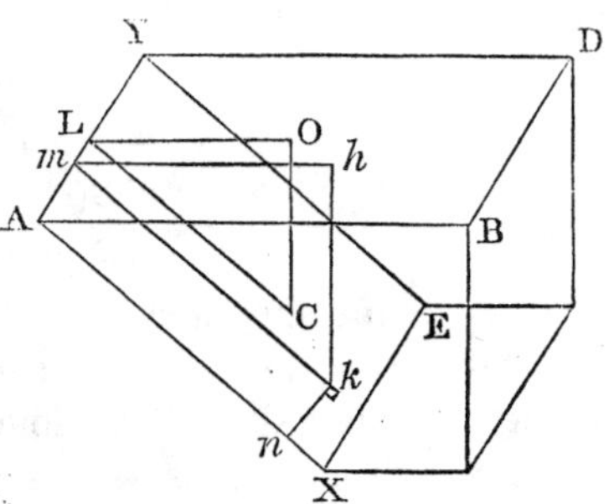

Then the normal pressure p on k will be

$$p=g\rho hk.$$

The pressure on each element will be of the same form, and their sum or resultant R will be

$$\mathrm{R}=\Sigma.p.$$

The moment of each partial pressure, in reference to the plane through AY perpendicular to AX (*Art.* 46), will be px, and the moment of each, in reference to the plane through AX perpendicular to AY, will be py. Hence, if $\bar{x}$ and $\bar{y}$ are the co-ordinates of the point of application C of the resultant R, we have (*Art.* 43)

$$R=\Sigma.p,\ R\bar{x}=\Sigma px,\ R\bar{y}=\Sigma.py.$$

If θ be the inclination of the side of the vessel or immersed surface to the surface of the fluid, we have

$$h=x\ \sin.\ \theta.$$

Hence $$p=g\rho xk\ \sin.\ \theta,$$

and $$R=\Sigma.g\rho xk\ \sin.\ \theta=g\rho\ \sin.\ \theta.\Sigma.xk,$$

$$R\bar{x}=\Sigma.g\rho x^2k\ \sin.\ \theta=g\rho\ \sin.\ \theta.\Sigma.x^2k,$$

$$R\bar{y}=\Sigma.g\rho yxk\ \sin.\ \theta=g\rho\ \sin.\ \theta.\Sigma.xyk.$$

$$\therefore\ \bar{x}=\frac{\Sigma.x^2k}{\Sigma.xk}, \quad (143)$$

and $$\bar{y}=\frac{\Sigma.xyk}{\Sigma.xk}. \quad (144)$$

When the upper boundary is below the surface of the fluid and at a distance a from it, then, since the pressures p are limited to the immersed surface, $h=(a+x)\ \sin.\ \theta$, and we shall have

$$\bar{x}=\frac{\Sigma(ax+x^2)k}{\Sigma(a+x)k}, \quad (145)$$

$$\bar{y}=\frac{\Sigma(a+x)yk}{\Sigma(a+x)k}. \quad (146)$$

Further, if the axis of x is so taken that it will bisect every horizontal line of the immersed surface, the pressures on opposite sides of this axis will obviously be equal, and the center of pressure will lie in this axis or $\bar{y}=0$.

It will be observed that the numerator of (143) is the moment of inertia of the surface, and that the denominator is the statical moment. Hence the denominator is equal to the area of the surface multiplied by the depth of its center of gravity (29). Hence, if A be the area and $\bar{x}_1$ this depth,

$$\bar{x}=\frac{\Sigma.x^2k}{A\bar{x}_1}, \quad (147)$$

$$\bar{y}=\frac{\Sigma.xyk}{A\bar{x}_1}. \quad (148)$$

306. Prop. *To find the center of pressure of a rectangular* *surface vertically immersed, and having one side in the surface of the fluid.*

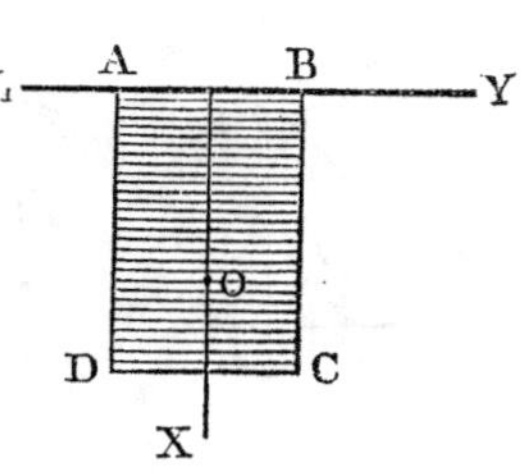

Let YY_1 be the line of intersection of the immersed surface ABCD with the surface of the fluid, and conceive ABCD divided into n rectangles by horizontal lines, n being a very large number. Let $AB=b$ and $AD=h$, and draw OX bisecting the rectangle. The center of pressure is obviously in OX and $\bar{y}=0$. To find $\bar{x}$, we have, since the height of each small rectangle is $\frac{h}{n}$, for the area of each,

$$k=\frac{bh}{n}.$$

The distances of these rectangles from the surface are

$$\frac{h}{n}, \frac{2h}{n}, \frac{3h}{n}, \&c.;$$

their squares, $$\frac{h^2}{n^2}, \frac{2^2h^2}{n^2}, \frac{3^2h^2}{n^2}, \&c.$$

Hence $$\Sigma.x^2k=\frac{bh^3}{n^3}+\frac{bh^3.2^2}{n^3}+\frac{bh^3.3^2}{n^3}+, \&c.,$$

$$=\frac{bh^3}{n^3}(1^2+2^2+3^2+ \ldots\ldots n^2).$$

But when n is very large, the sum of the mth powers of the natural numbers 1, 2, 3, &c., to n is $\frac{n^{m+1}}{m+1}$.

$$\therefore\ \Sigma.x^2k=\frac{bh^3}{n^3}.\frac{n^3}{3}=\tfrac{1}{3}bh^3.$$

Also, $$A\bar{x}_1=bh.\tfrac{1}{2}h=\tfrac{1}{2}bh^2.$$

$$\therefore\ \bar{x}=\frac{\frac{1}{3}bh^3}{\frac{1}{2}bh^2}=\tfrac{2}{3}h,$$

or the center of pressure is two thirds the height of the rectangle below the surface of the fluid.

397. Prop. *To find the center of pressure when the immersed*

surface is a triangle, having one side horizontal and the opposite vertex in the surface.

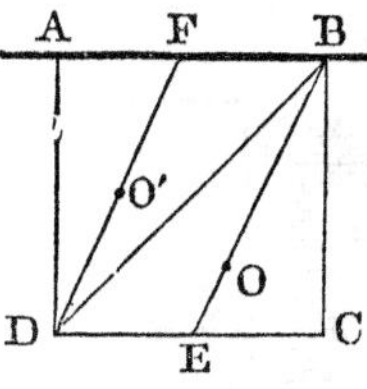

Let BDC be the triangle, h its height, and b its base, and suppose the triangle divided as before into n horizontal divisions.

The heights of the successive divisions are $\frac{h}{n}$;

the lengths, $\frac{b}{n}, \frac{2b}{n}, \frac{3b}{n}$, &c.;

the depths, $x_1=\frac{h}{n}, x_2=\frac{2h}{n}, x_3=\frac{3h}{n}$, &c.

Their areas, considering the horizontal sides of the successive trapezoids as differing insensibly from each other, are

$$k_1=\frac{bh}{n^2}, k_2=\frac{2bh}{n^2}, k_3=\frac{3bh}{n^2}, \text{ \&c.}$$

Multiplying these by the squares of their depths, and taking their sum, we have

$$\Sigma.x^2k=\frac{bh^3}{n^4}+\frac{2^3bh^3}{n^4}+\frac{3^3bh^3}{n^4}+, \text{ \&c.} \ldots \frac{n^3bh^3}{n^4},$$

$$=\frac{bh^3}{n^4}(1^3+2^3+3^3+, \text{ \&c.} \ldots n^3),$$

$$=\frac{bh^3}{n^4}\cdot\frac{n^4}{4}=\tfrac{1}{4}bh^3.$$

Also, $\mathrm{A}=\frac{1}{2}bh$, and $\bar{x}_1=\frac{2}{3}h$.

$\therefore \ \mathrm{A}\bar{x}_1=\frac{1}{3}bh^2$.

Hence $\bar{x}=\dfrac{\frac{1}{4}bh^3}{\frac{1}{3}bh^2}=\frac{3}{4}h$.

If now BE be drawn bisecting the base, the center of pressure will be in BE, and at a distance from the surface $\mathrm{BO}=\frac{3}{4}\mathrm{BE}$.

398. Prop. *To find the center of pressure when the base of the triangle lies in the surface.*

In this case the heights of the successive divisions of the triangle ADB (*Fig., Art.* 397) are $\frac{h}{n}$;

the lengths, $b-\frac{b}{n}, b-\frac{2b}{n}, b-\frac{3b}{n}$, &c.;

the depths, $x_1=\frac{h}{n}$, $x_2=\frac{2h}{n}$, $x_3=\frac{3h}{n}$, &c.:

the areas, $k_1=\frac{bh}{n}-\frac{bh}{n^2}$, $k_2=\frac{bh}{n}-\frac{2bh}{n^2}$, $k_3=\frac{bh}{n}-\frac{3bh}{n^2}$, &c.,

which, multiplied by the squares of the depths, and their sum taken, give

$$\Sigma.x^2k=\frac{bh^3}{n^3}-\frac{bh^3}{n^4}+\frac{2^2bh^3}{n^3}-\frac{2^3bh^3}{n^4}+\frac{3^2bh^3}{n^3}-\frac{3^3bh^3}{n^4}+, \&c.,$$

$$=\frac{bh^3}{n^3}(1^2+2^2+3^2+\ldots n^2)-\frac{bh^3}{n^4}(1^3+2^3+3^3+\ldots n^3)$$

$$=\frac{bh^3}{n^3}.\frac{n^3}{3}-\frac{bh^3}{n^4}.\frac{n^4}{4}=\tfrac{1}{3}bh^3-\tfrac{1}{4}bh^3=\tfrac{1}{12}bh^3,$$

and $A\bar{x}_1=\frac{1}{2}bh.\frac{1}{3}h=\frac{1}{6}bh^2$.

$$\therefore\ \Sigma.x^2k=\frac{\frac{1}{12}bh^3}{\frac{1}{6}bh^2}=\tfrac{1}{2}h.$$

Draw DF bisecting AB, and take FO′$=\frac{1}{2}$FD. O′ is the center of pressure of DAB.

399. Prop. *To find the center of pressure when a rectangle is immersed vertically, having a side parallel to the surface of the fluid and at a given distance below it.*

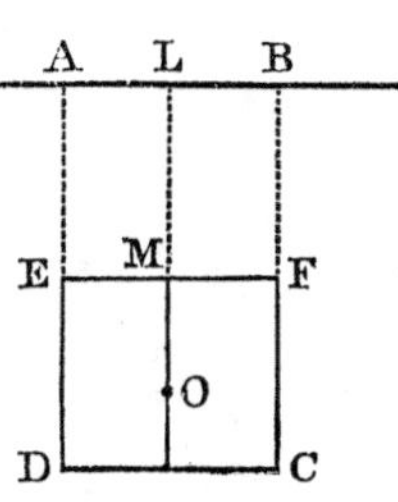

Let a be the distance of EF, the upper side of EFCD from AB the surface of the fluid, and conceive the rectangle divided as before into n divisions whose heights are $\frac{h}{n}$, and areas $\frac{bh}{n}$. Using (145), and reckoning x from EF, the origin of the surfaces, the depths are

$$a+\frac{h}{n},\ a+\frac{2h}{n},\ a+\frac{3h}{n},\ \&c.$$

Hence $\Sigma.(a+x)xk=\frac{abh^2}{n^2}+\frac{bh^3}{n^3}+\frac{2abh^2}{n^2}+\frac{2^2bh^3}{n^3}+\frac{3abh^2}{n^2}+\frac{3^2bh^3}{n^3}+, \&c.$

$$=\frac{abh^2}{n^2}(1+2+3+\ldots.n)+\frac{bh^3}{n^3}(1^2+2^2+3^2+\ldots n^2)$$

$$=\frac{abh^2}{n^2}.\frac{n^2}{2}+\frac{bh}{n^3}.\frac{n^3}{3}=\tfrac{1}{2}abh^2+\tfrac{1}{3}bh^3.$$

And $\Sigma(a+x)k=\text{A}\bar{x}_1=bh(a+\frac{1}{2}h)=abh+\frac{1}{2}bh^2.$

$$\therefore\ \bar{x}=\frac{\frac{1}{2}abh^2+\frac{1}{3}bh^3}{abh+\frac{1}{2}bh^2}=\frac{1}{3}.\frac{2h+3a}{h+2a}=\text{MO}.$$

Hence $\text{LO}=a+\bar{x}=\frac{1}{3}h.\dfrac{2h+3a}{h+2a}+a=\frac{2}{3}.\dfrac{d^3-a^3}{d^2-a^2},$

in which $d=\text{AD}=a+h$.

400. Prop. *To determine the conditions of equilibrium of fluids of different density.*

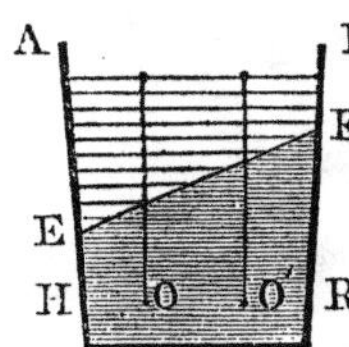

When fluids of different densities which do not mix or unite chemically are contained in the same vessel they will arrange themselves in the order of their densities, the most dense taking the lowest position. This is obvious from the ready displacement of their particles and *Art.* 390, *Cor.* 1.

The limiting surfaces of each will likewise be horizontal; for, in order to equilibrium, any horizontal stratum must sustain in every part of it the same pressure. But if the common surface, as EF, of two of them be inclined, then a horizontal stratum, as HR, will sustain at different points, as O and O′, superincumbent columns of fluid of different weights, and the equilibrium will not subsist.

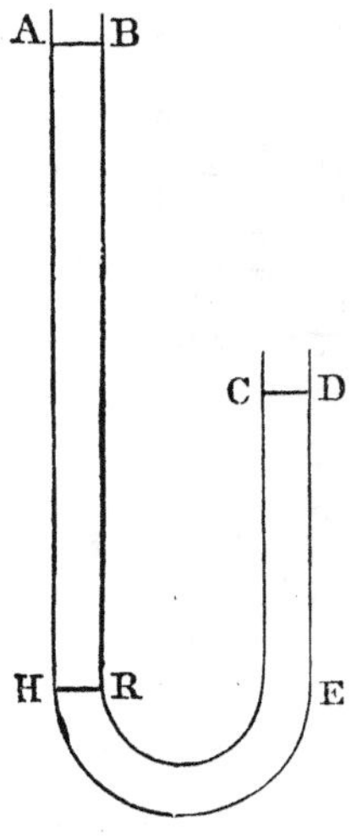

401. Prop. *In communicating tubes, the heights at which fluids of different densities will stand above their common base when in equilibrium, are inversely as their densities.*

In the bent tube AED, let one fluid occupy the portion AR, and the other the portion RED, the fluids having the common base HR. Let h be the height of the fluid in AR above HR, and ρ its density, h' the height of the surface CD above HR, ρ' its density, and let the common base HR=A. The pressure of the fluid in AR on A is $g\rho h\text{A}$, and that of the fluid in RED on the same base $g\rho' h'\text{A}$. In equilibrium these pressures will be equal.

Therefore $$g\rho h A = g\rho' h' A,$$

and $$\frac{h}{h'} = \frac{\rho'}{\rho}. \qquad (149)$$

402 EXAMPLES.

Ex. 1. A cubical vessel, each side of which is ten feet square, is filled with water, and a tube thirty-two feet long is fitted to an aperture in it, whose area is one square inch. If the tube be vertical, of the same diameter as the aperture, and filled with water, what is the pressure on the interior surface of the vessel, neglecting the weight of the water it contains?

Since the weight of a cubic foot of water is 1000 oz., the weight of one cubic inch is $\frac{1000}{1728} = .5787$ oz., and the weight of $32 \times 12 = 384$ cubic inches is $384 \times .5787 = 222.2208$ oz. $=$ 13.8888 lbs. This is the pressure on the aperture, or one square inch of the surface of the water in the vessel.

The number of square inches in the interior surface of the vessel is $6 \times 10^2 \times 144 = 86{,}400$.

Then, *Art.* 381, the pressures being in the ratio of the areas pressed, the whole pressure on the interior surface will be

$$P = 13.8888 \times 86399 = 1{,}200{,}000 \text{ lbs. nearly.}$$

Ex. 2 What is the pressure on the bottom of the vessel in *Ex.* 1 when the weight of the water in the vessel is taken into account; 1°, without the vertical tube, and 2°, with it.

1°. The base of the vessel is 100 square feet, and its depth 10 feet. Hence, *Art.* 385,

$$P_1 = 10 \times 100 \times 62.5 = 62{,}500 \text{ lbs.} = \text{the weight of the water.}$$

2°. The pressure transmitted to each square inch of the base by the water in the tube is 13.8888 lbs. Hence the whole transmitted pressure $P_2 = 13.8888 \times 10^2 \times 144 = 200{,}000$ lbs. nearly, and the entire pressure on the base is $P = 262{,}500$ lbs.

Ex. 3. What is the pressure on each vertical side of the vessel without the tube?

The depth of the center of gravity of each side of the vessel is 5 feet. Hence, *Art.* 386,

$$P_3 = 5 \times 10^2 \times 62.5 = 31{,}250 \text{ lbs.},$$

equal one half the pressure on the base.

Thus we have the whole pressure on the interior surface of the vessel, including that produced by the water in the tube,

$$
\begin{aligned}
P+P_1+4P_3 &= 1,200,000 \\
&+ \quad 62,500 \\
&+ \ 125,000 \\
&= 1,387,500 \text{ lbs.}
\end{aligned}
$$

Ex. 4. Required the pressure on the base of a conical vessel filled with water, the radius of the base being $r=5$ feet, and the altitude $a=10$ feet.

By (137), $P=62.5hA$ lbs., in which $h=a$, and $A=\pi r^2$.

$$\therefore\ P=62.5\pi r^2 a=62.5\times 3.1416\times 25\times 10=49,087.5 \text{ lbs.}$$

Ex. 5. Required the normal pressure on the concave surface of the cone of *Ex.* 4.

The slant height of the cone is $\sqrt{a^2+r^2}$, and the concave surface is $2\pi r\dfrac{\sqrt{a^2+r^2}}{2}$. The vertical depth of the center of gravity of the concave surface is $\dfrac{2a}{3}$.

$$\text{Hence (137), } P_1=62.5\times\frac{2a}{3}\pi r\sqrt{a^2+r^2}$$

$$=62.5\times\tfrac{2}{3}\times 10\times 3.1416\times 5\sqrt{125}=73,175.1 \text{ lbs.}$$

Ex. 6. Required the vertical pressure on the concave surface of the same cone.

Since the normal to the side of the cone makes the same angle with the axis of the cone that the side of the cone makes with the base, the cosine of the inclination of the normal to the axis is $\dfrac{r}{\sqrt{a^2+r^2}}$.

Hence the vertical pressure (*Art.* 387)

$$P_2=P_1\frac{r}{\sqrt{a^2+r^2}}=62.5\times\tfrac{2}{3}a\pi r\sqrt{a^2+r^2}\cdot\frac{r}{\sqrt{a^2+r^2}}=62.5\times\tfrac{2}{3}\pi r^2 a,$$

$$=\tfrac{2}{3}P\ (Ex.\ 4)=32,725 \text{ lbs.}$$

Ex. 7. Required the resultant of all the pressures on the interior surface of the same cone.

By *Art.* 388, the resultant of the horizontal pressures is zero.

and the resultant of the vertical pressures is equal to the excess of the downward pressure over the upward pressure.

$$\therefore \ R = P\ (Ex.\ 4) - P_2\ (Ex.\ 6) = 16{,}362.5 \text{ lbs.},$$

and this is the weight of the water in the cone.

Ex. 8. A rectangular parallelogram, whose sides a and b are 26 feet and 14 feet respectively, is immersed in water with the side b in the surface, and is inclined to the surface at an angle $\phi = 56°.35'$. Required the pressures P_1 and P_2 on the parts into which the parallelogram is divided by its diagonal.

Ans. $P_1 = 82286.5$ lbs.
$P_2 = 164573.0$ lbs.

Ex. 9. When $a = 30$ feet, $b = 20$ feet, and $\phi = 59°.38'$, required the pressures on the equal parts into which the parallelogram is divided by a line parallel to the horizon.

Ans. $P_1 = 121332\frac{2}{3}$ lbs.
$P_2 = 363998$ lbs.

Ex. 10. When the parallelogram of *Ex.* 9 is vertical, how far from the surface must the dividing line be drawn that the pressures on the two parts may be equal?

Ans. $x = 21.213$ feet.

Ex. 11. A sphere 10 feet in diameter is filled with water. Required the entire pressure on the interior surface, and the weight of the water.

Ex. 12. A vessel in the form of a paraboloid, with vertex downward, is filled with water, and revolves uniformly on its axis. Required the time of one revolution when the angular velocity is just sufficient to empty it.

Ex. 13. The concave surface of a cylinder filled with fluid is divided by horizontal sections into n annuli in such a manner that the pressure on each annulus is equal to the pressure on the base. Given the radius of the cylinder. Required its height, and the breadth of the mth annulus.

Ex. 14. A regular tetrahedron, with one face horizontal, is filled with water. Compare the pressures on the base, and on the other sides with the weight of the water.

Ex. 15. An iron vessel 40 feet long, every transverse section of which is an isosceles triangle whose base is 16 feet and altitude 20 feet, floats with its vertex downward. If a cubic foot of iron weighs 487.5 lbs., required the depth to which the vessel will sink when the sides and ends are one quarter of an inch thick and there is no deck.

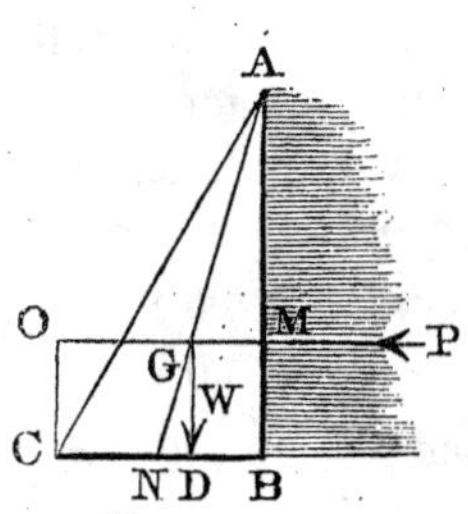

Ex. 16. An embankment of brick-work, of which ABC is a section, weighs 120 lbs. to the cubic foot. Its height AB is 14 feet, and its base BC is 6 feet. Find whether or not the embankment will be overthrown by the pressure of water on the surface AB.

Draw AN bisecting CB, and take NG$=\frac{1}{3}$NA. G will be the center of gravity of the section and the weight W will act at G. Take AM$=\frac{2}{3}$AB, and M will be the center of pressure (*Art.* 396). The resultant P of the pressure of the water will act at M and will pass through G. The moment of P to turn the embankment over C will be P.CO, while the moment of W to resist the overturn will be W.CD. If P.CO$>$W.CD, the embankment will be overturned.

Since the embankment is uniform throughout its length, as also the pressure on it, we may determine the stability by taking one foot in length.

Now $\quad W=1\times14\times6\times\frac{1}{2}\times120=5040$ lbs.,

and $\quad P=1\times14\times7\times62.5=6125$ lbs.,

$OC=BM=\frac{1}{3}\times14$ and $CD=\frac{2}{3}\times6=4$.

$\therefore\ W.CD=20160,$

and $\quad P.CO=28583\frac{1}{3}$.

The latter being greater than the former, the embankment will be overturned.

Ex. 17. A wall of masonry, a section of which is a rectangle, is 10 feet high, 3 feet thick, and each cubic foot weighs 100 lbs. Find the greatest height of water it will sustain without being overturned.

Ex. 18. If the height of the wall be 8 feet, its thickness 6 feet, and each cubic foot weigh 180 lbs., find whether it will stand or fall when the water is on a level with the top.

CHAPTER II.

SPECIFIC GRAVITY.

403. Def. The *specific gravity* of a body is the ratio of the weight of the body to the weight of an equal volume of some other body taken as the standard of comparison, and whose specific gravity, therefore, is taken as the unit.

Water is generally employed as the standard of comparison for solids and liquids, and atmospheric air for aeriform fluids.

Cor. If v, ρ, σ, and w be the volume, density, specific gravity, and weight respectively of one body, and v_1, ρ_1, σ_1, w_1 the same of another body; then, since $v=v_1$,

$$\frac{\sigma}{\sigma_1}=\frac{w}{w_1}=\frac{g.\rho.v}{g.\rho_1 v_1}=\frac{\rho}{\rho_1}, \qquad (150)$$

or the ratio of the specific gravities of two bodies is equal to that of their densities.

404. Prop. *To find the specific gravity of a body more dense than water.*

If the body be immersed in water it will descend (*Cor., Art.* 390). Let w be the absolute weight of the body, and w_1 its weight or the force with which it will descend when immersed in water. Then the loss of weight in water is $w-w_1$, and this is equal to the upward pressure of the water, or to the weight of a volume of water equal to that of the solid. Hence

$$\sigma=\frac{w}{w-w_1}=\frac{\text{absolute weight}}{\text{loss of weight}}. \qquad (151)$$

The absolute weight as well as the loss of weight is ascertained by the hydrostatic balance, which differs from the ordinary balance only in having a hook appended beneath one of the scale pans, to which the body may be suspended by a fine thread and allowed to sink in a vessel of water beneath it. The body is first placed in the scale pan and counterpoised by a weight w, and then suspended to the hook, and when immersed in the water its counterpoise w_1 again determined.

405. Prop. *To find the specific gravity of a body less dense than water.*

Since the body A is less dense than water, it will not descend in the water by its own gravity. Let a more dense body B be attached to it, and call the compound body C.

Let w =absolute weight of A,

$w'=$ " " " B, and w'_1=its weight in water

$w''=$ " " " C, and $w''_1=$ " " " "

Then $w'-w'_1$=loss of weight of B,

$w''-w''_1=$ " " " " C,

and $(w''-w''_1)-(w'-w'_1)$=loss of weight of A, the upward pressure of the water on A, and therefore equal to the weight of a volume of water the same as that of A.

But $$w''=w'+w,$$

and, by substitution,

$$(w''-w''_1)-(w'-w'_1)=w'+w-w''_1-w'+w'_1=w+w'_1-w''_1.$$

$$\therefore \ \sigma=\frac{w}{w+w'_1-w''_1}. \qquad (152)$$

Hence, add to the absolute weight of the body the difference of the weights of the more dense and compound bodies in water, and divide the absolute weight of the body by the sum.

406. Prop. *To find the specific gravity of a liquid.*

Let a body whose weight is w be weighed both in water and in the liquid, the weight in the former being w_1, and in the latter w_2. Then $w-w_1$ and $w-w_2$ are the weights of equal volumes of the two liquids. Hence

$$\sigma=\frac{w-w_2}{w-w_1}; \qquad (153)$$

or, if an empty bottle whose weight is w, weighs when filled with water w_1, and when filled with the liquid w_2, then

$$\sigma=\frac{w_2-w}{w_1-w}. \qquad (154)$$

407. Prop. *To find the weights of the constituents in a mechanical composition when the specific gravities of the compound and constituents are known.*

Let w, w_1, w_2 be the weights of the compound and constituents respectively.

σ, σ_1, σ_2 their respective specific gravities,

v, v_1, v_2 their volumes.

In all merely mechanical combinations,

$$v=v_1+v_2 \ (a), \text{ and } w=w_1+w_2 \ (b).$$

But
$$v=\frac{w}{g\rho},\ v_1=\frac{w_1}{g\rho_1}, \text{ and } v_2=\frac{w_2}{g\rho_2}.$$

$$\therefore (a)\ \frac{w}{\rho}=\frac{w_1}{\rho_1}+\frac{w_2}{\rho_2};$$

or, since their densities are as their specific gravities (150),

$$\frac{w}{\sigma}=\frac{w_1}{\sigma_1}+\frac{w_2}{\sigma_2}.$$

Substituting in succession the values of w_1 and w_2, obtained from (b), we have

$$w_1=w\left(\frac{1}{\sigma}-\frac{1}{\sigma_2}\right)\div\left(\frac{1}{\sigma_1}-\frac{1}{\sigma_2}\right)=\frac{(\sigma_2-\sigma)\sigma_1}{(\sigma_2-\sigma_1)\sigma}.w, \quad (155)$$

$$w_2=w\left(\frac{1}{\sigma}-\frac{1}{\sigma_2}\right)\div\left(\frac{1}{\sigma_2}-\frac{1}{\sigma_1}\right)=\frac{(\sigma_1-\sigma)\sigma_2}{(\sigma_1-\sigma_2)\sigma}.w. \quad (156)$$

408. *Hydrometer.* Instruments for determining the specific gravity of fluids are called Hydrometers or Areometers. They are made of glass, brass, &c., and are of two kinds: one in which the weight is constant, the other in which the volume is constant. In its simplest form the areometer consists of a hollow globe, one of whose diameters is prolonged in a flat or cylindrical stem of uniform size, and to the other extremity is attached a smaller globe loaded with mercury or shot, that it may float in a vertical position.

1°. Areometer of a constant weight.

The annexed figure may represent this form of the instrument. To graduate the stem, suppose it to sink in distilled water, at a given temperature, to a point s, and let the distance sr be divided into any number of equal parts, and continued

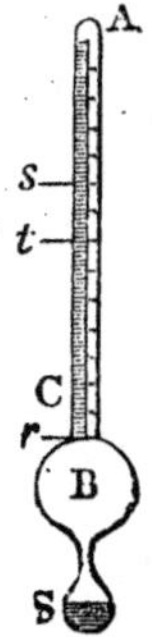

upward from s. When immersed in another fluid suppose it to sink to t, distant from s, x divisions. Let V be the volume of the portions immersed in water, v the volume included between any two divisions of the stem, and σ the specific gravity of the second fluid. Then ρ and ρ_1 being the densities of this fluid and water, the weight of the water displaced by the instrument will be $g\rho_1 V$, the weight of the other fluid displaced will be $g\rho(V-vx)$, and, since each is equal to the weight of the instrument,

$$g\rho_1 V = g\rho(V-vx).$$

$$\therefore\ \sigma = \frac{\rho}{\rho_1} = \frac{V}{V-vx} = \frac{\frac{V}{v}}{\frac{V}{v}-x}. \qquad (a)$$

If σ be previously known, and x be observed, we can determine the value of

$$\frac{V}{v} = x.\frac{\sigma}{1-\sigma}.$$

Substituting in (a) this value of $\frac{V}{v}$, and putting for x, 1, 2, 3 &c., the corresponding values of σ for each division of the scale will be known. These may be marked on the instrument, or the divisions may be numbered, and a table of their values formed to accompany the instrument.

Cor. From (a) we have

$$x = \frac{V}{v}\left(1-\frac{1}{\sigma}\right).$$

Giving to σ a small increment $d\sigma$, the corresponding increment of x is

$$x'-x = dx = -\frac{V}{v}.\frac{1}{\sigma^2}.d\sigma\,;$$

that is, for any small increment of the specific gravity of the fluid the corresponding change in the depth of immersion of the instrument varies as $\frac{V}{v}.\frac{1}{\sigma^2}$, which may be considered a measure of the susceptibility of the instrument.

2°. Areometer of a constant volume.

The principal obstacle to the use of the areometer of a constant weight is the inconvenience and difficulty of calculating and marking against the different divisions of the stem of each instrument a different scale of specific gravity, and constructing the stem of that perfectly uniform thickness which is necessary to the accuracy of the observations.

To obviate these difficulties, Fahrenheit conceived the idea of sinking the instrument always to the same depth by means of weights to be placed in a cup at the end of the stem.

Let W be the weight of the instrument, w_1 and w the weights respectively necessary to sink it to the same point a of the stem in water, and in the fluid whose specific gravity σ is required, V the constant volume of the portion immersed, ρ_1 and ρ the densities of the water and the fluid. Then

$$g\rho V = W + w \text{ and } g\rho_1 V = W + w_1.$$

$$\therefore \ \sigma = \frac{\rho}{\rho_1} = \frac{W+w}{W+w_1}.$$

Cor. Differentiating, we get

$$dw = (W+w_1)d\sigma = g\rho_1 V d\sigma;$$

that is, for any given small variation $d\sigma$ in the specific gravity of the fluid, the variation of w is as V, or the susceptibility of the instrument is as the volume of the portion immersed.

409. *Nicholson's Hydrometer.* This is a modification of the preceding, to adapt it to the determination of the specific gravity of solids as well as liquids. For this purpose, a metallic basket is attached to the lower extremity, in which the body may be placed, and its weight in water ascertained. The basket admits of reversal, so that the body may be retained under water when specifically lighter.

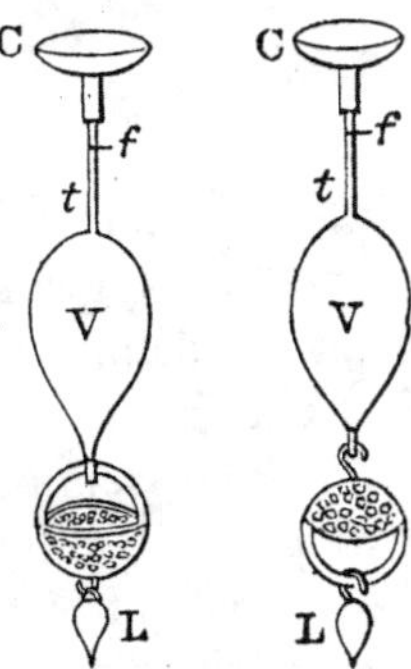

Let w be the weight in C necessary to sink the instrument to f. Replacing w by the body, let w_1 be the weight which must be *added* to sink the instrument to the same depth f. Removing the body to the basket beneath, let w_2 be the weight in C requisite to sink the instrument a third time to f.

Calling W the weight of the instrument, W_1 that of the water displaced by it when immersed to f, V the volume of the body, ρ its density, that of water being ρ_1.

Then $$W+w=W_1, \qquad (a)$$
$$g\rho V+W+w_1=W_1, \qquad (b)$$
and $$g\rho V+W+w_2=W_1+g\rho_1 V. \qquad (c)$$

Subtracting (*a*) from (*b*), and (*b*) from (*c*),

$$g\rho V=w-w_1,$$
$$g\rho_1 V=w_2-w_1.$$
$$\therefore\ \sigma=\frac{\rho}{\rho_1}=\frac{w-w_1}{w_2-w_1}.$$

410. EXAMPLES.

Ex. 1. A piece of wood weighs 12 lbs., and when annexed to 22 lbs. of lead, and immersed in water, the whole weighs 8 lbs. The specific gravity of lead being **11**, required that of the wood.

The specific gravity of lead being 11, if v be its volume, $11v=22$ or $v=2$. But this is the volume of water displaced by the lead, and the specific gravity of water being 1, its weight will be 2. Therefore the loss of weight in water is 2 lbs., and the actual weight in water 20 lbs.

Hence, *Art.* 405, $w=12$, $w''_1=8$, and $w'_1=20$.

$$\therefore\ \sigma=\frac{w}{w+w'_1-w''_1}=\frac{12}{12+20-8}=\frac{1}{2}.$$

Ex. 2. A diamond ring weighs $69\frac{1}{2}$ grains, and when weighed in water $64\frac{1}{2}$ grains. The specific gravity of gold being $16\frac{1}{2}$, and that of diamond $3\frac{1}{2}$, what is the weight of the diamond?

Let v, v' be the volumes of the gold and diamond respectively, and λ the weight of a unit of volume of water. Then $v\lambda$ is the weight of a volume of water equal in bulk to the gold, and $\frac{33}{2}v\lambda$ the weight of the gold. In the same manner, $\frac{7}{2}v'\lambda$ is the weight of the diamond. Hence

$$69\tfrac{1}{2}=\tfrac{33}{2}v\lambda+\tfrac{7}{2}v'\lambda, \qquad (a)$$

and $$64\tfrac{1}{2}=\tfrac{33}{2}v\lambda+\tfrac{7}{2}v'\lambda-(v+v')\lambda. \qquad (b)$$

Subtracting (*b*) from (*a*), we have

$$5=\lambda v+\lambda v'.$$

∴ from (*a*), $$69\tfrac{1}{2}=\tfrac{33}{2}(5-\lambda v')+\tfrac{7}{2}\lambda v',$$

$$139=165-26\lambda v',$$

$$\lambda v'=1,\ \tfrac{7}{2}\lambda v'=3\tfrac{1}{2},$$

or the weight of the diamond is three and a half grains.

Ex. 3. A body A weighs 10 grains in water, and a body B weighs 14 grains in air, and A and B connected together weigh 7 grains in water. The specific gravity of air being .0013, required the specific gravity of B, and the number of grains of water equal to it in bulk.

Let λ', λ'' be the number of grains in the volumes of water equal to the volumes of A and B respectively, and σ', σ'' their specific gravities. Then, by the conditions of the question,

$$(\sigma'-1)\lambda'=10, \qquad (a)$$

$$(\sigma''-.0013)\lambda''=14, \qquad (b)$$

$$(\sigma'-1)\lambda'+(\sigma''-1)\lambda''=7. \qquad (c)$$

From (*a*), (*b*), and (*c*), we have

$$10+\frac{14(\sigma''-1)}{\sigma''-.0013}=7,$$

$$14(1-\sigma'')=3(\sigma''-.0013).$$

$$\therefore\ \sigma''=.8237.$$

Hence, also, from (*b*),

$$\lambda''=\frac{14}{\sigma''-.0013}=\frac{14}{.8224}=17.023 \text{ grains.}$$

Ex. 4. When 73 parts by weight of sulphuric acid, the specific gravity of which is 1.8485, are mixed with 27 parts of water, the resulting dilute acid has a specific gravity equal to 1.6321. Required the amount of condensation which takes place by the mixture.

Let λ' be the number of parts by weight in a quantity of water equal in volume to that of the sulphuric acid, and λ'' in a quantity of water equal in volume to that of the mixture.

Then $$1.8485\lambda'=73,$$

or $$\lambda'=39.4915,$$

and $$1.6321\lambda''=73+27=100$$
or $$\lambda''=61.2707.$$

Now the condensation of the mixture will be expressed by the ratio of the diminution of the volume of water and acid, when mixed, to their united volume before mixture.

Hence $$\text{condensation}=\frac{\lambda'+27-\lambda''}{\lambda'+27}$$
$$=\frac{5.2208}{66.4915}=0.0785.$$

Ex. 5. Two fluids, the volumes of which are v and v', and specific gravities σ and σ', on being mixed, contract $\frac{1}{n}$th part of the sum of their volumes by mutual penetration. Required the specific gravity of the mixture.

Let σ'' be the specific gravity of the mixture; then the sum of their weight before and after mixture will be equal.

Hence $$(v\sigma+v'\sigma')\lambda=\left(1-\frac{1}{n}\right)(v+v')\sigma''\lambda,$$

which gives $$\sigma''=\frac{n}{n-1}\cdot\frac{v\sigma+v'\sigma'}{v+v'}.$$

Ex. 6. A body whose weight in a vacuum is 73.29 grains loses 24.43 grains by immersion in water. Required its specific gravity. *Ans.* $\sigma=3$.

Ex. 7. Required the specific gravity of a body which weighs 65 grains in a vacuum, and 44 grains in water.

Ans. $\sigma=3.0952$.

Ex. 8. An areometer sinks to a certain depth in a fluid whose specific gravity is .8, and when loaded with 60 grains it sinks to the same depth in water. What is the weight of the instrument? *Ans.* $w=240$ grains.

Ex. 9. A compound of gold and silver weighing $w=10$ lbs. has a specific gravity $\sigma=14$, that of gold being $\sigma'=19.3$, and that of silver being $\sigma''=10.5$. Required the weight w' and w' of the gold and silver in the compound.

Ans. $w'=5.483$,
$w''=4.517$

CHAPTER III.

COMPRESSIBLE OR AERIFORM FLUIDS.

411. Prop. *To find the tension of the atmosphere or other compressible fluid.*

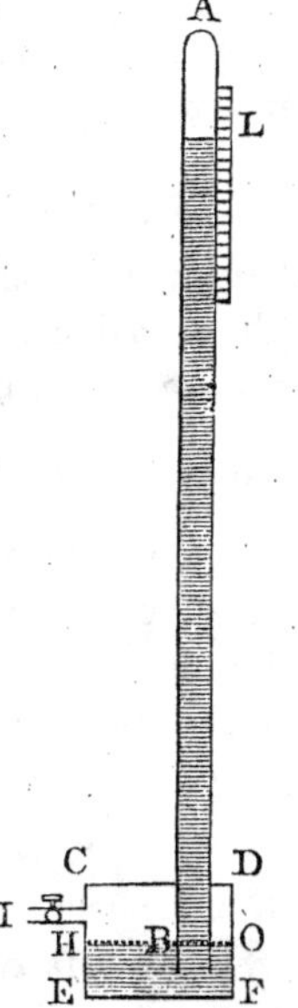

Let a glass tube AB, open at one end and closed at the other, be filled with mercury. Retain the mercury in the tube by the pressure of the finger, let it be inverted and the open end immersed beneath the surface of the mercury in the vessel CDFE. It will now be found that a column of mercury, as BL, will occupy a portion of the tube, while the remaining portion AL will be void. This column of mercury is sustained in the tube by the pressure of the atmosphere on the surface of the mercury in the cistern CF, and as there is no pressure on the mercury at L, the equilibrium is due to the equality of pressures of the atmosphere and mercury on the common base B (*Art.* 401). If we now suppose the cistern covered, so as to separate the air without from that within the cistern, the pressure of the external air can not be communicated to that within, and the mercurial column must be sustained by the expansive force or tension of the inclosed air, and be a measure of it.

If the tension of any other elastic fluid inclosed in a vessel be required, let the tube I from the cistern CEFD be fitted to an aperture in the vessel, and a communication be thus established between the inclosed fluid and the mercury in the cistern. The mercury in the tube will then rise or fall till an equilibrium takes place between the expansive force of the fluid and the weight of the mercurial column. The height at which the mercury stands in the tube above the surface of mercury in the cistern is ascertained by a graduated scale attached to the tube.

412. Schol. The mean pressure of the atmosphere at or near the level of the sea is generally employed in mechanics as the unit of pressure, and other expansive forces are compared with this and expressed in atmospheres. It has been ascertained by the barometer that the mean pressure of the atmosphere, at a temperature of 50°, is equivalent to a column of mercury 30 inches in height; or, the specific gravity of mercury being 13.598, to a column of water 13.598×2.5 feet$=34$ feet; or, the specific gravity of air, at a temperature of 50°, being $\frac{1}{810}$, to a column of air of uniform density, 27,540 feet $=5.2$ miles in height.

The tension is also measured by the pressure of the atmosphere on a unit of surface. Now 30 cubic inches of mercury is equal to 13.598×30 cubic inches of water$=407.94 \times \frac{1}{1728}$ cubic feet of water$=0.23607 \times 1000$ oz.$=14.75$ lbs. on the square inch.

The instrument above described involves the essential parts of a barometer, a more particular description of which belongs to Physics. Generally, instruments employed to determine the tension of elastic fluids are called *manometers*.

413. Prop. *To show that the tension of an aeriform fluid is inversely as its volume.*

Let AB be a vertical tube which communicates with the cylindrical vessel DEO_1H_1, closed at the top, and mercury be carefully introduced into the tube, so as to fill the horizontal part H_1O_1BC, and leave the air in DO_1 of the same density as the exterior air. The inclosed air will then have a tension of one atmosphere. Take $DH_2=\frac{1}{2}DH_1$, $DH_3=\frac{1}{3}DH_1$, &c., and draw the lines $H_2O_2R_2$, $H_3O_3R_3$, &c. Now if mercury be poured into the tube AB until it stands at H_2O_2 in the cylinder, it will be found to stand in the tube at the height

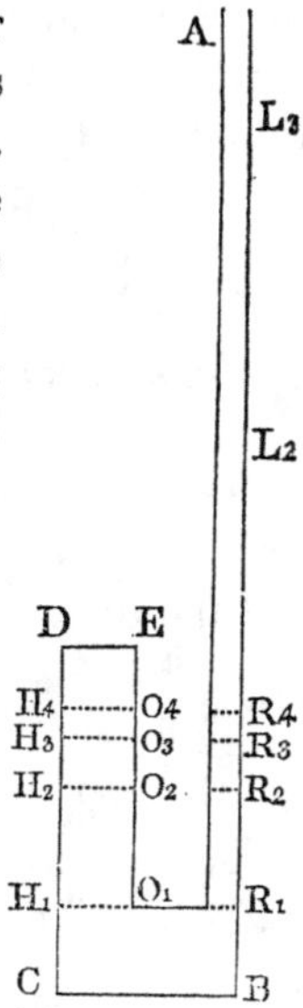

$L_2R_2=30$ inches above H_2O_2; or, when the air in the cylinder is reduced to half its volume, its tension or expansive force is two atmospheres. When the mercury rises in the cylinder to the height H_3O_3, it will be found to stand at the height $L_3R_3=60$ inches above H_3O_3; or, when the air is reduced to one third of its volume, it has a tension of three atmospheres, or, generally, the tension is inversely as the volume. The truth of the law has been tested experimentally as far as 26 atmospheres, and for all fractions of an atmosphere.

If V and V′ be the volumes of a given mass of air, p and p' the corresponding tensions or pressures on a unit of surface, then

$$\frac{p}{p'}=\frac{V'}{V}, \text{ or } pV=p'V'. \qquad (157)$$

This law is called the law of Mariotte, from the discoverer

414. Cor. Let ρ and ρ' be the densities of any mass of air corresponding to the volumes V and V′. Then, since the density is inversely as the volume,

$$\frac{\rho}{\rho'}=\frac{V'}{V}=\frac{p}{p'}, \qquad (158)$$

or the elastic force is directly as the density.

415. Prop. *To estimate the effect of heat on the volume and tension of atmospheric air.*

When air is inclosed in a vessel and heat is applied, its elastic force is increased, as may be shown by the method indicated in *Art.* 411, and by the same method the increase in the tension for a given increase of temperature may be ascertained. Experiment indicates that the tension of a given volume of dry air increases, by being heated from the freezing to the boiling point, 0.367 of its original value, and therefore, if the tension remains the same, the volume will increase 36.7 per cent. Let v be the increment of volume, the tension remaining the same, for one degree of Fahrenheit's thermometer; then

$\nu=0.002039$, and for an increase of t degrees of temperature, the increase of volume will be $\nu t=0.002039t$.

If now V_0 be the original volume of a given mass of air, and it be heated t_1 degrees, the tension remaining the same, the new volume V_1 will be

$$V_1=(1+\nu t_1)V_0;$$

and when heated t_2 degrees the corresponding volume will be

$$V_2=(1+\nu t_2)V_0.$$

$$\therefore \frac{V_1}{V_2}=\frac{1+\nu t_1}{1+\nu t_2}=\frac{1+0.002309t_1}{1+0.002309t_2}. \qquad (159)$$

But the densities ρ_1 and ρ_2 are inversely as the volumes.

Hence $$\frac{\rho_1}{\rho_2}=\frac{V_2}{V_1}=\frac{1+\nu t_2}{1+\nu t_1}.$$

If, also, a change take place in the tensions at the same time, let p_0 be the tension at 32°, p_1 the tension at $32°+t_1$, and p_2 that at $32°+t_2$. Then, since the tension is inversely as the volume,

$$V_1=(1+\nu t_1)\frac{p_0}{p_1}V_0 \text{ and } V_2=(1+\nu t_2)\frac{p_0}{p_2}V_0.$$

$$\therefore \frac{V_1}{V_2}=\frac{(1+\nu t_1)p_2}{(1+\nu t_2)p_1},$$

and $$\frac{\rho_1}{\rho_2}=\frac{(1+\nu t_1)}{(1+\nu t_2)}\cdot\frac{p_1}{p_2}=\frac{(1+\nu t_1)}{(1+\nu t_2)}\cdot\frac{b_1}{b_2}, \qquad (160)$$

in which b_1 and b_2 represent the measures of the tensions p_1 and p_2, or the corresponding heights of the barometer.

416. Prop. *To find the density of the air at different temperatures and under different pressures.*

By accurate experiments the weight of a cubic foot of air at a temperature of 32°, when the barometer stands at 30 inches, is found to be $\rho=0.08112$ lbs. avoirdupois. Hence, for the temperature $32°+t°$.

$$\rho_1=\frac{\rho}{1+\nu t}=\frac{0.08112}{1+0.002039t}\text{ lbs.} \qquad (161)$$

If, also, the barometer, instead of $b=30$ inches, should stand at some other height, as b_1 inches, the density will be expressed by

$$\rho=\frac{\rho}{1+\nu t}\cdot\frac{b_1}{b}=\frac{0.08112}{1+0.00204t}\cdot\frac{b_1}{30}=\frac{0.002704b_1}{1+0.00204t}\text{ lbs.} \qquad (162)$$

Whenever the elasticity of the air is expressed by the pressure $p=14.75$ lbs. on the square inch, instead of the barometric height b, the density for any other tension p_1 will be

$$\rho_1=\frac{\rho}{1+\nu t}.\frac{p_1}{p}=\frac{0.08112}{1+0.00204t}.\frac{p_1}{14.75}=\frac{0.0055p_1}{1+0.00204t}\text{ lbs.} \quad (163)$$

The density of steam is five eighths of the density of atmospheric air for the same temperature and tension. Therefore we have, for steam,

$$\rho_1=\frac{0.00344p_1}{1+0.00204t}\text{ lbs.}$$

417. Prop. *To determine the height corresponding to a given density of the atmosphere, and, conversely, the density in terms of the height.*

Since the density of the atmosphere at the surface of the earth is due to the pressure of the superincumbent portions o it, the density must decrease as the height increases.

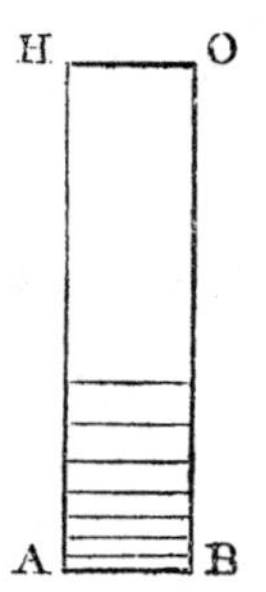

Let HOBA be a vertical column of air whose base AB is one square foot. Conceive it divided into portions of equal weights w, and heights x_0, x_1, x_2, &c., beginning at AB, so small that the density of each may be regarded as uniform. Then $x_0+x_1+x_2$, &c. . . . $x_n=\Sigma.x=h$ is the height of the nth stratum. Let p_0 be the tension of the lowest stratum, and p_n that of the nth, ρ_0 and ρ_n their densities respectively. Then the weight of the nth stratum is

$$w=1.x_n\rho_n=\frac{x_n\rho_0 p_n}{p_0}. \quad (158).$$

$$\therefore\ x_n=\frac{p_0}{\rho_0}.\frac{w}{p_n}. \quad (a)$$

But the tension of the $(n-1)$th stratum is p_n+w, and the height, therefore,

$$x_{n-1}=\frac{p_0}{\rho_0}.\frac{w}{p_n+w}.$$

In like manner, $$x_{n-2}=\frac{p_0}{\rho_0}\cdot\frac{w}{p_n+2w},$$

. . . .

$$x_0=x_{n-n}=\frac{p_0}{\rho_0}\cdot\frac{w}{p_n+nw}=\frac{p_0}{\rho_0}\cdot\frac{w}{p_0}$$

since $p_n+nw=p_0$.

Now, in the exponential series,

$$e^y=1+y+\frac{y^2}{1.2}+\frac{y^3}{1.2.3}+, \&c.;$$

when y is very small, we have

$$e^y=1+y,$$

or $$y=yl.e=l.(1+y)=2.30258\text{L}.(1+y), \qquad (b)$$

in which l denotes the Naperian logarithm, and L the common logarithm of a number.

If in (b), for y, we put $\frac{w}{p_n}$, we get

$$\frac{w}{p_n}=l.\left(1+\frac{w}{p_n}\right)=l.\left(\frac{p_n+w}{p_n}\right)=l.(p_n+w)-l.p_n.$$

$\therefore$ from (a), we have

$$x_n=\frac{p_0}{\rho_0}\{l.(p_n+w)-l.p_n\}.$$

Now, substituting p_n+w successively for p_n, we obtain

$$x_n\ =\frac{p_0}{\rho_0}\{l.(p_n+w)-l.p_n\},$$

$$x_{n-1}=\frac{p_0}{\rho_0}\{l.(p_n+2w)-l.(p_n+w)\},$$

$$x_{n-2}=\frac{p_0}{\rho_0}\{l.(p_n+3w)-l.(p_n+2w)\},$$

. . . .

$$x_0=x_{n-n}=\frac{p_0}{\rho_0}\{l.(p_n+nw)-l.(p_n+(n-1)w)\}.$$

By taking the sum of these equations, the terms in the brackets will all cancel, except $l.(p_n+nw)=l.p_0$ in the last and $l.p_n$ in the first, and we shall have

$$\Sigma.x=h=\frac{p_0}{\rho_0}.\{l.p_0-l.p_n\}=\frac{p_0}{\rho_0}l.\frac{p_0}{p_n}. \qquad (164)$$

To find p_n when h is given, we have

$$l.\frac{p_0}{p_n}=\frac{\rho_0 h}{p_0}.l.e,$$

or

$$\frac{p_0}{p_n}=e^{\frac{\rho_0 h}{p_0}}.$$

Hence

$$p_n=p_0.e^{-\frac{\rho_0 h}{p_0}}, \qquad (165)$$

where $e=2.71828$, the base of the Naperian system of logarithms.

418. Schol. Formula (164) may be adapted to the determination of heights by the barometer. To this end let b_0 and b_n be the heights of the barometrical columns at the lower and upper stations respectively. Then, since $\frac{p_0}{p_n}=\frac{b_0}{b_n}$,

$$h=2.30258\frac{p_0}{\rho_0}.\text{L}\frac{b_0}{b_n}. \qquad (a)$$

The value of $\frac{p_0}{\rho_0}$ may be determined from the consideration that p_0 expresses the weight or pressure of a column of the atmosphere on a unit of surface as one square foot, and $\frac{p_0}{\rho_0}$ must express the height of this column on the supposition that its density is uniform. Now a cubic foot of air at 32° weighs 0.08112 lbs., and a cubic foot of water at the same temperature weighs 62.37917 lbs., and therefore the specific gravity of water referred to air is $\frac{62.37917}{0.08112}=769$. Hence the height of a homogeneous atmosphere at a temperature of 32° is $34\times769=26146$ feet.

$$\therefore\ 2.30258\frac{p_0}{\rho_0}=2.30258\times26146=60204 \text{ feet.}$$

It is here assumed, however, that the lower station is at or near the level of the sea, and no account is taken of the variation of gravity at different elevations. From numerous observations made at different elevations above the sea, and at known differences of height, this coefficient is found to be 60345 feet at a temperature of 32°. But the actual temperature of the air at both the lower and upper stations will, in general, differ from the standard temperature of the formula, and, since the density of air varies uniformly with the temperature, we may use the mean of the temperatures of the air at the two stations. Let t_0 and t_n be the indications of Fahrenheit's thermometer; then the mean temperature will be $\frac{1}{2}(t_0+t_n)$, and the deviation t from the standard will be

$$t=\tfrac{1}{2}(t_0+t_n)-32.$$

The expansion of dry air is 0.00204 for a change of 1°; but when the atmosphere contains vapor, it is found, by comparing the rates of expansion of vapor and of dry air, and assuming a certain mean humidity for the air, that the rate is expressed by 0.00222. Incorporating this correction in (*a*), and using the coefficient determined by observation, we have

$$h=60345(1+.00222t)\mathrm{L}.\frac{b_0}{b_n}. \qquad (166)$$

It is further obvious that a change in the length of the mercurial column will be produced by a change of temperature of the mercury. Let τ_0 and τ_n be the temperatures of the mercury, as shown by a thermometer *attached* to the cistern of the barometer; then, since mercury expands at the rate of 0.0001 for each degree,

$$b_n=b'_n(1+.0001)(\tau_0-\tau_n),$$

where b'_n is the *observed* height of the barometer at the upper station. Using this value of b_n, the difference of elevation between two stations or the height of a mountain may be determined with considerable accuracy.

In the determination of the constant coefficient, the variation of gravity at different elevations is allowed for in the assumption that this coefficient is that which belongs to the mean height above the sea at which observations are usually made, and to the latitude of 45°. When the latitude differs from this, it will be necessary to multiply the result by

$$(1+.002837 \cos. 2\psi),$$

ψ being the latitude at which the observations are made.

419. EXAMPLES.

Ex. 1. A cylindrical tube 40 inches long is half filled with mercury, and then inverted in a vessel of mercury. How high will the mercury stand in the tube, the pressure of the external air being equivalent to 30 inches?

Let l be the length of the tube, a the length of the portion occupied by the air before it was inverted, h the height of mercury due to the pressure of the external air, x the height of the mercury after the tube is inverted, and h' the column of mercury equivalent to the tension of the air in the tube.

Then (157), $ha=h'(l-x)$.

But $h=h'+x$ or $h'=h-x$.

$\therefore\ ha=(h-x)(l-x)$,

whence we get $x=\frac{1}{2}(l+h)\pm\frac{1}{2}\sqrt{(l-x)^2+4ah}$.

Using the data of the question, $l=40$, $a=20$, and $h=30$, we get

$$x=10 \text{ or } 60.$$

The first value is that which pertains to the specific question.

Ex. 2. A tube 30 inches long, closed at one end and open at the other, was caused to descend in the sea with the open end downward until the inclosed air occupied only one inch of the tube. How far did it descend?

Ex. 3. A spherical air-bubble having risen from a depth of 1000 feet in water, was one inch in diameter when it reached the surface. What was its diameter at the bottom?

Ex. 4. Required the equation of the curve described by the extremities of a horizontal diameter of the air-bubble of *Ex.* 3, supposing its center to move in a vertical line.

Ex. 5. What number of degrees must a given volume of air be heated to double its elasticity?

Ex. 6. The following barometrical observations were made at the White Rocks, on the bank of the Connecticut River some two miles below the city of Middletown:

	Barometer.	Det. Ther.	At. Ther.
At the base,	$b_0=30.09$ in.	$t_0=83.0$	$\tau_0=84.5$
On the summit,	$b'_n=29.65$ in.	$t_n=85.0$	$\tau_n=83.5$

Hence $t=\frac{1}{2}(t_0+t_n)-32°=52°$ and $1+.00222t=1.11544$.
Also, $\tau_0-\tau_n=1°$ and $b_n=b'_n(1+.0001\times1)=29.653$ in.

$$\therefore h=60345\times1.11544.\times L\frac{30.09}{29.653}.$$

30.09	log.	1.4784222		
29.653 a.c.	"	$\bar{8}.5279314$		
		0.0063536	log.	$\bar{7}.8030190$
		1.11544	"	0.0474462
		60345.	"	4.7806413
		427.67 feet	"	2.6311074.

Ex. 7. The following observations were made by Humboldt at the Mountain of Quindiu, New Grenada, in lat. 5°:

	Barometer.	Det. Ther.	At. Ther.
At the level of the Pacific,	30.036 in.	77°.54	77°.54
On the summit,	20.0713	65°.75	68°.00

Required the height of the mountain.

Ans. 11500 feet nearly.

HYDRODYNAMICS.

---

420. PROP. *The velocity of a fluid in a tube of variable diameter, kept constantly full, is in different transverse sections inversely as the areas of the sections.*

Since the tube is supposed constantly full, and the fluid incompressible, the same quantity of fluid must pass through every section in a unit of time. But admitting the fluid to have the same velocity in every part of the same section, the quantity which flows through any section in a unit of time will be the product of the area of the section by the velocity. If, therefore, k and k' be the areas of any two sections, and v and v' the velocities at each respectively,

$$kv=k'v',$$

and

$$v : v'=k' : k. \qquad (167)$$

In reality, the velocity is diminished by the sides of the tube, and is therefore in any section least near the sides of the tube and greatest near the central portions.

421. PROP. *The velocity with which a fluid issues from a small orifice in the bottom of a vessel kept constantly full, is equal to that acquired by a body falling freely through the height of the fluid above the orifice.*

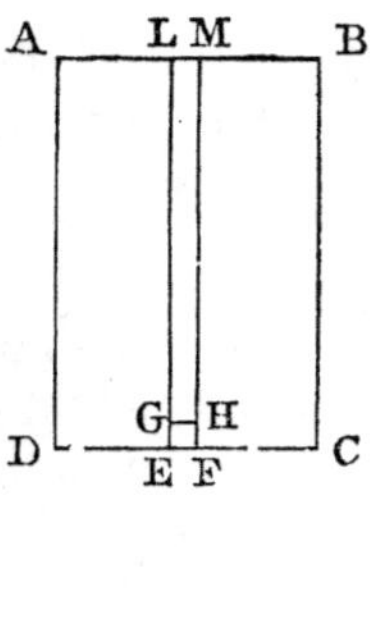

Let EF represent a very small orifice in the bottom of the vessel ABCD, filled with a fluid to the level of AB, GF a stratum whose thickness FH$=h'$ is indefinitely small, and FM$=h$ the whole height of the column vertically above the orifice.

A L M B
D G H C
E F

If now the stratum GF fall by its own weight through HF$=h'$, the velocity will be

$$v=\sqrt{2gh'}.$$

But if the stratum be urged by its own weight and the weight of the column LH, calling the force in this case g', the velocity v' will be

$$v'=\sqrt{2g'h'}.$$

But the forces g and g' are as the weights of the columns GF and LF, or as their heights h' and h. Hence

$$\frac{g'}{g}=\frac{h}{h'} \text{ or } g'=\frac{gh}{h'}.$$

Substituting this value of g' in that of v', we have

$$v'=\sqrt{2gh}, \qquad (168)$$

which is the velocity of a body falling freely through the height h.

Cor. If the orifice be made in the side of the vessel, and a tube be inserted so as to direct the current *obliquely, horizontally*, or *vertically upward*, the velocity of efflux will be the same, since the pressure of fluids is the same in every direction.

In the first case, its path will be a parabola whose equation is (64).

In the second case, the angle of elevation $a=0$, which reduces the equation to

$$x^2=4hy,$$

the equation of a parabola whose axis is vertical and origin of co-ordinates at the vertex or orifice.

In the last case, if all obstructions are removed, the fluid will rise to the height of the surface of the fluid in the vessel.

422. Prop. *To determine the horizontal distance to which a fluid will spout from an orifice in the vertical side of a vessel.*

Let the vessel ABCD be filled to the level AB. If the fluid issue horizontally from the orifice O, the equation of its path is

$$x^2=4hy,$$

in which h=OB, the height of the fluid above the orifice, is the impetus or height due to the velocity.

To determine the range, let y=OC=a. Hence

$$x = CP = 2\sqrt{hy} = 2\sqrt{ha} = 2.OM,$$

or the horizontal distance is equal to twice the ordinate at the orifice, in a semicircle whose diameter is the vertical distance from the surface of the fluid to the horizontal plane.

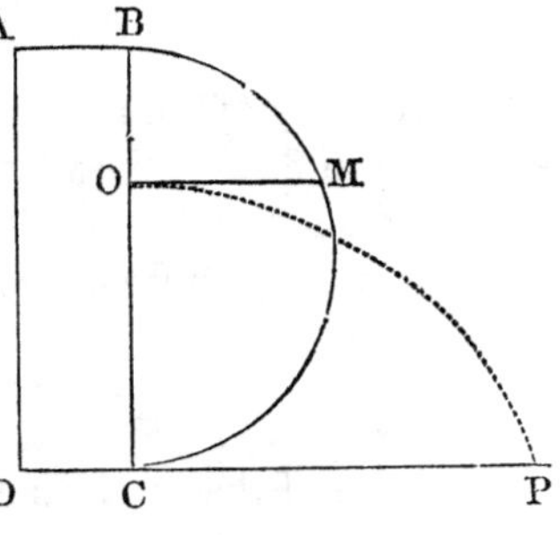

Cor. When the orifice is at the middle of BC, the range is a maximum, and at equal distances above and below the middle the range will be the same.

423. Prop. *To find the quantity of fluid discharged in a given time from a small orifice in the bottom of a vessel when the fluid is maintained at the same constant height.*

Let h be the constant height of the fluid in the vessel, t the given time, k the area of the orifice, and Q the quantity.

Then $$Q = kvt = kt\sqrt{2gh}. \qquad (169)$$

If the orifice be circular, and r its radius, then $k = \pi r^2$, and

$$Q = \pi\sqrt{2g}.tr^2\sqrt{h},$$
$$= 25.195.tr^2\sqrt{h}.$$

The time being in seconds, and r and h in feet, Q will be in cubic feet. If the weight W be required,

$$W = 62.5Q.$$

Cor. 1. The time required for the efflux of a given quantity is

$$t = \frac{Q}{\pi r^2\sqrt{2gh}}.$$

Cor. 2. Since t and g are constant in (169),

$$Q \propto k\sqrt{h}.$$

Hence the quantity discharged in the same time from orifices differing in size and distance from the surface, varies as the size of the orifice and square root of its depth.

If the orifices are the same, $Q \propto \sqrt{h}$. In order, therefore, that the discharge from one orifice may be twice that from another, its depth must be four times as great.

424. Prop. *To determine the time in which a cylindrical vessel filled with water will empty itself by a small orifice in its base.*

The velocity of efflux at any instant is $v=\sqrt{2gh}$, h being the height of the fluid in the vessel. But since the velocities v and v', at the orifice and at any transverse section of the vessel, are inversely as their areas, k and k' (167),

$$v'=\frac{k}{k'}v=\frac{k}{k'}\sqrt{2gh}\propto\sqrt{h}.$$

The surface of the fluid, therefore, in the vessel descends with a velocity proportional to the square root of the space over which it must pass, and the circumstances of its motion are precisely the same as those of a body projected vertically upward. But the velocity of a body projected upward is such that if it continued uniformly it would move over twice the space through which it must move in the same time to acquire or expend that velocity. Hence the time of descent of the surface of the fluid to the base is twice that required for the descent of the same superficial stratum when the vessel is kept full. In the latter case, if Q be the whole quantity of water in the vessel (169),

$$t=\frac{Q}{k\sqrt{2gh}}.$$

Therefore, in the former,

$$t=\frac{2Q}{k\sqrt{2gh}}. \qquad (170)$$

If r' be the radius of a tranverse section of the vessel, r that of the orifice,

$$Q=\pi r'^2h \text{ and } k=\pi r^2.$$

$$\therefore\ t=\frac{2r'^2}{r^2\sqrt{2g}}\sqrt{h}.$$

425. Schol. The preceding deductions are founded on the hypothesis that the fluid particles descend in straight lines to the orifice, and all issue with a velocity due to the height of the fluid surface. Experiment shows that this is true only of

those vertically above the center of the orifice, that those situated about the central line of particles, take a curvilinear course as they approach the orifice, being turned inward toward this line or spirally around it, and this deviation from a vertical rectilinear path is the greater the more remote they are from the central line. This deviation will necessarily occasion a diminution of vertical velocity, and therefore a diminution of total discharge. The oblique direction of the exterior particles gives to the vein of issuing fluid, when the orifice is circular, the form nearly of a conic frustum, whose larger base is the area of the orifice. This diminution of a section of the issuing fluid is called the *contraction of the vein*, and the vein itself, from the orifice to the smallest section, the *vena contracta*, or *contracted vein*.

The results of most experiments agree in making the length of the contracted vein, when the orifice is circular and horizontal, equal to the radius of the orifice. There is a greater discrepancy in the results of experiments for determining the ratio of the diameters of the two ends of the contracted vein. When water flows through orifices in thin plates it is found to be about 0.8. The ratio of the areas of the two ends, called the *coefficient of contraction*, is therefore 0.64.

The actual discharge is found by experiment to differ slightly from the theoretical for other causes. The ratio of the former to the latter is found to be about 0.97, and is called the coefficient of velocity. The product of the coefficients of velocity and contraction, called the coefficient of efflux, is 0.62.

Hence, for the actual velocity of discharge through orifices in thin plates, we have

$$v_1 = 0.62v = 0.62\sqrt{2gh},$$

and for the quantity in a unit of time (169),

$$Q = kv_1 = 0.62k\sqrt{2gh}.$$

When the orifice or pipe through which the discharge is made has the length and form of the vena contracta, the velocity will be

$$v_1 = 0.97v = 0.97\sqrt{2gh},$$

and the quantity

$$Q = kv_1 = 0.97k\sqrt{2gh},$$

k being the smallest section of the contracted vein.

By the use of cylindrical or conical adjutages, the quantity of the discharge is increased. More seems to be gained by the adhesion of the fluid particles to the sides of the tube, in preventing the contraction of the vein, than is lost by the friction. The discharge is greater when the adjutage is conical and the larger end is the discharging orifice.

426. Prop. *To determine the quantity of water which will flow from a rectangular aperture.*

1°. When one side of the aperture coincides with the surface of the water.

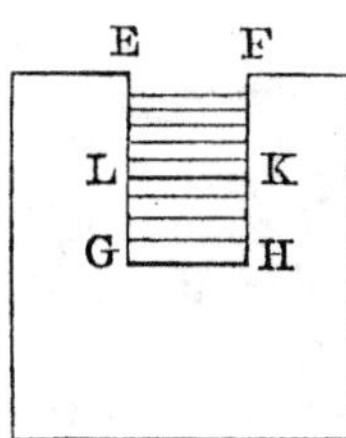

Let h be its height and b its breadth, and conceive the aperture to be divided horizontally into a very large number n of equal divisions so narrow that the velocity of the fluid in every part of each may be regarded as the same. The larger n is, the more nearly will the hypothesis be satisfied.

The depth of these successive divisions below the surface will be

$$\frac{h}{n}, \frac{2h}{n}, \frac{3h}{n}, \text{\&c.}$$

The velocities in each will be

$$\sqrt{2g\frac{h}{n}}, \sqrt{2g\frac{2h}{n}}, \sqrt{2g\frac{3h}{n}}, \text{\&c.};$$

and since the areas of these divisions are each $b\frac{h}{n}$, the quantities discharged by each in a unit of time will be (169)

$$\frac{bh}{n}\sqrt{2g\frac{h}{n}}, \frac{bh}{n}\sqrt{2g\frac{2h}{n}}, \frac{bh}{n}\sqrt{2g\frac{3h}{n}}, \text{\&c.},$$

and their sum, or the whole quantity Q discharged, will be

$$Q=\frac{bh\sqrt{2gh}}{n\sqrt{n}}(1^{\frac{1}{2}}+2^{\frac{1}{2}}+3^{\frac{1}{2}}+, \text{\&c.} \ldots n^{\frac{1}{2}}).$$

But $$1^{\frac{1}{2}}+2^{\frac{1}{2}}+3^{\frac{1}{2}}+, \text{\&c.} \ldots n^{\frac{1}{2}}=\frac{n^{\frac{1}{2}+1}}{\frac{1}{2}+1}=\tfrac{2}{3}n^{\frac{3}{2}}.$$

$$\therefore\ Q=\tfrac{2}{3}bh\sqrt{2gh}=\tfrac{2}{3}b\sqrt{2gh^3}.$$

Or, if $v=\frac{bh}{Q}$ be the mean velocity.

$$v=\tfrac{2}{3}\sqrt{2gh}.$$

2°. When the upper side of the rectangular aperture is not coincident with the surface of the fluid.

Let its depth $EL=h_1$, and the distance $EG=h$, as before. Then the quantity which issues from the aperture LH will be equal to the quantity which would issue from EH, diminished by the quantity which would issue from EK, or

$$Q=\tfrac{2}{3}b\sqrt{2gh^3}-\tfrac{2}{3}b\sqrt{2gh_1^3}=\tfrac{2}{3}b\sqrt{2g}(h^{\frac{3}{2}}-h_1^{\frac{3}{2}});$$

and if $Q=b(h-h_1)v$,

$$v=\tfrac{2}{3}\sqrt{2g}.\frac{h^{\frac{3}{2}}-h_1^{\frac{3}{2}}}{h-h_1}.$$

427. Prop. *To determine the quantity of water which will flow from a triangular aperture.*

1°. When the vertex of the triangle is in the surface.

Let the base $GH=b$, and the height $HF=h$, and let the triangle be divided into n parts, as before, of equal but very small heights. The altitudes of these elements are each $\frac{b}{n}$, and the lengths, beginning at the vertex of the triangle, are $\frac{b}{n}, \frac{2b}{n}, \frac{3b}{n}$, &c. Their areas, regarding them as parallelograms, which we may do, since n is indefinitely large, are

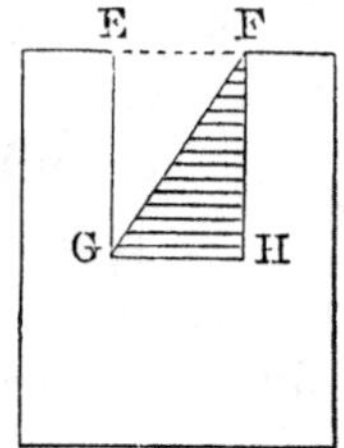

$$\frac{b}{n}.\frac{h}{n},\ \frac{2b}{n}.\frac{h}{n},\ \frac{3b}{n}.\frac{h}{n},\ \text{\&c.}$$

Hence the discharges through each are

$$\frac{bh}{n^2}\sqrt{2g\frac{h}{n}},\ \frac{2bh}{n^2}\sqrt{2g\frac{2h}{n}},\ \frac{3bh}{n^2}\sqrt{2g\frac{3h}{n}},\ \text{\&c.},$$

and the whole discharge will be

$$Q = \frac{bh\sqrt{2gh}}{n^2\sqrt{n}}(1^{\frac{3}{2}} + 2^{\frac{3}{2}} + 3^{\frac{3}{2}} +, \&c. \ldots n^{\frac{3}{2}})$$

$$= \frac{bh\sqrt{2gh}}{n^2\sqrt{n}}\left(\frac{n^{\frac{3}{2}+1}}{\frac{3}{2}+1}\right) = \tfrac{2}{5}.bh\sqrt{2gh} = \tfrac{2}{5}b\sqrt{2gh^3}.$$

Also, $\quad v = \frac{Q}{\frac{1}{2}bh} = \frac{4}{5}\sqrt{2gh}.$

2°. When the base of the triangle is in the surface.

In this case the quantity which will flow through EFG will equal the discharge through EFGH minus that through FGH or

$$Q = \tfrac{2}{3}bh\sqrt{2gh} - \tfrac{2}{5}bh\sqrt{2gh} = \tfrac{4}{15}bh\sqrt{2gh},$$

and $\quad v = \frac{8}{15}\sqrt{2gh}.$

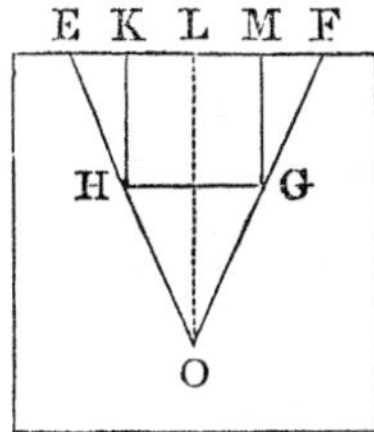

Cor. If the aperture be a trapezoid, as EFGH, whose upper base $EF = b_1$ lies in the surface, lower base $HG = b_2$, and altitude $KH = h$, we may find the discharge by regarding the aperture as made up of the rectangle KHGM, and the two triangles EKH and MGF. Hence

$$Q = \tfrac{2}{3}b_2h\sqrt{2gh} + \tfrac{4}{15}(b_1 - b_2)\sqrt{2gh} = \tfrac{2}{15}(2b_1 + 3b_2)h\sqrt{2gh}.$$

Also, through the triangle HOG, whose base $HG = b_2$, and depth of vertex $LO = h_1$, the quantity of the discharge will be

$$Q = \tfrac{4}{15}b_1h_1\sqrt{2gh_1} - \tfrac{2}{15}(2b_1 + 3b_2)h\sqrt{2gh}.$$

In a similar manner, rectilinear orifices of other forms may be divided into triangles, trapeziums, &c., and the discharge determined.

428. Prop. *To determine the velocity with which an elastic fluid will issue from a small orifice into an unlimited void when urged by its own weight.*

By reference to *Art.* 421, it will be seen that the result there obtained is independent of the density of the fluid, and will therefore be true of all fluids, whatever be their density. For the forces g and g' are as the weights of the stratum at the orifice, and of the column vertically above it; that is,

$$g : g' = w : w' = gh : gh' = h : h'.$$

Consequently, $v = \sqrt{2gh'}$

expresses the velocity with which every fluid of uniform dens ity will issue from a small orifice.

COR. The velocity with which air will rush into a vacuum is that which a heavy body would acquire in falling from the height of a homogeneous atmosphere. The height of a homogeneous atmosphere (*Art.* 418), when the temperature is 32° and the barometer stands at 30 in., is $h'=26146$ feet, which gives $v=1297$ feet. But for a temperature of 60° the height h_1 will be (159)

$$h_1 = h'(1+\nu t),$$

in which $t=60°-32°=28°$, and the value of ν for a mean state of humidity of the air will be 0.00222 instead of 0.002039 (*Art.* 412).

$$\therefore\ v=\sqrt{2gh_1}=\sqrt{2gh'(1+0.00222\times 28)}$$
$$=\sqrt{2\times 32\tfrac{1}{6}\times 26146\times(1+0.00222\times 28)}=1336 \text{ feet}$$

MOTION OF FLUIDS IN LONG PIPES.

429. The subject of the conveyance of water in pipes is one of considerable practical importance. But the motion of fluids in long pipes is so much affected by adhesion and friction in the interior, by the resistance occasioned by curvature, and by the disengagement of air, which remaining stationary when the pipes are laid along a level surface, or collecting in the higher portions of them, where they are curved, opposes the flow of the fluid, that theoretical results are of little practical value; besides, investigations conducted on hypotheses involving all the causes which affect the motion of the fluid are too difficult for an elementary work.

The experiments of Bossut, in 1779, are those which are usually relied on for information on this subject. Water was allowed to flow through pipes whose diameters were $1\frac{1}{3}$ in. and 2 in., and lengths from 30 to 180 feet. They were chiefly of tin, and inserted in the side of a reservoir filled to a constant height, either one foot or two feet above the center of the pipe The following principles were established:

1°. The discharges in given times with pipes of the same

length, and with the same head of water, are proportional to the squares of the diameters.

2°. For pipes of different lengths and of the same diameter, the discharge was inversely proportional to the square roots of the lengths.

To determine the supply which may be expected from a pipe of given dimensions, it was found that a pipe 30 feet long and $1\frac{1}{3}$ inch in diameter would discharge at its extremity about one half of that which would issue from a simple orifice, or short pipe of the same diameter.

Couplet, in 1730, found that a pipe of stone or iron, 600 fathoms in length and 12 inches in diameter, with a head of 12 feet of water, discharged $\frac{1}{10}$th, and a pipe of the same diameter and 2340 fathoms in length, with a head of 20 feet, discharged $\frac{1}{10}$th of that which would have been obtained from a simple orifice

To determine the quantity which flows through a section of a natural stream, it is usual to measure the breadth and the depth at different points of a transverse section, and find the area of the section. The velocity at various points of the section is measured by the hydraulic quadrant, or rheometer of some kind. The mean velocity, multiplied by the area, gives the quantity which flows through the section in a second.

430. Prop. *To determine generally the velocity with which a fluid will issue from an orifice of any size in the bottom of a cylindrical vessel.*

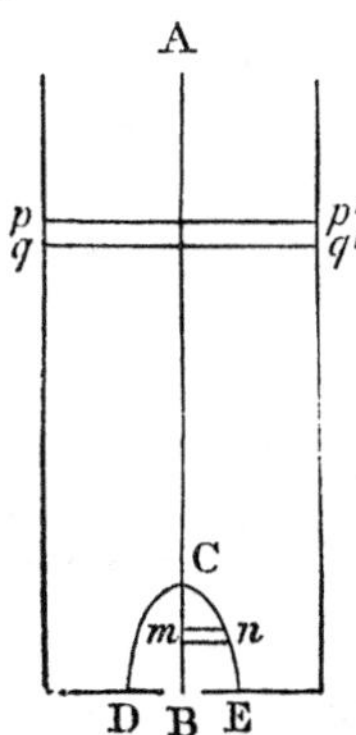

We shall adopt the usual hypothesis of a division of the fluid into thin laminæ, and that in their descent their parallelism is preserved. Let the distance of the lamina $pp'qq'$ from the surface A be x, and K the area of its surface. Then, if dx be its thickness and ρ its density, $\rho \mathrm{K} dx$ will be its mass, and $g\rho \mathrm{K} dx$ its weight, or the force with which it will descend if free. Now the resistance with which it meets in its descent is the difference of pressures on its lower and upper surfaces. Let p represent the pressure on a unit of the upper surface by the water above it; then $p+dp$ will be the upward pressure on a unit of the lower surface, or the resistance of the water below it, and the difference $-dp$, multiplied by K, will be the resistance experienced by the whole lamina. Hence the moving force will be

$$g\rho \mathrm{K}dx-\mathrm{K}dp,$$

and the acceleration, [VI.] and [IX.],

$$\phi=\frac{g\rho \mathrm{K}dx-\mathrm{K}dp}{\rho \mathrm{K}dx}=\frac{g\rho dx-dp}{\rho dx}=\frac{d^2x}{dt^2}. \qquad (a)$$

Let v be the velocity of discharge at B, k the area of a section of the issuing fluid, and dt the element of time in which the surface K descends a distance equal to dx. Then (167) we have

$$\mathrm{K}dx=kvdt \text{ or } \frac{dx}{dt}=\frac{kv}{\mathrm{K}}. \qquad (b)$$

By differentiation, regarding dt as constant, we get

$$\frac{d^2x}{dt^2}=\frac{kdv}{\mathrm{K}dt}.$$

$$\therefore \frac{kdv}{\mathrm{K}dt}=\frac{g\rho dx-dp}{\rho dx}.$$

or $$g\rho dx-dp=\frac{\rho kdv}{\mathrm{K}}.\frac{dx}{dt}.$$

Hence (b) $$g\rho dx-dp=\rho\frac{k^2}{\mathrm{K}^2}.vdv.$$

Integrating, $$g\rho x-p=\rho\frac{k^2}{\mathrm{K}^2}.\frac{v^2}{2}+\mathrm{C}.$$

Let now $x=h=\mathrm{AB}$ and $k=\mathrm{K}$; then

$$g\rho h-p=\rho\frac{v^2}{2}+\mathrm{C};$$

and subtracting the preceding from this, we have

$$g(h-x)=\frac{v^2}{2}\left(1-\frac{k^2}{\mathrm{K}^2}\right),$$

or $$v=\sqrt{\frac{2g(h-x)}{1-\frac{k^2}{\mathrm{K}^2}}}. \qquad (171)$$

If k be very small, $v=\sqrt{2g(h-x)}$ nearly; and if the vessel be kept constantly full, $x=0$ and $v=\sqrt{2gh}$ (168).

If $k=K$, $v=\infty$, or the velocity must be infinite. From which we infer that a section of the issuing fluid can never be equal to a section of the vessel. If a cylindrical tube be vertical and filled with a fluid, the portion of the fluid at the lower extremity, being urged by the pressure of all above it, will necessarily have a greater velocity than those portions which are higher, and therefore (167) a section of the issuing fluid is necessarily less than a section of the tube.

431. PROP. *To determine the time in which a cylindrical vessel will exhaust itself by a small orifice in the base.*

Since k is very small, we have $v=\sqrt{2g(h-x)}$ for the velocity at the end of the time t when the surface of the fluid has descended through the space x. Let this velocity be supposed constant during the indefinitely small time dt. Then the quantity discharged in this time will be

$$dQ=kdt\sqrt{2g(h-x)}=Kdx,$$

dx being the descent of the surface in the time dt.

Hence $$dt=\frac{K}{k\sqrt{2g}}.(h-x)^{-\frac{1}{2}}dx.$$

Integrating, $$t=-\frac{2K}{k\sqrt{2g}}.(h-x)^{\frac{1}{2}}+C.$$

If $x=0$, $t=0$, and $C=\frac{2K}{k\sqrt{2g}}\sqrt{h}$.

$$\therefore\ t=\frac{2K}{k\sqrt{2g}}(\sqrt{h}-\sqrt{h-x}), \qquad (172)$$

and when $x=h$, or the whole is discharged,

$$t=\frac{2K\sqrt{h}}{k\sqrt{2g}}=\frac{2Kh}{k\sqrt{2gh}}=\frac{2Q}{k\sqrt{2gh}}\ (170). \qquad (173)$$

The time is therefore twice as great as that which is required to discharge the same quantity when the vessel is kept constantly full.

432. PROP. *To determine the quantity which will issue from an aperture of any size and form in the side of a prismatic vessel.*

Let the *Fig.*, *Art.* 430, represent one face of the vessel, and DCE the aperture.

The element of the area of the orifice will now be ydx.

$$\therefore\ dQ=tydx\sqrt{2g(h_1+x)},$$

in which $h_1=$AC, $x=$Cm, and $y=mn$.

This, integrated between the limits $x=0$, and $x=h-h_1=$CB will give the quantity discharged in the time t.

If the aperture be rectangular, $y=$a constant$=b$. Then

$$dQ=bt\sqrt{2g(h_1+x)}dx,$$

and $$Q=\tfrac{2}{3}bt\sqrt{2g}(h_1+x)^{\frac{3}{2}}+C,$$

and, between the limits above named,

$$Q=\tfrac{2}{3}bt\sqrt{2g}(h^{\frac{3}{2}}-h_1^{\frac{3}{2}}). \qquad (174)$$

If the orifice extend to the top of the vessel, then $h_1=0$, and

$$Q=\tfrac{2}{3}bt\sqrt{2g}.h^{\frac{3}{2}}=\tfrac{2}{3}bht\sqrt{2gh},$$

or $$v=\frac{Q}{bht}=\tfrac{2}{3}\sqrt{2gh},$$

which is the velocity due to the depth of the center of pressure of the aperture below the surface of the fluid.

In the foregoing the vessel is supposed to be kept constantly full, or the surface of the fluid to be very large compared with the aperture.

433. EXAMPLES.

Ex. 1. A vessel, formed by the revolution of a semi-cubical parabola about its axis, which is vertical, is filled with a fluid till the radius of its surface is equal to its distance from the vertex; to find the time in which the fluid will be discharged through a small hole at the vertex.

The equation of the semi-cubical parabola is $ay^2=x^3$.

Hence (167) $$kvdt=-Kdx,$$

or $$k\sqrt{2gx}dt=-\pi y^2dx=-\frac{\pi x^3dx}{a}.$$

$$\therefore\ dt=-\frac{\pi}{ka\sqrt{2g}}.x^{\frac{5}{2}}dx.$$

Taking the integral between the limits $x=0$ and $x=a$, we have

$$t=\frac{\pi}{ka\sqrt{2g}}.\tfrac{2}{7}a^{\frac{7}{2}}=\frac{\pi\sqrt{2a^5}}{7k\sqrt{g}}.$$

Ex. 2. To find the time in which a paraboloid of revolution whose altitude is h and parameter p, full of fluid, will empty itself through a small orifice at its vertex, its axis being vertical.

Ans. $t=\dfrac{2\pi ph^{\frac{3}{2}}}{3k\sqrt{2g}}$.

Ex. 3. A conical vessel, the radius of whose base is r and altitude h, is filled with a fluid. Required the time in which the surface of the fluid will descend through half its altitude, the orifice being at the vertex and the axis vertical.

Ans. $t=\dfrac{\pi r^2h^{\frac{1}{2}}(2^{\frac{5}{2}}-1)}{20kg^{\frac{1}{2}}}$.

Ex. 4. Find the times in which a fluid contained in a vessel, formed by the revolution of the curve whose equation is $y^4=a^3x$ about the axis of x, will descend through equal distances h, supposing a small orifice at the vertex, and the axis vertical.

Ans. Each $t=\dfrac{\pi a^{\frac{3}{2}}h}{k\sqrt{2g}}$.

Ex. 5. The horizontal section of a cylindrical vessel is 100 square inches, its altitude is 36 inches, and it has an orifice whose section is one tenth of a square inch. In what time, if filled with a fluid, will it empty itself, allowing for the contraction of the vein?

Ans. $t=11^m 36^s.5$.

HYDROSTATIC AND HYDRAULIC INSTRUMENTS

MARIOTTE'S FLASK.

433. This piece of apparatus was devised to illustrate several successive positions of equilibrium of air and water, and variations in the velocity of discharge. In the annexed section of this flask, A, B, C are apertures so small that when open and water is issuing the air can not enter, and the reverse. The apertures at first are all closed, the vessel is filled with water, and the tube DF is inserted at M and filled with water to the height D.

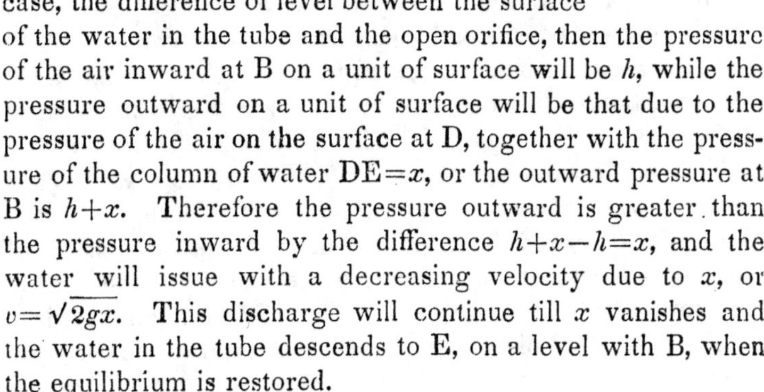

1°. Let the aperture B be opened.

If h be the height of a column of water equivalent to one atmosphere, and x represent, in every case, the difference of level between the surface of the water in the tube and the open orifice, then the pressure of the air inward at B on a unit of surface will be h, while the pressure outward on a unit of surface will be that due to the pressure of the air on the surface at D, together with the pressure of the column of water $DE=x$, or the outward pressure at B is $h+x$. Therefore the pressure outward is greater than the pressure inward by the difference $h+x-h=x$, and the water will issue with a decreasing velocity due to x, or $v=\sqrt{2gx}$. This discharge will continue till x vanishes and the water in the tube descends to E, on a level with B, when the equilibrium is restored.

2°. Let B be now closed and A be opened.

Now the pressure downward on the surface of the water in the tube at E is h. The pressure upward on the same surface is that due to the pressure of the air at A, and to the weight of a column of water equal in height to AB=HE=x. or the upward pressure on E is $h+x$. The resultant upward pressure on E is therefore $h+x-h=x$, and the surface of the water in the tube will rise till x vanishes, or till it arrives at H, when the equilibrium will again be restored.

In the mean time, however, the air has entered at A and risen to the upper part of the vessel, where it occupies the space above ho. Since an equilibrium now exists, the expansive force of the air above ho, together with a column of water equal in height to LH, equilibrates the pressure h of the air on the water in the tube at H. If h' be the height of a column of water equivalent to the tension of the air above ho, and y=LH represent the difference of level of the surfaces of the fluid in the tube and flask, we have

$$h'+y=h \text{ or } h'=h-y.$$

3°. Let A and B be closed and C opened.

The pressure at C inward is h, outward is $h+x$, representing by x, as before, the vertical distance of the surface of the water in the tube from the open orifice C. Therefore the resultant of the pressures at C is $h+x-h=x$, and the water will flow from C with a velocity due to the height x. As the surface of the fluid in the tube descends, x diminishes, and the velocity of discharge also, till the surface in the tube arrives at F, at which time x becomes equal to the height of F above C. In this position of the fluid, the equation $h'=h-y$ still obtains, y being now the distance LF. But an equilibrium not being established, air will enter the vessel at F and rise to ho. This operation will continue till $y=o$ and $h'=h$. During this entrance of the air, and the descent of ho to F, the value of x has remained unchanged, and therefore the velocity of discharge has remained constant. After this the value of x will diminish, and the velocity of discharge also, till the water in the flask descends to the level of C.

BRAMAH'S HYDROSTATIC PRESS.

434. In the annexed section of this instrument L and H are vertical cylindrical cavities, in a solid mass of metal or other

strong material. The diameter of H is considerably less than that of L, and they communicate through an aperture N, in which is a valve opening into L. A is a strong piston or solid cylinder of iron fitting closely to the surface of the hollow cylinder, and movable in it. IH is a piston similarly applied to the other cavity H, and movable by means of a lever KP, whose fulcrum is at P. At O is a valve opening upward, and below it the cylinder H is continued downward to a reservoir of water. The lever KI being raised, the water ascends, as in the common pump, from the reservoir G into the cavity H. The lever being then pressed down, the valve O closes, and the water is forced through the valve N into L, and, acting on the piston A, communicates a pressure to any substance, as books, included between its upper surface and a strong cross bar BC firmly connected with the solid cylinder DE. When the water from H is forced into L, its reaction closes N, and the same operation is repeated.

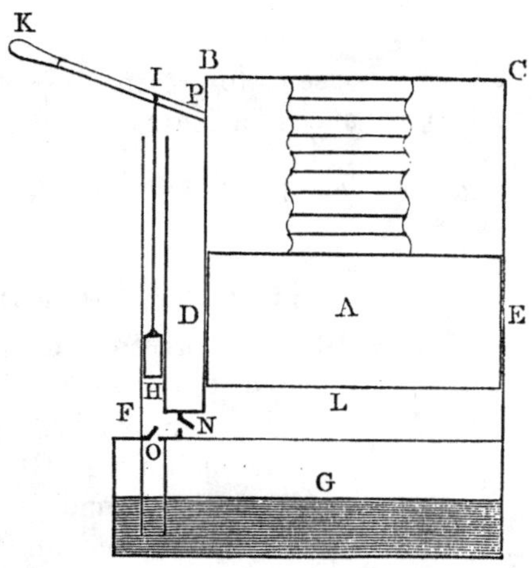

The pressure on the base of A (134) is to the force impressed by H as the area of the base of A is to that of H.

Let R and r be the radii of the pistons A and H, L and l the longer and shorter arms of the lever KP, p the power applied at K, and P the resulting pressure on the base of A Then the force p', applied to the piston H, will be

$$p'=\frac{pL}{l}.$$

But (134),

$$\frac{P}{p'}=\frac{\pi R^2}{\pi r^2}.$$

$$\therefore\ P=p'\frac{R^2}{r^2}=p.\frac{L}{l}.\frac{R^2}{r^2}.$$

This press seems to present the simplest and most effective of all contrivances for increasing human power.

If $p=100$ lbs., $\frac{L}{l}=10$, and $\frac{R}{r}=10$, then $P=100,000$ lbs

HYDROSTATIC BELLOWS.

435. This instrument presents an illustration of what is termed the Hydrostatic Paradox: that fluids, however unequal in quantity, may be made to equilibrate. It consists of two circular boards, A and B, firmly connected by a cylindrical coating of leather or other flexible material. CM is a tube communicating with the lower portion of the cylinder. Water being poured into the tube CM, the boards A and B will separate, B will rise, and a weight W, very large compared with that of the water in CM, may be supported on B. The fluid in CM which counterpoises W is that above P, the level of B.

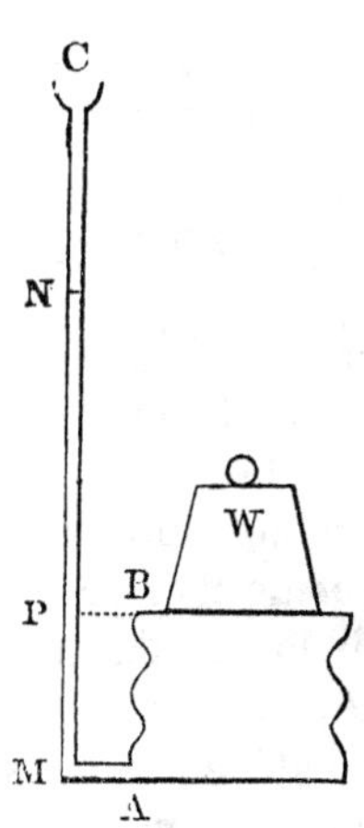

Let k be the area of a horizontal section of the tube, K that of a section of the cylinder, or that of B, and p the pressure of the fluid NP above P on k.

Then (134) $$\frac{p}{W}=\frac{k}{K}.$$

If σ be the specific gravity of the weight W, V its volume, and $NP=h$, then

$$\frac{g\rho_1 hk}{g\rho V}=\frac{k}{K}.$$

$$\therefore\ h=\frac{\rho V}{\rho_1 K}=\sigma.\frac{V}{K}.$$

If V′ be the whole volume of the fluid in the instrument, and H the height of the cylinder AB, then

$$V' = H(K+k) + hk = H(K+k) + \sigma V\frac{k}{K}.$$

$$\therefore\ H = \frac{V'K - \sigma Vk}{K(K+k)}.$$

Let now the quantity of fluid be increased by v, and the corresponding increment $\triangle H$ in the height of W will be

$$\triangle H = \frac{K(V'+v) - \sigma Vk}{K(K+k)} - \frac{KV' - \sigma Vk}{K(K+k)} = \frac{v}{K+k}.$$

DIVING BELL.

436. The diving bell is commonly a hollow cylinder or parallelopiped, one end of which is closed. It is immersed with the open end downward, weights being added to sink and keep it steady in its descent. As the vessel descends the fluid continually exercises a greater pressure on the contained air, condenses it, and occupies a greater portion of the vessel. If the form be as above supposed, to find the height x of that portion of the bell free from water, when the top is sunk to the depth H, let h be the height of a homogeneous atmosphere, a the height of the bell, ρ' the density of the external air, ρ that of the air in the bell, and ρ_1 that of water. Then the pressure p, on a unit of surface of air in the bell, will be

$$p = g\{h\rho' + (H+x)\rho_1\}.$$

But the unit of pressure arising from the elasticity of the internal air is $g\rho h$.

$$\therefore\ g\{h\rho' + (H+x)\rho_1\} = g\rho h.$$

The quantity of air in the bell being constant, and a horizontal section of it the same throughout (155) and (156),

$$\rho x = \rho' a, \text{ or } \rho = \rho'.\frac{a}{x}.$$

$$\therefore\ h\rho' + H\rho_1 + x\rho_1 = \rho' a\frac{h}{x},$$

or

$$\frac{\rho'}{\rho_1}hx + Hx + x^2 = \frac{\rho'}{\rho_1}ah,$$

and σ being the specific gravity of air,

$$\sigma hx + Hx + x^2 = \sigma ah.$$

$$\therefore\ x = -\tfrac{1}{2}(H+\sigma h) \pm \sqrt{\tfrac{1}{4}(H+\sigma h)^2 + \sigma ah}.$$

SEA GAGE.

437. The vessel AB is perforated with holes, and has within it, fixed in a vertical position, a glass tube, having one end closed and the other immersed in a cup of mercury. A is a hollow sphere whose buoyancy is sufficient to raise the instrument when the weight W suspended at the bottom is detached. The instrument is allowed to sink in the water whose depth is to be determined, and by a mechanical contrivance the weight is detached when it strikes the bottom, so that the gage will ascend by the buoyancy of the ball. The height MP to which the mercury has risen in the tube is marked on the interior of the tube, by the adhesion of oil or other viscid substance placed on its surface.

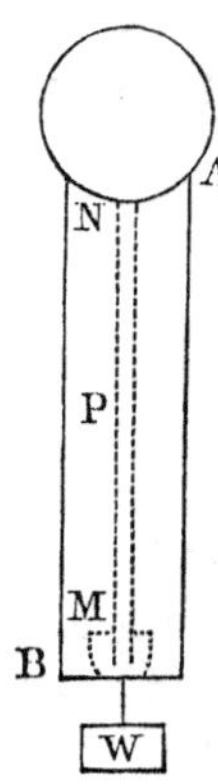

Let h be the height of the barometer at the surface, l the length of the tube above the surface of the mercury in the cup, x the depth of the water, and $MP=h_1$. If h_2 be the height of a column of mercury which would be sustained by the elasticity of the air in NP, then (155)

$$h_2(l-h_1)=hl \text{ or } h_2=\frac{hl}{l-h_1}.$$

Now the elasticity of the air in NP+ the weight of the column MP= the pressure of the atmosphere on the surface of the fluid + the weight of a column of water extending to the bottom: or, ρ being the density of mercury, and ρ_1 that of the water,

$$\rho h_2+\rho h_1=\rho h+\rho_1 x.$$

$$\therefore\ x=\frac{\rho}{\rho_1}(h_2+h_1-h)=\sigma h_2+\sigma(h_1-h)$$

$$=\sigma\frac{hl}{l-h_1}+\sigma(h_1-h)$$

$$=\frac{\sigma h}{1-\frac{h_1}{l}}+\sigma(h_1-h).$$

in which σ is the ratio of the specific gravity of mercury to that of the water of the ocean.

SIPHON.

439. The siphon is an instrument for transfe ring fluids from one vessel V to another V′ in which the surface of the fluid is lower. It consists of a bent tube with one branch longer than the other. Suppose this tube to be filled with the fluid from the vessel V, and to have its extremities immersed in the fluids in the two vessels.

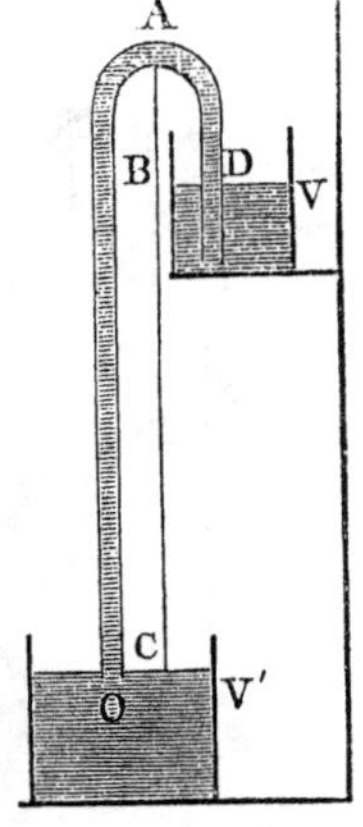

Let $CB=a$ be the difference of level of the fluids in the vessels, ρ_1 the density of the fluid, ρ the density of the surrounding air, and k the area of a section of the tube. The pressure on the surface of the fluid in V′ without the tube by the atmosphere will be greater than that on the surface of the fluid in V by $g\rho ak$. This excess of pressure acting at O upward will urge the fluid from V′ to V. At the same time, the pressure exerted by the liquid in the siphon at the level of the fluid in V will be greater than that at the level of the fluid in V by $g\rho_1 ak$. This excess of pressure will urge the fluid from V to V′. Hence the resultant pressure is equal to $gak(\rho_1-\rho)$, and in the direction DAO. The fluid will therefore flow from V to V′, and, since ρ is small, it will move with a force nearly equivalent to a column of the fluid whose height is a and base k. It is not necessary that the longer branch should be immersed in the fluid of V′.

If h be the height of a column of the fluid in equilibrium with the atmosphere, and $AB=x$, then the upward pressure on k in the shorter branch, at the level of the fluid, is that due to h, while the downward pressure is that due to x. Therefore the fluid is urged in the direction DAO, at A, by a force represented by $h-x$. As long as $x<h$ the fluid will rise to A and pass into the longer branch.

THE COMMON PUMP.

440. AB and BC are cylinders connected together as in the figure, the former called the body or barrel of the pump, and the latter the suction pipe. At B is a valve opening upward. M is a piston accurately fitting the barrel, and movable by means of a rod EV and a lever EF. In the piston is a valve V, which opens upward. The suction pipe BC is immersed in the fluid to be raised.

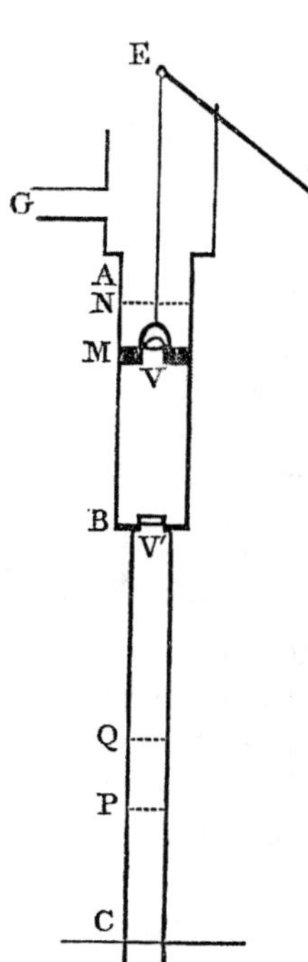

To understand the action of the pump, conceive the piston M to be depressed to B. The air in MB being thus condensed, will open the valve V and escape. If now the piston be raised to A, the pressure of the external air will keep the valve V closed, the air in BC, by its elasticity, will open the valve V′ and diffuse itself uniformly through CA. The pressure on the surface of the water in the tube at C being thus diminished, the pressure of the air on the water without the tube will cause it to rise to some point P in the tube, till the weight of PC and the elasticity of the air in AC together shall produce a pressure on a section of the water in the tube at C equal to the upward pressure, occasioned by the external air on the water of the reservoir; that is, until the tension of the air in AP, together with the weight of PC, equal one atmosphere. The valve V′ being now equally pressed on both sides, will close by its own weight. Let now the piston descend again to B. The air in AB being condensed, will again open V and escape. By a second ascent of the piston to A, the air in BP will expand through PA, and the volume of water PC will increase in length till the pressure on the section at C shall again be one atmosphere. By repeated strokes of the piston the water will rise to B and pass through V'. When this is effected, a portion will pass through V at each descent of the piston, and be lifted by its ascent to the reservoir above A, and be discharged through the spout G.

To determine the distance BP, and the elasticity of the air in BP after n strokes, let h_n be the height of a column of water which will measure the elasticity of the air in BP, and $\text{BP}=x_n$; also, let h_{n+1} and x_{n+1} represent the same quantities after the $(n+1)$th stroke. Let h be the height of a column of water equivalent to one atmosphere, $\text{AB}=a$ the length of a stroke of the piston, $\text{BC}=b$ the distance of the lower valve from the surface of the water, K be a section of the barrel AB, and k a section of the suction pipe BC. Then, since the tension of the air in BP, together with the weight of the column of water PC, is equal to one atmosphere,

$$h_n+(b-x_n)=h \text{ and } h_{n+1}+(b-x_{n+1})=h.$$
$$\therefore\ h_n-x_n=h-b \quad \text{and } h_{n+1}-x_{n+1}=h-b.$$

Now P being the position of the surface of the water after the nth stroke, and Q after the $(n+1)$th, h_n and h_{n+1} measure the densities of the air in BP and BQ; and since, by the $(n+1)$th stroke, the air in BP is expanded through AQ (155),

$$h_{n+1}(\text{K}a+kx_{n+1})=kh_nx_n,$$

and substituting the value of $x_{n+1}=h_{n+1}+b-h$, we have

$$h_{n+1}(\text{K}a+k(h_{n+1}+b-h))=kh_nx_n.$$

$$\text{Hence } h_{n+1}=-\tfrac{1}{2}\left\{\frac{\text{K}a}{k}+b-h\right\}\pm\tfrac{1}{2}\sqrt{\left\{\frac{\text{K}a}{k}+b-h\right\}^2+4h_nx_n},$$

$$\text{and} \qquad x_{n+1}=\tfrac{1}{2}\left\{\frac{\text{K}a}{k}+h-b\right\}\pm\tfrac{1}{2}\sqrt{\left\{\frac{\text{K}a}{k}+b-h\right\}^2+4h_nx_n}.$$

The length of the pipe BC is necessarily less than h, otherwise the weight of PC, before it becomes equal to BC, would be equal to one atmosphere, and, could a vacuum be produced in MP, no further ascent of the fluid would follow.

To find the pressure on the piston when the water is raised to the point N above the piston, let it be represented by a column of water whose height is x. The pressure upward against the base of M is that of a column of water equal in height to

$h-$MC. The pressure on the upper surface is $h+$NM. Therefore the resultant pressure is the difference of these, or

$$x=h+\text{NM}-h+\text{MC}=\text{NM}+\text{MC}=\text{NC}.$$

441. Sometimes, by not preserving a proper relation between the distances AB and AC, the water will cease to rise, even though BC$<h$. To show this, let the barrel and suction pipe be equal, or K$=k$, and the lower valve be at some point below B, as at C. When the water has risen to X and the piston is raised to A, the tension of the air in AX, together with the weight of the column XC, is equal to one atmosphere. If now, when the piston descends to B, the tension of the air in BX does not become greater than that of the external air, the valve in the piston will not be opened, no air will escape from BX, and the water will rise no further. To find the point X when the air in AX, reduced to the volume of BX, becomes *equal* in tension to one atmosphere, let AX$=x$, AB$=a$, AC$=l$. Then BX$=x-a$. Now when the piston is raised to A, the air in BX being expanded through AX, the tension becomes (155) $\frac{h(x-a)}{x}$.

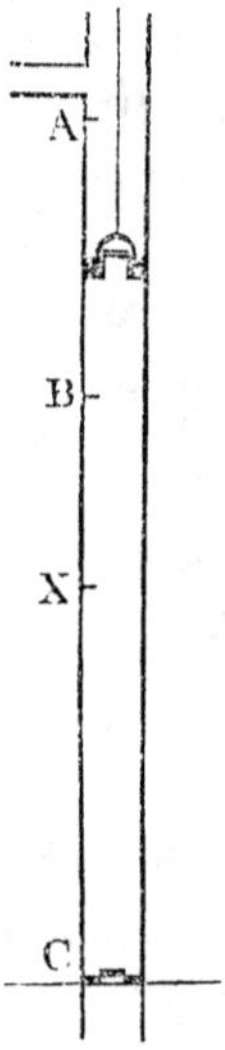

$$\therefore\ \frac{h}{x}(x-a)+l-x=h,$$

and

$$x=\tfrac{1}{2}l\pm\tfrac{1}{2}\sqrt{l^2-4ah}.$$

If $l^2<4ah$, the value of x is imaginary, and no point of stoppage exists. When $l^2=4ah$, $x=\frac{1}{2}l$, and there is one point at half the distance from A to C; and when $l^2>4ah$, there will be two points.

If $a=10$, $l=36$, and $h=32$,

$$x=20 \text{ and } 16.$$

AIR PUMP.

442. B and B′ are cylinders or barrels, commonly of the same size, in which pistons P and P′, with valves V and V′ opening upward, are movable by means of rack-work, the one ascending as the other descends. At the lower extremity of the barrels there are apertures with valves opening upward, and communicating by the pipe *ac* with the receiver R, from which the air is to be exhausted.

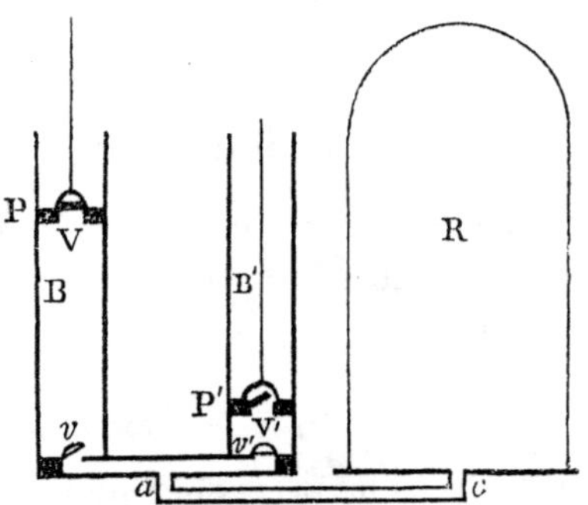

The ascent of the piston P tends to produce a vacuum in B, the valve V being closed by the pressure of the external air. The air in R, by its elastic force, opens the valve *v* and fills the barrel B. When P descends, the elasticity of the air in B closes *v* and opens V, through which the contents of B escape into the external air. The action of both pistons is manifestly the same, and thus for each descent of either piston a volume of air equal to that of either barrel is expelled from the machine.

Let V be the volume of the receiver pipes and one barrel, V_1 the capacity of either barrel, and ρ_n the density of the air in the machine after the *n*th stroke. Now the quantity of air in the machine after the *n*th stroke is $\rho_n V$, and by the $(n+1)$th stroke the volume V_1, or quantity $\rho_n V_1$, is expelled. There will remain then in the machine after the $(n+1)$th stroke the quantity $\rho_n V - \rho_n V_1$, and this being diffused through the space V, we have, by (155) and (156),

$$\rho_{n+1} V = \rho_n V - \rho_n V_1 = \rho_n (V - V_1),$$

or

$$\rho_{n+1} = \rho_n \left(1 - \frac{V_1}{V}\right).$$

If ρ_0 be the initial density of the air by making *n* equal to 0, 1, 2, &c., successively, we have

$$\rho_1=\rho_0\left(1-\frac{V_1}{V}\right)$$

$$\rho_2=\rho_1\left(1-\frac{V_1}{V}\right)=\rho_0\left(1-\frac{V_1}{V}\right)^2$$

. . . .

$$\rho_n=\rho_0\left(1-\frac{V_1}{V}\right)^n.$$

It is obvious that ρ_n can never become zero as long as n is finite, and therefore, even theoretically, perfect exhaustion is impossible.

CONDENSER.

443. If in the preceding figure the valves V, V′, v, v', were made to open downward, it would represent a condenser. By each descent of the piston a volume of air equal to that of the barrel will be forced into R, and will by its reaction close the valves v and v', and be retained there.

To find the density of the air in the receiver after n strokes of the piston, let ρ_n and ρ_{n+1} be the densities of the air in **R** after the nth and $(n+1)$th strokes, V the volume of the receiver R and the pipe ac, V_1 that of either barrel, and ρ_0 the density of the external air. Then

$$\rho_{n+1}V=\rho_n V+\rho_0 V_1.$$

$$\therefore\ \rho_{n+1}=\rho_n+\rho_0\frac{V_1}{V}.$$

If n be made successively equal to 0, 1, 2, 3, &c., we have

$$\rho_1=\rho_0\left(1+\frac{V_1}{V}\right)$$

$$\rho_2=\rho_1+\rho_0\frac{V_1}{V}=\rho_0\left(1+\frac{2V_1}{V}\right)$$

. . . .

$$\rho_n=\rho_0\left(1+\frac{nV_1}{V}\right).$$

CLEPSYDRA.

444. The clepsydra, or water clock, is a contrivance for measuring time by the descent of the surface of a fluid which flows through a small aperture in the base of the vessel containing it.

Suppose the vessel a prism. It is required to determine what scale must be marked on its side, that the coincidence of the descending surface with the successive lines of division may mark equal successive intervals of time. Let a be the altitude of the prism, x the distance of the surface from the base of the vessel at the end of the time t, from the beginning of the motion.

Equation (172) gives, using x instead of $h-x$,

$$x=a-\frac{k\sqrt{2ga}}{K}t+\frac{k^2g}{2K^2}t^2.$$

Let Δx and Δt be corresponding increments of x and t; then

$$x+\Delta x=a-\frac{k\sqrt{2ga}}{K}(t+\Delta t)+\frac{k^2g}{2K^2}(t+\Delta t)^2.$$

Subtracting the preceding equation from this, and we have

$$\Delta x=-\frac{k^2g}{K^2}\left\{\frac{K}{k}\sqrt{\frac{2a}{g}}-t-\tfrac{1}{2}\Delta t\right\}\Delta t.$$

The time t is in seconds, and, to determine the divisions corresponding to successive *minutes* of time, put $\Delta t=60''$, and give to t the successive values 0, 60'', 120'', &c.

In a vessel of the form supposed in *Ex.* 4, *Art.* 433, the vertical distance of the successive divisions will all be equal, or the surface will descend equal distances in equal times.

THE END.

www.ingramcontent.com/pod-product-compliance
Lightning Source LLC
LaVergne TN
LVHW020220110826
845151LV00003B/772

9781425532321